Science, Technology and Innovation Studies

Science, technology and innovation (STI) studies are interrelated, as are STI policies and policy studies. This series of books aims to contribute to improved understanding of these interrelations. Their importance has become more widely recognized, as the role of innovation in driving economic development and fostering societal welfare has become almost conventional wisdom. Interdisciplinary in coverage, the series focuses on the links between STI, business, and the broader economy and society. The series includes conceptual and empirical contributions, which aim to extend our theoretical grasp while offering practical relevance. Relevant topics include the economic and social impacts of STI, STI policy design and implementation, technology and innovation management, entrepreneurship (and related policies), foresight studies, and analysis of emerging technologies. The series is addressed to professionals in research and teaching, consultancies and industry, government and international organizations.

More information about this series at http://www.springer.com/series/13398

Dirk Meissner • Leonid Gokhberg •
Ozcan Saritas
Editors

Emerging Technologies for Economic Development

Springer

Editors
Dirk Meissner
National Research University
Higher School of Economics
Moscow, Russia

Leonid Gokhberg
National Research University
Higher School of Economics
Moscow, Russia

Ozcan Saritas
National Research University
Higher School of Economics
Moscow, Russia

ISSN 2570-1509 ISSN 2570-1517 (electronic)
Science, Technology and Innovation Studies
ISBN 978-3-030-04368-1 ISBN 978-3-030-04370-4 (eBook)
https://doi.org/10.1007/978-3-030-04370-4

Library of Congress Control Number: 2019934100

This Springer imprint is published by the registered company Springer Nature Switzerland AG.
The registered company address is: Gewerbestrasse 11, 6330 Cham, Switzerland

Contents

Editors and Contributors

About the Editors

Dirk Meissner is Deputy Head of the Laboratory for Economics of Innovation at HSE ISSEK and Academic Director of the Master Program "Governance for STI." Dr. Meissner has 20 years of experience in research and teaching technology and innovation management and policy. He has strong background in policy making and industrial management for STI with special focus on foresight and roadmapping, funding of research and priority setting. Prior to joining HSE, Dr. Meissner was responsible for technology and innovation policy at the presidential office of the Swiss Science and Technology Council. Previously, he was management consultant for technology and innovation management with Arthur D. Little. He is member of OECD Working Party on Technology and Innovation Policy. He is Associate Editor of *Technological Forecasting and Social Change*, *IEEE Transactions on Engineering Management*, *Journal of the Knowledge Economy*, member of Editorial Review Board at Small Business Economics and Journal of Knowledge Management. He guest edited *Special Issues in Industry and Innovation* journal, *Journal of Engineering and Technology Management*, *Technological Analysis and Strategic Management* among others.

Leonid Gokhberg is First Vice-Rector of the HSE and also Director of HSE ISSEK. His area of expertise is statistics and indicators on STI as well as foresight and policy studies in this area. He has authored over 400 publications in Russian and international peer-reviewed journals, monographs, and university textbooks. Prof. Gokhberg has coordinated dozens of national and international projects funded by public agencies, businesses, and international organizations. He has served as a consultant of the OECD, Eurostat, UNESCO, and other international and national agencies. Leonid is also a member of the Global Innovation Index Advisory Board, the OECD Government Foresight Network, and OECD and Eurostat working groups and task forces on indicators for S&T as well as steering committees of various prestigious international and national initiatives. Prof. Gokhberg is Editor-in-Chief of the Scopus-indexed scientific journal *Foresight and STI Governance* and editor of the Springer academic book series *Science, Technology, and Innovation Studies*, and participates on the editorial boards of several other influential journals. He holds Ph. D. and Dr. of Sc. degrees in Economics.

Ozcan Saritas is Professor of Innovation and Strategy at the HSE and Editor-in-Chief of *Foresight*—the journal of future studies, strategic thinking and policy. He worked as Senior Research Fellow at the Manchester Institute of Innovation Research, the University of Manchester. His research focuses on innovation and policy research with particular emphasis on socio-economic and technological foresight. With a Ph.D. from the "Foresight and Prospective Studies Program," he introduced the "Systemic Foresight Methodology," and has produced a number of publications on the topic. Dr. Saritas has extensive work experience with the international organizations, including the United Nations, OECD, and the European Commission. He has been involved in large-scale national, multinational, and corporate research and consultancy projects on sectors including energy, climate change, agriculture, food, water, transportation, and ICT among others; published a number of articles in respected journals; and have delivered keynote speeches in more than

50 countries across the world. Besides his research and publication activities, he designs and delivers academic and executive education courses on Foresight and Strategic Planning. He has recently co-authored a book, entitled *Foresight for Science, Technology and Innovation* published by Springer, which has become one of the key readings in the field.

Contributors

Pavel Bakhtin Information and Analytical Systems Unit, HSE ISSEK, Moscow, Russia

Serhat Burmaoglu Faculty of Economics and Administrative Sciences, Izmir Katip Celebi University, Izmir, Turkey

Michael Clark Portland State University, Portland, OR, USA

Tugrul Daim Portland State University, Portland, OR, USA

Boris Ermolenko National Research University, Higher School of Economics, Moscow, Russia

Georgy Ermolenko National Research University, Higher School of Economics, Moscow, Russia

Leonid Gokhberg National Research University, Higher School of Economics, Moscow, Russia

Qin Guo Portland State University, Portland, OR, USA

Aharon Hauptman Tel Aviv University, Tel Aviv, Israel

Andrew James Manchester Institute of Innovation Research, The University of Manchester, Manchester, UK

Elena Khabirova Unit for Digital Economy Studies, HSE ISSEK, Moscow, Russia

Marina Klubova National Research University, Higher School of Economics, Moscow, Russia

Ilya Kuzminov National Research University, Higher School of Economics, Moscow, Russia

Lubov Matich National Research University, Higher School of Economics, Moscow, Russia

Dirk Meissner National Research University, Higher School of Economics, Moscow, Russia

Liliana Proskuryakova National Research University, Higher School of Economics, Moscow, Russia

Vitaly Roud National Research University, Higher School of Economics, Moscow, Russia

Pavel Rudnik National Research University, Higher School of Economics, Moscow, Russia

Vladimir Salun National Research University, Higher School of Economics, Moscow, Russia

Ozcan Saritas National Research University, Higher School of Economics, Moscow, Russia

Mitsutaka Shirasaki Genentech, Beaverton, OR, USA

Alexander Sokolov National Research University, Higher School of Economics, Moscow, Russia

Anton Timofeev National Research University, Higher School of Economics, Moscow, Russia

Natalia Veselitskaya National Research University, Higher School of Economics, Moscow, Russia

Konstantin Vishnevskiy National Research University, Higher School of Economics, Moscow, Russia

Haydar Yalcin Faculty of Economics and Administrative Sciences, Izmir Katip Celebi University, Izmir, Turkey

Faculty of Humanities and Social Sciences, Izmir Katip Celebi University, Izmir, Turkey

Abbreviations

Terms

A&F	Agriculture and Forestry
EHS	Environment, Health and Safety
FTA	Future Oriented Technology Assessment
GC	Grand Challenges
GVC	Global Value Chain
HDM	Hierarchical Decision Model
ICT	Information and Communication Technologies
iFORA	Intelligent Foresight Analytics (HSE)
IoT	Internet of Things
PET	Privacy Enhancing Technology
PPP	Public Private Partnership
R&D	Research and Development
RRI	Responsible Research and Innovation
S&T	Science and Technology
SME	Small and Medium Sized Enterprises
STEEP	Social, Technological, Economic, Environmental and Political
STEEPV	Social, Technological, Economic, Environmental, Political and Value
STI	Science, Technology and Innovation
SWOT	Strengths, Weaknesses, Opportunities, Threats
URS	User Requirement Specification

Institutions

EC	European Commission
FAO	Food and Agricultural Organization of the United Nations
FDA	Federal Drug Agency
HSE	National Research University Higher School of Economics

IEA	International Energy Agency
ISSEK	Institute for Statistical Studies and Economics of Knowledge, National Research University Higher School of Economics
NISTEP	National Institute of Science and Technology Policy
OECD	Organization for Economic Cooperation and Development
UNESCO	United Nations Educational, Scientific and Cultural Organization
WEF	World Economic Forum

What Do Emerging Technologies Mean for Economic Development? 1

Dirk Meissner, Leonid Gokhberg, and Ozcan Saritas

Economic development of countries and regions is commonly perceived as associated to innovation which in turn is thought of to be driven by technology (ies) (Camagni and Capello 2013). In this understanding it follows that technologies are crucial for economic development which is precondition for employment creation and social welfare of regions and countries. However, generating economic and societal value from technologies is not a simple linear process which can be established easily. The potential values from technologies for different interest groups closely correlate with the technology development, e.g. life cycle, stage (Lundvall and Borrás 1997; Proskuryakova et al. 2017; Gokhberg and Meissner 2016). It has been recognized that mature technologies in the late phases of their life cycle provide less potential for regional and country economic development than technologies at the beginning of the life cycle. This however holds true only while considering the long-term value and impact generated; in selected cases where technologies are mature but take the form of platform technologies with related impact on industry standards, the economic impact might be reasonably strong and also sustainable, and in other cases of mature technologies, a short-term impact might appear possible based on the actual technology with its respective applications and the technologies' perception as standard setting throughout industries, e.g. becoming platform technology (Gertler 2003; Gokhberg et al. 2016). Besides what has already emerged and near market, it is also very important to recognize the 'technological emergence'. Recently, more and more efforts are dedicated for technological emergence to get ready and capitalize on what is likely to emerge.

D. Meissner (✉) · L. Gokhberg · O. Saritas
National Research University, Higher School of Economics, Moscow, Russia
e-mail: dmeissner@hse.ru; lgokhberg@hse.ru; osaritas@hse.ru

D. Meissner et al. (eds.), *Emerging Technologies for Economic Development*, Science, Technology and Innovation Studies,
https://doi.org/10.1007/978-3-030-04370-4_1

Although it is hard to predict their emergence, once detected more recently big data and analytics are used to track technologies with potential transformational impacts. Investigating in emerging and cutting-edge technologies can inform policy and strategy and reveal new opportunities for social and economic development. In this respect, opportunities created by emerging and new technologies are numerous.

Whether emerging or already emerged, technologies in the early stages of life cycles with already tested and proven applications and market awareness created appear more promising enablers of sustainable development (Huggins and Johnston 2009; Ketels and Memedovic 2008). Obviously emerging technologies inherit a strong application and sometimes multiple use potential, but still they are featured by reasonable uncertainty for completion and broad diffusion, hence providing risks which cannot be neglected (Laur et al. 2012; Carayannis et al. 2018). However, frequently the chances and opportunities provided by emerging technologies are considered to outweigh the respective risks, which is why investment in emerging technologies is on the rise for long including public investment.

A reasonable number of leading countries have rather precise view of the most important socio-economic and science and technology (S&T) priorities, which often include emerging technologies like nanotechnologies and biotechnologies, among many others. These technologies are often seen as instruments for enabling new technologies with further socio-economic impacts and addressing grand challenges. For instance, the development of the 'graphene' has brought a number of potential applications for the development of battery technologies for energy storage. Fast-charging batteries enabled by graphene can revolutionize how energy is stored and can be used in a number of socio-economic sectors like information and communication technologies, transport, agriculture and health. Therefore, when emerging technologies are focused, their alignment with broader societal, environmental and all other demand side issues, like price and affordability, equity and accessibility, among the others, should be taken into consideration. Beyond these, when setting priorities for new technologies at the national or corporate levels, there are also other issues to be addressed such as risks and uncertainties associated to those technologies (Klofsten et al. 2015; Meissner et al. 2017). Thus, technology assessment and also technology and economic foresight need to take the latter much more into account than previously done (Kuzminov et al. 2018). Therefore, emerging technologies need to be considered from a holistic point of view, e.g. as systems of innovation. The systems perspective on innovation, in this book with a strong focus on emerging technologies, provides a more solid assessment of opportunities and risks while at the same time requires broader skills and competences by those assessing the technologies from the different perspectives.

The book gives a comprehensive overview on the variety of emerging technologies and approaches to assessing emerging technologies and estimating their contribution to economic development, by considering them in technical terms as well as their prospective applications. Chapters in the book identify and describe existing and potential markets for emerging technology-based applications. Overall, the book stresses the interdisciplinary nature of emerging technologies. Complementary to the technology and market descriptions, methods like integrated roadmapping are introduced and explained through real case examples. Moreover, the book shows integrated roadmaps for nanotechnologies and develops policy measures and recommendations for companies to further develop these technologies. These measures are illustrated with the use of policy measures and roadmaps for implementation.

In the book different trends and scenarios are discussed for the development of emerging technologies based on a number of existing and emerging macroeconomic trends of the global, and the national, economies. The book gives valuable insights in the emerging technology development and their applications in different sectors. It is structured along major application fields for emerging technologies across five major sections. Following the introduction, Part I sheds light on emerging technologies in materials and manufacturing. Part II highlights the applications in the energy- and transport-related technologies. Then, Part III focuses on water and agriculture, while Part IV outlines human- and security-oriented applications. The concluding part focuses on measuring emerging technologies and the meaning of science, technology and innovation (STI) policies for leveraging the impact of emerging technologies.

Part I on **materials and manufacturing** starts with an analysis of nanotechnologies and new materials, especially carbon fibres. *Saritas, Sokolov and Vishnevski* find that recent advancements have led to the development of new materials with improved specifications and reduced dimensions. Cutting-edge metals, foams and other substances make buildings, vehicles and gadgets more energy efficient and environmentally friendly. Due to their strategic importance, new materials are at the radar of national policies. In recent decades, a number of developed and developing countries have conducted numerous studies to determine STI-based development prospects for the sphere of converging technologies with a particular focus on new materials, nanotechnologies and their production. All carbon fibre production technologies involve in one way or another pyrolysis of raw material. No 'revolutionary' technologies for carbon fibre production are expected to emerge in the near future. The carbon fibre industry is currently changing from custom production (e.g. for the aerospace industry) to general mass market-oriented production. The most important aspect is believed to be increasing capacities of individual production lines.

Nanotechnology for high-tech industries especially for light-emitting diodes is described by *Roud, Sokolov and Meissner* in the following chapter. They argue that enhancing energy efficiency has been one of the top policy goals in many countries in the last decade. Innovative lighting solutions and light-emitting diodes (LED) in particular are among the very much promising opportunities to increase energy efficiency. LED technologies are becoming dominant in a number of application segments. Demand for economic and energy security makes the development of the LED industry one of the national priorities in many countries including Canada, the USA, Japan, China and European countries, among others. The key areas of the LED industry's development envisage designing materials with unprecedented characteristics and using nanoscale components. At the same time, the technology level of semiconductor industries as such to a large extent is determined by chips processing—the key stage of LED production chain.

Another application field for nanotechnologies is found in traditional industries. *Meissner and Rudnik* analyze catalysts for petroleum refining, namely, the potential for development and the application of nanotechnologies in catalytic oil refining processes. Catalysts play a key role in most oil refining processes, so improving their properties and developing new catalyst types are seen as high-priority technological objectives, the achievement of which would have a major effect on the efficiency of production in the fuel-and-energy complex. Such technologies have the potential for deeper oil processing; accelerated development of motor, diesel and other fuels' production; and production of petrochemical materials—provided advanced, high-performance equipment, innovative efficient catalysts and sorbents are used and the principle of combining production processes at a single high-capacity installation is observed. The authors argue that Russia offers significant opportunities for expanding into external markets which requires efforts to acquire knowledge and experience related to the application of innovative international technologies and processes to achieve breakthrough in domestic production. Along this way Russia has a chance to become an international technology development hub, offering conditions for creating and disseminating far beyond its borders the most advanced industrial and research technologies, attracting international intellectual capital and high-tech companies and—most importantly—achieving competitive positions to secure advantages in a specific segment of the technology chain, to successfully integrate into the international division of labour.

Part II focuses on emerging technologies in the **energy and transport domains**. The part begins with an analysis of measures to increasing the share of renewables in the energy mix and assessing the technological potential for evidence-based policy-making. The authors, *Ermolenko, Ermolenko and Proskuryakova*, find that waves of innovations in energy technologies led to the paradigm shifts in the industrial and societal development. They argue that the centuries-long dominant position of fossil fuels as economic drivers has led to the establishment of major players—multinational companies that would like to preserve their business as long as possible. The chapter analyzes recent technical developments of wind power and solar power especially and their contribution to the changing energy mix and environmentally friendly energy production. Approaches taken in Russia are used to illustrate concerted political initiatives to promote environmentally friendly energy production in a country which is used to fossil fuels-based energy production to a reasonable extent.

A roadmap for fuel cell electric vehicle (FCEV) Global Market Creation is presented by *Sokolov, Saritas and Meissner*. They find that in parallel with the technological development, recent discussions about global warming and climate change caused by carbon dioxide emissions brought public support for emission-free vehicles, which are perceived as an asset. Despite of advancements and public support, the introduction of FCEVs is still not at the desirable levels. Car manufacturers frequently announce the near-time launch of FCEVs, which are postponed with the same frequency. Saritas, Meissner and Sokolov aim to make an attempt of a systemic analysis for a broader and more holistic analysis, which may portray the bigger picture, and help to understand the industrial dynamics better.

Their key argument is that the stakeholder base in the transportation industry and thus for FCEVs is broader than usually thought, and there are a number of other issues to be addressed for a successful and widespread launch of FCEVs. The chapter illustrates the links between the key technologies for FCEVs, the consumer properties of existing and advanced FCEVs, the most promising products and their respective market shares, volumes and growth rates. It highlights the structure of potential demand for innovative products and outlines their most prospective markets. The roadmap also provides an assessment of technical capabilities required for manufacturing of products with the most preferable consumer properties, which would allow generating the significant competitive advantages for FCEVs.

In their contribution on emerging technologies in the aircraft and shipbuilding industries *Klubova, Veselitskaya, Matich* and *Salun* investigate emerging technologies in aircraft and shipbuilding industries aimed at addressing common grand challenges. For this purpose, the authors propose a framework for global trends and technology identification in two fields of the transport sector based on scenario and roadmapping approaches. The suggested framework supports policy-makers, companies and other interested parties set priorities, select innovation projects and implement them based on a vision of a desirable future.

Living systems and environment are the main subjects of **Part III**. In the first chapter, *Saritas and Vishnevski* view nanotechnology as a response to grand challenges with special emphasis on the case of water treatment and purification. Their chapter sheds light on the potential uses of nanotechnologies for supplying clean water. It shows that new technologies are needed for effective and resource-efficient water treatment, and nanotechnologies may offer affordable solutions. Overall, the chapter advocates that supplying drinking water to billions of people, while at the same time protecting water resources, is the best strategic guideline for the water supply industry and for applying new techniques and materials including nanotechnology. Developing new technologies for deep purification of water with nanotechnology will increase healthy water supply by removing both visible and invisible impurities hazardous to human and animal health. Thus, the chapter discusses nanotechnology solutions for water treatment and purification. How nanotechnologies can significantly increase the efficiency of certain traditional water purification processes such as coagulation, sorption and flotation is discussed. Next, technological, market and institutional aspects are considered regarding nano-technology solutions. The chapter is concluded with future scenarios and strategic steps to be taken for the implementation of nano-based water treatment and purification.

Gokhberg, Kuzminov, Bakhtin, Timofeev and Tochilina introduce an approach towards identifying emerging technologies. They exemplify the methodology in the case of agriculture and food. The paper discloses a new approach to sectoral emerging technology identification and their future development analysis. Their proposed research strategy relies on a combination of traditional foresight methods with big data-augmented science, technology and innovation (STI) landscape mapping. On the first step, the results of text-mining analysis present the ontology of currently emerging technologies in global agriculture and food (A&F) sector. On the second step, these technologies with the usage of text-mining techniques were aggregated: (a) future technological market forecasts and (b) parameters of their potential to bring answers to sectoral and national challenges. Based on this big data-augmented analysis, prospective science and technology (S&T) development areas for the Russian A&F sector were highlighted. The comparative benefits of the proposed approach for evidence-based STI policy and corporate strategic planning are shortly summed up in terms of its objectivity and comprehensive information source coverage.

The part concludes with a chapter on technology assessment, trends and wild cards in human enhancement. First, *Guo, Clark, Shirasaki and Daim* propose a solid approach towards assessing container closure integrity testing technology for the biotech industry. They develop a framework for assessing technologies in the biotechnology sector. The proposed framework is applied for evaluating several container closure integrity testing technology alternatives. Then, in the chapter *Saritas* describes emerging technologies, trends and wild cards in human enhancement. He argues that the key question is whether human will be at the centre of change as it has historically been or will be put aside by smarter machines. Machines then will become the principal creator of new technologies and race far ahead of humanity. While the machines are advancing, emerging technologies also offer new possibilities for humans to remain competitive against machines through the enhancement of physical and mental capacities. The development and use of, such as, neuroscience, silicon chips and smart technologies offer new opportunities. A desirable future is that there would be an ecosystem of humans and machines, where both complement each other rather than competing with each other. Machines would continue to support humanity's well-being and quality of life and offer new possibilities for advanced 'human-machine', 'human-human', 'human-work' and human-environment interactions. Following the exploration of technologies, this chapter reviews and discusses the implications of human enhancement on the future of work, where the socio-economic impacts of emerging technologies can be well observed. These technologies are expected to make revolutionary changes in working environment as people will be able to work harder, longer and smarter. The chapter will also address some of those ethical issues associated with human enhancement technologies.

Another promising field for applying emerging technologies is the **security sector** which is the core of **Part IV**. *Hauptman* illuminates the 'dark side' of emerging technologies. He postulates that almost every new technology developed for the benefit of society has a potential 'dark side', manifested, for example, by security or privacy threats. His chapter presents selected findings of two European projects, which by employing foresight methods tried to shed light on these dark sides: one project assessed the threats of potential abuse (by criminals or terrorists) posed by selected emerging technologies, while the second focused on the impacts of emerging technologies on privacy. The chapter pays special attention to new, even surprising possibilities opened by the convergence of technologies and also deals with related policy issues. The foresight processes, which help to explore evolving security or privacy aspects of new technologies, reflect a need for continuous analysis of the unfolding technology landscape for potential, sometimes surprising, implications.

New issues and impacts for defence and security are discussed by *James*. He illustrates how after 09/11 foresight studies shifted in security thinking away from a focus on state-centric threats towards a much broader view of security risks recently. This expanded perspective includes risks presented by the vulnerability of the European society to the failure of critical infrastructure, to pandemics, environmental change and resource-based conflicts. The chapter places a particular emphasis on the treatment of technological change in these defence and security foresight studies and argues that the growing importance of dual-use technologies is likely to mean that defence will play a declining role as a sponsor and lead user of advanced technologies in the future.

Defence 4.0 by means of Internet of Things in Military is analyzed by *Burmaoglu, Saritas and Yalcin*. They show how scientific and technological developments have been influencing military concepts and practice, particularly following the inception of the scientific revolution in the late sixteenth century. Their argumentation starts from the fact that a number of technologies have been developed for defence, found their civilian applications and vice versa. Wherever the boost for change comes from, the nature of warfare has changed radically both due to S&T advancements and changing socio-economic and geopolitical contexts. Despite of the barriers due to strict organizational culture, armies have adapted themselves into changing characteristics of warfare through new concepts and instruments. Among S&T developments, recent advancements in information and communication technologies (ICTs) bring enormous opportunities as well as challenges for defence. One of the recent phenomena emerged with the rapid development of ICTs is the Internet of Things which affects every aspect of life with a growing number of devices communicating with each other. While the possibilities introduced by the Internet of Things have been providing immense benefits, the increasing number of connections makes the system ever more complex and vulnerable because of the difficulty of securing huge networks.

To conclude, **Part V** summarizes the previous chapters and looks especially on **challenges to science, technology and innovation policy**. *Meissner, Gokhberg and Saritas* consider how to stimulate emergence and convergence of technologies. They argue that technology development and application are increasingly taking place under ever more challenging conditions. In their view, technology features such as complexity, interdisciplinarity and budget as well as time constraints and also application features such as increased solutions' reliability, multiple application potential and direct and immediate economic impact create additional challenges for technology developers. Especially emerging technologies are frequently considered to demonstrate features which fulfil these expectations and requirements. Therefore, companies and policy-makers often wish to support the development of such emerging technologies and leverage their potential for achieving commercial impact and regional economic development. This requires that aims and goals of technology support are formulated in a more flexible form as currently practised, and no definite fixed indicators and deliverables are described. In doing so governments are asked to obtain a more entrepreneurial attitude which is not expressed in standardized public announcements but which is filled with live by the public sector itself. The authors conclude that emerging technologies provide significant opportunities for companies and research institutions; however, for regional economic development, it requires policy-makers looking beyond the existing policy measures. This means especially concerted—e.g. consistent and coherent—STI policy approaches solving the routine policy-maker dilemma which is often to develop new STI policy measures instead of rethinking and streamlining the existing STI policy mix.

Acknowledgements The book chapter was prepared within the framework of the Basic Research Program at the National Research University Higher School of Economics (HSE) and supported within the framework of the subsidy by the Russian Academic Excellence Project '5-100'.

References

Camagni R, Capello R (2013) Regional innovation patterns and the EU regional policy reform: toward smart innovation policies. Growth change 44(2):355–389

Carayannis EG, Grigoroudis E, Campbell DF, Meissner D, Stamati D (2018) The ecosystem as helix: an exploratory theory-building study of regional co-opetitive entrepreneurial ecosystems as quadruple/quintuple helix innovation models. R&D Manag 48(1):148–162

Gertler MS (2003) Local knowledge: tacit knowledge and the economic geography of context, or the undefinable tacitness of being (there). J Econ Geogr 3:75–99

Gokhberg L, Meissner D (2016) Seizing opportunities for national STI development. In: Deploying foresight for policy and strategy makers. Springer, Cham, pp 267–273

Gokhberg L, Meissner D, Sokolov A (2016) Foresight: turning challenges into opportunities. In: Deploying foresight for policy and strategy makers. Springer, Cham, pp 1–8

Huggins R, Johnston A (2009) Knowledge networks in an uncompetitive region: SME innovation and growth. Growth Change 40(2):227–259

Ketels CH, Memedovic O (2008) From clusters to cluster-based economic development. Int J Technol Learn Innov Dev 1(3):375–392

Klofsten M, Bienkowska D, Laur I, Sölvell I (2015) Success factors in cluster initiative management. Ind High Educ 29(1):65–77

Kuzminov I, Meissner D, Lavrynenko A, Tochilina E (2018) Technology classification for the purposes of futures studies WP BRP 78/STI/2018. National Research University Higher School of Economics

Laur I, Klofsten M, Bienkowska D (2012) Catching regional development dreams: a study of cluster initiatives as intermediaries. Eur Plann Stud 20(11):1909–1921

Lundvall B, Borrás S (1997) The globalizing learning economy—implications for technology policy at the regional, national and European level. In: Paper to the EU-TSER workshop on globalization and the learning economy: implications for technology policy. Brussels

Meissner D, Polt W, Vonortas NS (2017) Towards a broad understanding of innovation and its importance for innovation policy. J Technol Transf 42(5):1184–1121

Proskuryakova L, Meissner D, Rudnik P (2017) The use of technology platforms as a policy tool to address research challenges and technology transfer. J Technol Transf 42(1):206–227

Dirk Meissner is Deputy Head of the Laboratory for Economics of Innovation at HSE ISSEK and Academic Director of the Master Program "Governance for STI". Dr. Meissner has 20 years experience in research and teaching technology and innovation management and policy. He has strong background in policy making and industrial management for STI with special focus on Foresight and roadmapping, funding of research and priority setting. Prior to joining HSE Dr. Meissner was responsible for technology and innovation policy at the presidential office of the Swiss Science and Technology Council. Previously he was management consultant for technology and innovation management with Arthur D. Little. He is member of OECD Working Party on Technology and Innovation Policy. He is Associate Editor of Technological Forecasting and Social Change, IEEE Transactions on Engineering Management, Journal of the Knowledge Economy, member of Editorial Review Board at Small Business Economics and Journal of Knowledge Management. He guest edited Special Issues in Industry and Innovation journal, Journal of Engineering and Technology Management, Technological Analysis and Strategic Management among others.

Leonid Gokhberg is First Vice-Rector of the HSE and also Director of HSE ISSEK. His area of expertise is statistics and indicators on STI as well as foresight and policy studies in this area. He has authored over 400 publications in Russian and international peer-reviewed journals, monographs, and university textbooks. Prof. Gokhberg has coordinated dozens of national and international projects funded by public agencies, businesses, and international organizations. He has served as a consultant of the OECD, Eurostat, UNESCO, and other international and national agencies. Leonid is also a member of the Global Innovation Index Advisory Board, the OECD Government Foresight Network, and OECD and Eurostat working groups and task forces on indicators for S&T and, as well as steering committees of various prestigious international and national initiatives. Prof. Gokhberg is Editor-in-Chief of the Scopus-indexed scientific journal Foresight and STI Governance and editor of the Springer academic book series Science, Technology, and Innovation Studies, and participates on the editorial boards of several other influential journals. He holds PhD and Dr. of Sc. degrees in Economics.

Ozcan Saritas is Professor of Innovation and Strategy at the HSE and editor-in-chief of "Foresight"—the journal of future studies, strategic thinking and policy. He worked as Senior Research Fellow at the Manchester Institute of Innovation Research, The University of Manchester. His research focuses on innovation and policy research with particular emphasis on socio-economic and technological Foresight. With a PhD from the "Foresight and Prospective Studies Program," he introduced the "Systemic Foresight Methodology", and has produced a number of publications on the topic. Dr. Saritas has extensive work experience with the international organizations, including the United Nations, OECD, and the European Commission. He has been involved in large scale national, multinational and corporate research and consultancy projects on sectors including energy, climate change, agriculture, food, water, transportation, and ICT among others; published a number of articles in respected journals; and have delivered keynote speeches in more than 50 countries across the world. Besides his research and publication activities, he designs and delivers academic and executive education courses on Foresight and Strategic Planning. He has recently co-authored a book, entitled "Foresight for Science, Technology and Innovation" published by Springer, which has become one of the key readings in the field.

Part I

Materials and Manufacturing

New Materials: The Case of Carbon Fibres

2

Ozcan Saritas, Alexander Sokolov, and Konstantin Vishnevskiy

2.1 Introduction

New materials are one of the key pillars of the convergence in science, technology, and innovation (STI). Recent advancements in STI have led to the development of new materials with improved specifications and reduced dimensions. Cutting-edge metals, foams, and other substances make buildings, vehicles, and gadgets more energy efficient and environmentally friendly. Due to their strategic importance, new materials are at the radar of national STI policies. In recent decades, a number of developed and developing countries have conducted numerous studies to determine STI-based development prospects for the sphere of converging technologies with a particular focus on new materials, nanotechnologies, and their production (Kim et al. 2014).

A number of foresight studies have been undertaken at national (Saritas et al. 2007; Vishnevskiy and Yaroslavtsev 2017), regional (Battistella and Pillon 2016), and sectoral levels (Aydogdu et al. 2017) to investigate the future of new materials and nanotechnologies. Russia is one of the countries, which prioritised the domain of new materials and nanotechnologies for future STI development. Russia's future development agenda considers new materials and nanotechnologies among the key priority areas along with information and communication technologies; life sciences (biotechnology, medicine, and public health); rational uses of natural resources, transport, and space systems; and energy efficiency (Saritas 2015). In Russia, strategies are formulated to develop necessary competences in STI domains with a particular focus on nanotechnology, biotechnology, and power engineering sectors (Shmatko 2016).

O. Saritas (✉) · A. Sokolov · K. Vishnevskiy
National Research University, Higher School of Economics, Moscow, Russia
e-mail: osaritas@hse.ru; sokolov@hse.ru; kvishnevskiy@hse.ru

D. Meissner et al. (eds.), *Emerging Technologies for Economic Development*,
Science, Technology and Innovation Studies,
https://doi.org/10.1007/978-3-030-04370-4_2

At the regional level, Battistella and Pillon (2016) assert that new materials are one of the key technological components of regional development and policy. At the industrial level, new materials and nanotechnologies play a major role. For instance, Burmaoglu and Saritas (2017) underline the critical role of new materials for the defence sector. Saritas and Proskuryakova (2017) discuss the use of new materials and nanotechnologies for water purification. International organisations also focus on new materials and nanotechnologies due to their strategic importance. For instance, the OECD considers these technologies as a critical component towards the next production revolution along with the Internet of Things (IoTs), robotics, industrial biotechnology, and 3D printing (OECD 2017).

Among new materials, carbon fibres (CF) hold a prominent place and are defined as a nanostructured organic material containing between 92 and 99.99% of carbon (RCN 2010). CFs and composites based on them attract researchers' and manufacturers' attention due to their properties, such as good electroconductivity and nearly zero linear expansion coefficient, which make them indispensable for certain special applications. High binding energy allows CFs to retain their strength in a wide range of temperatures (up to 2200 °C). They are the most heat-resistant among all known fibrous materials, which means they can be used as thermal shields and heat-insulating materials.

The most important competitive advantages of CFs are believed to be their high modulus of elasticity and breaking point values, lightness, low friction coefficient, resistance to atmospheric impact, and chemical reagents. Certain specific features of CF materials allow them to be combined with other types of fibres, such as boron, glass, and organic ones (e.g. Kevlar 49, Armos, SVM). This enables products to be made that combine the advantages of the two initial materials. Such hybrid composites are already applied in aerospace, as well as in sporting equipment manufacturing.

Access to advanced technologies for the mass production of CFs and the scale of their industrial application are currently seen as indicators of national science, technology, and production potential levels and help to achieve sustainable development in various sectors of the economy. Currently CFs are most commonly used as reinforcing fillers in composites. They are considered to be a promising construction material for making various important products.

Thus, the present chapter examines CFs in detail and makes a future-oriented assessment. Long-term foresight future-oriented technology assessment (FTA) studies (Miles et al. 2016) and scenario analyses (Erdmann and Schirrmeister 2016; Amanatidou et al. 2016) have been undertaken to examine emerging trends in advanced science convergence (Vaseashta 2014). The present study first begins a detailed account of the CFs and CF composites in Sects. 2.2 and 2.3, respectively. Different types of CFs and composites are presented with their key advantages and uses. Next, their properties and uses are compared with a SWOT analysis. Competing technologies and the advantages of CFs and composites are discussed. Following the production and technology of CFs, Sect. 2.4 examines the demand side. A detailed market analysis and pricing forecasts are followed by future projections on demand. Here is a scenario approach employed as one of the most frequently used methods in foresight (Saritas and Burmaoglu 2015). The chapter is rounded off in Sect. 2.6 with a set of technology strategies to exploit emerging market opportunities and broader societal, environmental, and economic and other issues to consider when innovating in CF technologies.

2.2 Carbon Fibres (CFs)

Carbon fibre (CF) has increasingly been recognised as a revolutionary material with desirable properties including strength, rigidity and weight, and environmental friendliness. These properties made CFs to drive many industries such as wind power, aerospace, as well as automotive. Therefore, it is important to understand the nature of this highly demanded material, which is likely to open new horizons for many industries. The section below begins with an analysis of the types of carbon and a comparative analysis of them.

2.2.1 Types of CFs and a Comparative Analysis

CFs differ by precursors, which are the raw material used during the synthesis. Up to 90% of the CFs are produced from polyacrylonitrile (PAN), and the remaining are made of petroleum pitch or rayon. These materials are organic polymers, which consist of long strings of molecules bounded by carbon atoms. Four groups of CFs can be distinguished: (1) PAN, (2) pitch (coal, peat, wood, and oil tar distillation residue), (3) viscose, and (4) gas phase. These are briefly described below.

2.2.1.1 Polyacrylonitrile (PAN)-Based CFs

Fibres of this type are produced by sequential processing of primary PAN fibre (DoD 2005). An analysis of their strengths and weaknesses compared with other kinds of CF and with alternative products is given in Table 2.1.

2.2.1.2 Pitch-Based Carbon Fibres

There are several varieties of such fibres, the ones based on liquid-crystal (mesophase) and other regular (isotropic) pitches. Mesophase pitches can be used to make CFs with a high modulus of elasticity value, while isotropic pitches enable CFs to be produced relatively cheaply. A SWOT analysis of these types of CFs is presented in Table 2.2.

2.2.1.3 Viscose-Based CFs

Viscose is an artificial chemical fibre made of rayon by subjecting it to a chemical treatment to achieve structural modification. Cellulose is one of the most commonly used natural polymers, constantly reproduced on a major scale (see Table 2.3 for the SWOT analysis).

Table 2.1 SWOT analysis of PAN-based carbon fibres

Strengths	*Weaknesses*
Compared with other kinds of carbon fibre	
• Possibility to produce long unbroken fibres • Perfected technology	• Lower modulus of elasticity value than pitch-based CFs
Compared with alternative products	
• Favourable combination of high breaking point and modulus of elasticity values • Heat resistance	• High costs of primary PAN fibre • Low quality of Russian-made carbon fibres compared with the highest international level, which negatively affects export opportunities • The CFs' strength is limited by discrete defects in the polymer substrate and on the fibres' surface
Opportunities	*Threats*
• Application of CFs for energy saving and environmentally neutral production purposes • Production of fibres with specific physical and chemical properties (electroconductivity, high sorptive capacity) based on PAN bundles • Wide application in aerospace, defence industries, and nuclear power engineering	• Risk of pressure by environment protection organisations due to emission of cyan-hydric acid during production

Source: HSE

Table 2.2 SWOT analysis of pitch-based CFs

Strengths	*Weaknesses*
Compared with other kinds of CFs	
• High modulus of elasticity (up to 900 hPa) of mesophase pitch-based CFs • Low raw materials' costs of isotropic pitch-based CFs • Good electroconductivity	• Low bending strength
Compared with alternative products	
• High thermal conductivity	• High production costs • Low production volume • 'Dirty' production • Mismatched and unstable raw materials
Opportunities	*Threats*
• Used in centrifuge bowls, in combination with PAN-based fibres: pitch-based fibres provide axial reinforcement, PAN-based fibres—peripheral reinforcement • Thick monofilaments are used as cores for carborundum and other ceramic fibres • Used in space-based solar batteries and for spacecraft equipment manufacturing	• Russia lacks the raw materials—pitches suitable for making fibres • Russia lacks industrial extrusion (or other elongation) technologies for making pitch-based fibres • Limitations imposed by environment protection organisations due to pitch's carcinogenic properties

Source: HSE

Table 2.3 SWOT analysis of viscose-based CFs

Strengths	*Weaknesses*
• Relatively low costs of the raw material—viscose fibre • High graphitisability • Good textile properties • High electroconductivity • Heat insulation • Possibility to produce long fibres	• Low modulus of elasticity values • Low-strength properties, limiting the scope for application in composite production • High ash content • Mismatched raw materials
Opportunities	*Threats*
• Good prospects for medical applications • Application in aerospace industry—as heat insulation in rocket nozzles	• Lack of raw materials supplies

Source: HSE

Table 2.4 SWOT analysis of gas phase-based CFs

Strengths	*Weaknesses*
• A simple and inexpensive way to produce fibres • Maximum strength (close to monocrystals) • Easily adjustable fibre structure • Electroconductivity • Environmentally neutral production	• Hard to produce unbroken fibres • Insufficient reliability due to low cohesion of fibre layers
Opportunities	*Threats*
• Extremely high physical and chemical properties • Low production costs	• Technology is still at the laboratory prototypes stage

Source: HSE

2.2.1.4 Gas Phase-Based CFs

Gas phase-based fibres can be produced via catalytic pyrolysis of a hydrocarbon gas (methane, ethylene, acetylene, carbon monoxide, etc.) at 500–1500 °C. This technology is still under development and not used for mass production. Table 2.4 presents a SWOT analysis of the gas-based CFs.

The comparison of above precursors leads to a conclusion that PAN-based CFs have the biggest market, both for mass and specialised applications. PAN-based fibres are used as high-strength construction composites. Pitch-based fibres have a rather limited scope of application, mainly for special purposes (e.g. PAN- and pitch-based fibres can be used in combination with manufactured gas centrifuges).

Viscose-based CFs supplement replace graphite in high-temperature processes are used as a carbon material, primarily in medicine. Production costs of this material are determined by two factors: low product yield (10–15% of the primary viscose fibre mass) and a complex and expensive extraction process during graphitisation. It should be noted that since viscose-based CFs were first discovered and produced in Russia (in the 1970s and 1980s), the relevant production technologies and

Table 2.5 Carbon fibres' characteristics

	Carbon fibre type			
	PAN-based	Viscose-based	Pitch-based	Gas phase-based
Strength (hPa)	1.8–7.0	0.35–0.70	1.4–4.0	1.0–4.0
Modulus of elasticity (hPa)	200–600	20–60	140–930	200–300
Price (USD per kilo)	40	20	300	NA
Market volume (demand)	■■■	■□□	■□□	■■□
Technological development	■■■	■■□	□□□	□□□
Fibre yield from the raw material	■■■	■□□	■■□	□□□
Availability of raw materials and production capacities	■■■	■□□	□□□	□□□
Biocompatibility	□	■	□	□

'■'—presence of property. Levels: ■■■, high; ■■□, medium; ■□□, low; □□□, no such property
Source: HSE

equipment remained practically unchanged. Accordingly, the quality of the fibres currently produced remains quite low, with a relatively high costs and prices (USD60–100 per kilo).

The technology for making CFs from gas phase is still being developed, so these fibres are not yet on the market. However, gas phase-based CFs have many potential applications due to expected low production costs and relatively high properties. The competitive advantages of each type of CFs are described in Table 2.5.

2.2.2 Competing Technologies for CFs

CFs' growing popularity in various industries is limited by competition from alternative materials—both traditional (such as steel, aluminium, titanium, etc.) and new ones, particularly aramid, basalt, boron fibres, silicon carbide, and aluminium oxide fibres. On the whole, CF surpasses the other materials (see Table 2.6), but the latter are also applied in various industries.

2.2.2.1 Aramid Fibres

The main technical characteristics of aramid fibres, which determine the prospects and advantages of their application, include a low density (approximately 1400–1500 kg/m^3 compared to an average of 1700–1900 kg/m^3 for CFs); high tensile strength; higher impact strength compared with high-modulus CFs; and high static and dynamic load resistance. The high specific strength of materials based on aramid fibres means that the mass of a construction can be significantly reduced, while adhering to all relevant hardness and strength requirements. This feature of aramid fibre-based materials turns out to be quite efficient in terms of

Table 2.6 Properties of carbon fibres and alternative products

Material	Strength (hPa)	Modulus of elasticity (hPa)	Density (t/cm^3)	Diameter (microns)
High-strength CF	3.6–7.2	300	1.8	5–10
High-modulus CF	2.5–3.25	500–800	1.8–2.2	5–10
Glass fibre	3.5–4.6	72–110	2.5–2.9	3–25.8
Aluminium oxide fibre	2.2–2.4	385–420	3.95	10–25
Silicon carbide fibre	3.1–4.0	410–450	2.7–3.4	100–140
Boron fibre	3.45	400	2.6	100–200
Basalt fibre	1.8–3	90–115	2.2–2.6	10
Steel	2–3	200	7.8	
Aluminium	0.3–0.6	70–80	2.6–2.7	
Titanium	1–1.2	110	4.5	

Source: HSE

improving products' operational and economic characteristics including, for example, in the aerospace industry, fireproof clothing, sporting equipment, etc.

Using aramid fibres as the basis for composite materials can significantly increase their damage resistance properties. Such composites, like metals, have a plastic deformation potential which reduces their vulnerability to brittle fracture. On the other hand, a major flaw of aramid fibres—and therefore composite materials based on them—is very low compression and bending strength, which somewhat limit their application. In addition, such fibres are hard to colour, which can be seen as another drawback.

2.2.2.2 Basalt Fibres

Basalt fibres are made of natural minerals by melting and then transforming them into fibre without using any chemical additives. Currently, two major kinds of basalt fibres are manufactured: staple and unbroken fibres.

Depending on diameter, staple fibres are divided into several types with different thermal conductivity, sound absorption, density, and other properties, which determine their application areas:

- Microfine—less than 0.6 microns in diameter, used in super fine air-gas and liquid filters, for making thin paper and special products.
- Ultrafine—0.6–1.0 microns in diameter, used in superlight heat insulation and sound absorbent products, paper, fine air-gas and liquid filters.
- Superfine—1.0–3.0 microns in diameter. Special thermal processing of this kind means microcrystalline material can be produced with an acid resistance 2.5 times higher than regular fibres and with a hygroscopicity twice as low. The main advantage of this basalt fibre is zero shrinkage during exploitation. Microcrystalline fibre is used to make thermal-resistant heat insulation materials, plates, and filters for operation in aggressive high-temperature environments.

- Fine (9–15 microns) and thickened (15–25 microns) rock fibres are widely used as filtration base in hydraulic facilities' drainage systems.
- Rough fibres 50–500 microns in diameter are corrosion-resistant and can be used to strengthen construction materials.

Unbroken basalt fibre has such properties as resistance to high-temperature and aggressive environments, durability, mechanical strength, and low hygroscopicity. Due to its characteristics, basalt fibre is used in shipbuilding, wind energy generation, automobile industry, construction, etc.

2.2.2.3 Boron Fibres

The main advantage of boron fibres over carbon ones is high strength combined with high elasticity: boron fibre has tensile strength of 4300 mPa, modulus of elasticity of 380 hPa, and density of 2.63 g/cm^3. Moreover, in terms of hardness, boron fibre is second only to diamond. Boron fibres' mechanical properties are in effect the same as those of CF. However, boron fibre has a bigger cross-section than CF, with approximately 100 microns diameter (also manufactured are 140 and 200 microns boron fibres)—compared to a diameter of about 5–6 microns for CF. Boron fibres' increased diameter determines a higher productivity of their production process, which is based on chemical deposition over a tungsten wire core or CF.

Boron fibre's larger diameter has advantages and disadvantages. The advantages include simple operation, good matrix penetration of interfibrillar space due to small specific external surface, and high resistance to loss of stability under pressure. The disadvantages include poor flexibility of the fibre, which limits its application area. Another significant drawback of boron fibres and the composites based on them is their high costs, due to complex production technology—frequently an order of magnitude higher than that for other fibres.

2.2.2.4 Silicon Carbide Fibre

Silicon carbide fibre has several specific features and advantages. First, silicon carbide fibre can work without oxidation under quite high temperatures (carbon fibre starts to oxidise in an oxygen environment at 400 °C). In addition, from a technological point of view, making composite materials based on a metallic matrix and silicon carbide fibre is easier than producing CF-based metallic composites—because their reactivity to metals is low—but wetting of fibres' surface with melted metals is quite good.

Unlike electroconductive CF, silicon carbide fibre is a semiconductor, and its conductive properties are adjustable up to a point. The areas and scope of application of this kind of fibres and derived composites are determined by their technical characteristics, which in addition to high temperature resistance also include high shock resistance, high bending and tensile strength, and high durability.

2.2.2.5 Aluminium Oxide Fibres

The main mechanical properties of aluminium oxide fibres, including strength and modulus of elasticity, are similar to those of carbon fibres. Aluminium oxide fibres' advantages over CFs are better electrical insulation properties, colourlessness, and ability to retain properties under high temperature and during contact with melted metal in open-air environments. The latter characteristic permits the making of composite materials based on aluminium oxide fibres by casting, including shaping them into complex forms. On the other hand, a serious drawback of this kind of fibre is its relatively high density which somewhat limits its application areas. Aluminium oxide fibres are successfully applied for strengthening metals. Such composite materials retain good mechanical properties under high temperature and have high electric conductivity, etc. The above properties mean that these materials can be used for long periods of time in high-temperature environments, and they can also be pressed, rolled, or subjected to secondary shaping under temperatures close to metal matrix's melting temperature.

2.2.3 Advantages of CFs

CFs exceed all known composite fibre fills in terms of strength and modulus of elasticity. Accordingly, carbon-filled plastics based on CFs have much better elasticity and strength properties than aluminium or steel. At the same time, the specific weight of CF remains under 2 g/cm^3, which means that constructions can be twice as light as those made from aluminium and five times lighter than steel. Such materials are increasingly applied in aircraft construction and in products where the moment of inertia is crucial (e.g. centrifugal energy storage systems, high-speed centrifuges). Thus, as shown in Tables 2.7 and 2.8, CFs are superior to all other fibres overall.

Table 2.7 Comparative analysis of the properties of carbon fibres and alternative materials

Fibre material	Strength (hPa)	Modulus of elasticity (hPa)	Density (g/cm^3)	Diameter (microns)
Carbon high-strength	3.6–7.2	300	1.8	5–10
Carbon high-modulus	2.5–3.25	500–800	1.8–2.2	5–10
Glass	3.5–4.6	72–110	2.5–2.9	3–25.80
Aluminium oxide	2.2–2.4	385–420	3.95	10–25
Silicon carbide	3.1–4.0	410–450	2.7–3.4	100–140
Boron	3.45	400	2.6	100–200
Basalt	1.8–3	90–115	2.2–2.6	10
Steel	2–3	200	7.8	
Aluminium	0.3–0.6	70–80	2.6–2.7	
Titanium	1–1.2	110	4.5	

Source: HSE

Table 2.8 Comparison of CF and alternative materials

Type	Presence of "nanocomponent"	Maximum operational temperature, °C	Maximum Young's modulus value, hPa	Possible matrix types	Composite areas of application
Carbon fibre	Yes	3000	700	• Polymeric • Metallic • Ceramic • Carbon	1. Aerospace 2. Construction 3. Energy 4. Medicine 5. Sports 6. Oil and gas production 7. Industry
Boron	Yes	400	450	1. Polymeric 2. Metallic (A1)	Aerospace
SiC/W, SiC/C	Yes	800	420	Metallic (Ti)	Aerospace
SiC, fine	Yes	1300	450	1. Metallic 2. Ceramic	Aerospace, automobiles, high-voltage cables
Polycrystalline oxides	Yes	1200	350	1. Metallic 2. Ceramic	Aerospace, automobiles, high-voltage cables, stationary gas turbines
Monocrystalline oxides and oxide eutectics	Yes/No	1600	450	1. Metallic 2. Ceramic	Aerospace, stationary gas turbines

Source: HSE

2.3 CF Composites

CF composites are complex structures created by combining CFs (which are used as reinforcing elements) with a binding matrix. They have high values of three basic properties: strength, rigidity, and low specific weight. CF exceeds all known composite fibre fills in terms of strength and modulus of elasticity. Accordingly, carbon-filled plastics based on them have much better elasticity and strength properties than aluminium or steel. At the same time, the specific weight of CF remains under 2 g/cm^3, which means that CF constructions can be twice as light as aluminium and five times lighter than steel ones.

2.3.1 Types of CF Composites and a Comparative Analysis

CF composites consist of CFs and a binding matrix. The matrix, together with the reinforcing element, is the most important component of composite materials. It is the properties of the matrix which largely determine the resulting composite's properties. Material of the same kind as the reinforcing element is frequently used as the matrix material, which means that composites can be created using the maximum permissible parameters of the reinforcing fibre fill. Currently, ceramic and metallic matrices are widely used. In terms of *binding techniques* for connecting fibres with the matrix, all CF composites can be divided into three types:

- *Solid-phase* binding techniques for connecting fibres with the matrix are based on assembling blank packages, i.e. first combining reinforcing elements and the matrix and subsequently binding the components with each other to make the end product by hot pressing, forging, rolling, diffusion welding, extrusion, etc.
- *Liquid-phase* techniques are based on employing various methods of moulding melted metal matrices to impregnate the preliminary arranged system of fibres (in vacuum, under normal or increased pressure). An advantage of this technique is that it enables complex-configuration products to be made with minimum (or no) subsequent processing and with limited impact on fragile components.
- Creation of a metallic matrix using *deposition techniques* amounts to depositing alternating matrix and reinforcing layers on the substrate: metal layers are placed on the fibres using various methods (e.g. gas-phase deposition, chemical, electrolytic deposition, etc.) to fill the interfibrillar space.

Four types of composite materials, believed to have the best application prospects, are described below:

- Polymeric matrix CF composites (coal plastics)
- Carbon-carbonic composites
- Metallic matrix CF composites
- Ceramic matrix CF composites

Table 2.9 Main characteristics of carbon-carbonic composites

Property	Value
Density (g/cm^3)	>1.70
Thermal conductivity coefficient (kcal/m·h degree, at 50 °C):	
– by X, У axes – by Z axis	5–10 4–8
Linear thermal expansion coefficient by X, У, Z axes, α·10–6 1/°C in 20–100 °C temperature range	2.5–1.6
Specific heat, kcal/kg degree at 50 °C	0.19
Failure stress under stretching, kg-force/cm^2	
– by Z axis – by X, У axes	>200 >300
Failure stress under pressure, kg-force/cm^2	
– by Z axis – by X, У axes	>1100 >1200
Modulus of elasticity under stretching, E·10–3, kg-force/cm^2	
– by Z axis – by X, У axes	210 400
Modulus of elasticity under pressure, E·10–3, kg-force/cm^2	
– by Z axis – by X, У axes	210 230
Failure stress under shifting, kg-force/cm^2	
– by X, У axes	≥300

Source: HSE

2.3.1.1 Carbon-Carbonic Composite Materials

Carbon-carbonic composite material is a modular carbon-graphite material based on a 3D carbon structure made of CF and a pyrocarbon matrix. The choice of technologies applied to make carbon-carbonic composites is determined by the techniques employed to make the carbon matrix. Currently two techniques are used most frequently: carbonisation of a polymeric matrix of a previously moulded coal plastic blank and deposition of pyrocarbon from the gas phase into CF substrate pores. Frequently, a combination of the above techniques is used to give the composite the required properties. In turn, a pyrocarbon matrix is made using isothermal and non-isothermal (thermal gradient) techniques.

Carbon-carbonic composite materials are produced through a pyrocarbon saturation technological process in thermal gradient gas aggregates. Reinforcing the carbon matrix with CF enables materials with higher thermal characteristics than modular graphite and glass carbon to be produced. Carbon-carbonic composites increase products' strength and durability, reduce the weight of constructions, and do not require additional reinforcement by metal frameworks or similar, i.e. they are monocarbon materials. In addition, this material is heat-resistant: when exposed to high-temperature gas flows, its loss of mass is insignificant, and the product retains its original shape (see Table 2.9).

CF-based composite materials are widely used in nonferrous metallurgy to make fasteners for melting furnaces' sheathing. Due to their high biocompatibility,

Table 2.10 SWOT analysis of carbon-carbonic composites

Strengths	*Weaknesses*
• High elasticity and strength properties • High heat resistance • Chemical stability • Significant reserve of accumulated research from Soviet times; production technologies comparable with the leading international level	• Average thermal oxidation resistance • Labour-intensive, complex technology • Fragility
Opportunities	*Threats*
• Possibility to adjust functional properties by adding metals • Application in aerospace and the automobile industry (brake systems, taking into account the trend to increased speed) and to help components endure increased thermal load	• Highly specialised consumption

Source: HSE

resistance to biological environment's impact, nontoxic nature, and electroconductivity close to human tissues, these materials can be applied in medicine. In mechanical engineering they are used as a nonmetallic, self-lubricating material in heavy-load frictional units' slider bearings and in car brakes. Finally, carbon-carbonic composite materials are used to make tennis rackets.

In the near future, carbon-carbonic composites that are very high density (similar to the theoretical density of graphite) will be in demand. No radically new technologies would be required to meet this demand. The problem could be solved by combining already available production processes and certain other technological approaches (such as applying high pressure, combined matrices, etc.).

A SWOT analysis for the aforementioned fibre type is given in Table 2.10.

2.3.1.2 Metallic Matrix CF Composites

Metallic matrices of fibre composites are *light metals* (with density of 4–5 g/cm^3), such as aluminium, magnesium, and beryllium; *high-temperature metals* such as titanium, nickel, and niobium; and *various alloys*.

Aluminium alloys combining physical, mechanical, and technological properties are widely used as a matrix material. From a technological point of view, aluminium matrices can be divided into several types: deformable, casting, and powder. For magnesium matrices, MA2-1, MA5, and MA8 magnesium alloys are used and certain others. Titanium matrices are well suited for high-temperature work and welding, and they can retain their high strength properties (360–1050 mPa) under increased temperature (300–450 °C) for long periods of time.

Titanium and magnesium matrices remain highly resistant to deformation even under increased temperature, which makes it necessary to apply super plastic deformation modes to make composites with fragile fibres.

A SWOT analysis for the aforementioned fibre type is given in Table 2.11.

2.3.1.3 Ceramic Matrix CF Composites

Ceramic matrix materials' properties include high melting temperature, compression strength which is retained under sufficiently high temperatures, and resistance to

Table 2.11 SWOT analysis of metallic matrix CF composites

Strengths	*Weaknesses*
• High modulus of elasticity • Higher gas resistance compared with coal plastics • Can be used under high temperature	• Low resistance to melting of metal • Low corrosion resistance • Formation of aluminium carbide at the junction of fibres and metal, which destroys the material • Heavy weight compared with other CF composites • Not particularly good radiolocation properties • Specific recycling requirements
Opportunities	*Threats*
• Small-size design coupled with metallic elements (e.g. in truss-type load-bearing designs such as the ones used in space shuttles and Buran spacecraft) • Elements operating for long periods of time under increased temperature (high-pressure compressor blades, turbine blades) require 'metallic behaviour type'	• No consumer demand

Source: HSE

oxidation. New ceramics based on highly refractory thorium, aluminium, beryllium, zirconium, magnesium, and vanadium oxides are widely applied in equipment designed for extreme operational environments.

Along with refractoriness, ceramics have high tensile and impact strength and are resistant to vibration and thermal shock. Ceramic-based composite materials have a high modulus of elasticity combined with low plasticity, which requires an optimal balance of the properties of the matrix and reinforcing materials. In other words, it is important to select reinforcing materials with a higher modulus of elasticity than that of the matrix to make the material sufficiently strong.

A SWOT analysis for the aforementioned fibre type is given in Table 2.12.

2.3.1.4 Polymeric Matrix CF Composites (Coal Plastics)

Polymeric matrix CF composites (coal plastics) are a polymeric composite material made of interlaced CF threads placed into a polymeric resin matrix. Hardened epoxide, polyether and other thermosetting resins, and polymeric thermoplastic materials are used as matrices. The main advantages of polymeric matrix composites include a high specific strength and modulus of elasticity, resistance to aggressive chemical environments, low thermal and electrical conductivity, etc. In the course of making such materials, it is relatively easy to bind the reinforcing elements with the matrix under moderate temperature and pressure. Both traditional (such as moulding, contact vacuum, and autoclave forming) and specialised production processes (e.g. winding, pultrusion, etc.) are used when the material and the product are made at the same time.

A SWOT analysis for the aforementioned fibre type is given in Table 2.13.

Table 2.12 SWOT analysis of ceramic matrix CF composites

Strengths	*Weaknesses*
• Average mechanical properties • Chemical stability • Thermal stability • Increased strength	• Low resistance to thermal oxidation • Carbon fibre must be specially treated for the use with such matrixes (e.g. borated) • Fragility
Opportunities	*Threats*
• Application in components operating under high temperatures (internal combustion engines, rockets, etc.) • Possibility to make nonconducting materials conductive	• Highly specialised application areas

Source: HSE

Table 2.13 SWOT analysis of polymeric matrix CF composites

Strengths	*Weaknesses*
• High elasticity and strength properties under stretching • Low production costs • Relatively simple production technology • Thermal stability • Corrosion-resistant	• Average compression strength • Need to have specific ratio of fibres to achieve required technical properties • Vulnerable to 'precision' strikes (e.g. a carbon car bonnet can turn into a sieve if frequently hit by small pebbles) • Hard to restore the original look (unlike metal or glass fibre components); even after small damage, the whole component must be replaced • Carbon components are prone to fading under sunlight • Problems with recycling and reuse
Opportunities	*Threats*
• Possibility to adjust functionality by adding small amounts of carbon materials • Wide application in aviation and consumer goods	• Lagging far behind international producers

Source: HSE

Coal plastics are much superior to regular construction materials (aluminium, steel, etc.) in strength, modulus of elasticity, mass, etc. Coal plastics also display chemical and thermal stability, are corrosion-resistant, environmentally neutral, and do not require expensive utilisation.

Coal plastics' advantage over glass-fibre plastics is their low density and higher modulus of elasticity; coal plastics are both very light and strong materials. CFs and coal plastics have a practically zero linear expansion coefficient.

Black coal plastics are good electrical conductors, which somewhat limit their application sphere. Coal plastics are used in aircraft construction, rocket building, mechanical engineering, spacecraft and medical equipment, prostheses, in light bicycles, and other sporting goods.

Currently technologies for the production of carbon-carbonic and polymeric matrix CF composites are the most developed. Russia has sufficiently high

Table 2.14 Carbon fibre composites' characteristics

	Carbon fibre composites			
	Polymeric matrix (coal plastics)	Carbon-carbonic	Metal matrix	Ceramic matrix
Strength	0.9–3.5 hPa	■■□	■□□	■■□
Modulus of elasticity	■■■	■■□	■■□	■□□
Technological development	■■□	■■■	■□□	□□□
Simple technology	■■■	■□□	□□□	□□□
Chemical and thermal stability	■□□	■■□	■□□	■■□
Market volume (demand)	■■■	■■□	■□□	■□□

Source: HSE

technological potential for developing and producing carbon-carbonic composites and to make products with certain properties above average international values. At present these expensive materials—which rely on complex technologies for production—are only applied in defence, aerospace, and nuclear power industries. Regarding coal plastics, Russian production potential is significantly below international levels. However, the possibility to achieve high elasticity and strength characteristics as well as relatively simple production technologies combined with comparatively low costs suggests that this market should be promoted for application in numerous industries such as construction, energy, automobile industry, shipbuilding, sporting goods manufacturing, etc. (Table 2.14).

Coal fibre composites based on ceramic and metallic matrices have not yet been widely applied due to complex production processes and unstable characteristics (caused by low corrosion resistance and high fragility of ceramic matrices). Coal aluminium (a metallic matrix coal fibre composite) was developed as early as the beginning of the Soviet period. However, it was never used on an industrial scale because of insufficient research.

Thus coal plastics (polymeric matrix composites) have the biggest market potential due to their high physical and chemical properties as well as simple production technology. Carbon-carbonic composites could fill the specialised applications niche, especially in areas which already have the necessary technological potential (e.g. brake disks). Metal matrix-based composites could be used in aircraft construction, while ceramic matrix composites could be applied in high-temperature environments.

2.3.2 Competing Technologies for CF Composites

Along with CF composites, other materials can be used to accomplish similar objectives (see Table 2.15). Developing a strategy for new materials, one should take into account the specific features of competing technologies. Combining

Table 2.15 Main properties of coal plastics and alternative products

Properties	Coal plastic		Organic plastics	Glass-fibre plastics
	Ribbon	Bundle		
Breaking point, mPa	900	700	1600	600
Modulus of elasticity, hPa	150	130	80	50
Density, g/cm^3	1.5	1.53	1.35	2.2

Source: HSE

various kinds of composites would enable hybrid structures to be made, featuring all their advantages.

Currently scientists identify several materials which compete with CFs in various areas: glass, boron, organic, basalt plastics, composite-based silicon carbide fibres, powder-filled polymers, and textolites.

2.3.2.1 Glass-Fibre Plastics

Glass-fibre plastics are polymeric composite materials in which thermosetting synthetic resins are usually used as a matrix (e.g. phenolic, phenol-formaldehyde, epoxy, polyether, polyimide, organosilicon, and thermoplastic polymers such as polyamides, aliphatic polyamides, polyethylene, polystyrene, etc.). For reinforcement, glass fibres are used in the form of threads, bundles, glass fabrics, and cut fibres formed out of melted inorganic glass. The filler's characteristics largely determine the mechanical properties of glass-fibre plastics. For example, composites containing particularly positioned unbroken fibres, unidirectional and crossed, have the highest strength and rigidity.

The main technical characteristics which determine glass-fibre plastics' advantages are their high strength; relatively low density and thermal conductivity; good electrical insulation properties; transparency to radio waves; and resistance to aggressive environments, specifically water resistance and chemical stability.

Active application of glass-fibre plastics began in the mid-twentieth century, when this composite material began to be used to make 'blisters' (dome-like constructions housing locator antennas). It should be noted that initially, plastic fibre was introduced in reinforced glass fibre to neutralise the faults of fragile matrices, with a small number of fibres. With time the content of fibre in glass-fibre plastics has increased to 80% of the total composite's mass. The matrix's purpose also changed: it became simply a binding structure connecting strong fibres to each other.

Due to their technical properties and relatively low production costs, glass-fibre plastics are currently used quite widely in such areas as construction, shipbuilding, radio electronics, consumer goods production, sport goods, window frames, etc. Glass-fibre plastics are also used as construction and heat insulation material to make boats, launches, rocket engines, car bodies, cisterns, helicopter blades, and corrosion-resistant equipment and to build pipelines.

2.3.2.2 Boron-Filled Plastics

Boron-filled plastics are composite materials in which thermosetting resins are used as a matrix: epoxy, polyimide, and certain other polymers. The matrix's heat resistance defines boron plastics' thermal properties, which is reflected in the material's relatively low operational temperatures. As a filler, these composites contain boron fibres in the form of monothreads, bundles, and ribbons in which boron threads are interlaced with other threads. Mechanical properties of the resulting composites are defined by the filler's characteristics.

Boron fibres have high mechanical properties, including high fatigue strength, hardness, and the highest compression strength among all other fibres. Materials made of boron fibres are resistant to aggressive environments. A serious drawback is their fragility, which complicates the processing of boron plastic products and limits the range of their shapes. Moreover, due to certain specific features of boron fibres production technology (deposition of boron from chloride onto expensive tungsten substrate), their production costs may be very high at about USD400 per kilo.

Due to high production costs, boron plastics are not widely used. They are mostly applied in aircraft and spacecraft equipment to reduce the mass of high-load parts operating in aggressive environments (Rutkovskaya and Koledayev 2010).

2.3.2.3 Organoplastics

Organoplastics are composite materials in which thermosetting resins are used as a matrix, such as epoxy, polyether, phenolic, and polyimides. Synthetic fibres are used as filler (which accounts for 40–70% of such composites' mass) in the form of threads, bundles, fabrics, felt, paper, etc. Synthetic organic fibres are most commonly used, especially aramid ones. Natural or artificial fibres are applied more rarely. The mechanical properties of the resulting organoplastics are determined by the filler's characteristics. This means that the degree of the filler's macromolecular orientation plays a major role in improving them. For example, macromolecules of poly-paraphenylene terephthalamide (Kevlar) are mostly oriented towards the fabric's axis, which ensures a high tensile strength of the fibres.

Generally, organoplastics have a number of advantages over other materials; they are lighter than glass fibre and coal plastics due to their low density and have sufficiently high tensile density; high fatigue strength; and high resistance to impact and dynamic load. Furthermore, aramid fibre-based organoplastics can operate for a long time under increased temperature (about 180–200 °C). Their major weakness is low compression and bending strength.

Organoplastics are widely used in many industries such as aerospace, automobile, shipbuilding, mechanical engineering, chemical, radio electronics, sports goods manufacturing, etc. Kevlar-based materials are used to make bulletproof armour.

2.3.2.4 Basalt Plastics

Basalt plastics are composite materials in which basalt threads are used as a reinforcing element, while epoxy, epoxy-phenolic, phenol-formaldehyde, and polyamide resins are used as a binder. In terms of their basic technical characteristics, basalt plastics are certainly not inferior to glass fibre ones. They even surpass them by such parameters as modulus of elasticity, impact strength, and resistance to aggressive environments. One of the main advantages of basalt plastics is their high strength. On this criterion, basalt fibres are close to carbon fibres, so basalt plastic products are three times stronger and four times lighter than steel ones. In addition, they are highly durable and can work for 100–200 years.

Another obvious advantage of basalt plastics is their high dielectric properties and stability of their main qualitative characteristics during prolonged exploitation. Application of basalt plastics widens the range of special properties composite materials which enable them to operate under increased temperature and humidity.

Basalt fibres used to produce basalt plastics are made from the natural material basalt (which makes up 30% of the Earth's crust). According to experts, basalt plastics production has a short technological chain. Thus, due to their properties, basalt plastics can successfully compete with metal and glass-fibre plastic products, surpassing them by corrosion resistance including alkali resistance. Basalt plastics' main drawback is their relatively low compression and bending strength.

The areas of application of basalt plastics include electrical and radio engineering, production of acoustic materials, communication systems, mechanical engineering, industrial and civilian construction, and road building. Basalt plastics can be used to make road slabs, boat hulls, basalt plastic reinforcement rods, fencing panels for buildings, hot and cold water pipes, hydraulic facilities' coating, etc. In asbestos cement and concrete products, it is efficient to replace asbestos and metal components with basalt plastic reinforcing elements. Several types of basalt plastic construction reinforcement rods are also available, offering a number of advantages when building bridges of complex design, various other structures, dams, and subterranean installations.

The application of rovings made of integrated glass-basalt threads means that basalt plastic reinforcement elements with a filling degree of up to 80 weight percent can be made; basalt textolites based on basalt fabric have a filling degree of up to 70 weight percent; hardened basalt laminates based on basalt paper are also produced (Table 2.16). One kilo of basalt plastic reinforcing elements saves 9 kg of reinforcing steel.

Table 2.16 Characteristics of basalt plastics products

Property	Basalt plastic reinforcement elements	Basalt textolites	Hardened basalt laminates
Breaking point, mPa			
Stretching	1080–1380	140–220	
Bending	670–780	120–200	
Compression	460–490	–	
Modulus of elasticity under stretching, mPa	89,000–93,000	39,000–40,000	12,000–14,000
Water absorption in 24 h	0.01	0.02	0.02

Source: HSE

2.3.2.5 Powder-Filled Polymers

Adding various fillers significantly improves composites' mechanical properties (e.g. strength, elasticity, heat resistance, etc.) and adds the required visual features. Moreover, fillers are applied to reduce production costs. For example, calcium carbonate and kaolin (or china clay) are cheap fillers which have practically unlimited reserves. At the same time, their white colour is optimal for producing paintable materials.

Filled polymers were first produced in the USA where Dr. Leo H. Baekeland invented a way to synthesise phenol-formaldehyde resin (Bakelite), to which the scientist added wood flour to make it stronger. Bakelite-based powder-filled composite production technology amounts to an irreversible hardening, in a special form, of a mixture of partially hardened polymer and filler under pressure. Filled thermoreactive polymers are now widely used: there are tens of thousands of filled polymer brands available on the market.

Powder-filled polymers are used to make polyvinylchloride materials for production of sufficiently strong and elastic pipes, facing tiles, electric insulation materials, and polyether glass-fibre plastics (AIMS 2008). Adding talc to certain polymers such as polypropylene significantly improves their properties and increases their modulus of elasticity and heat resistance. Soot is mostly used as rubber filler but is also added to such polymers as polystyrene, polyethylene, and polypropylene. In addition, organic fillers such as wood flour, ground nutshells, and various plant and synthetic fibres are widely used.

2.3.2.6 Textolites

Textolites are laminated plastics based on various thermoreactive and thermoplastic polymers and thermoreactive resins, such as phenol-formaldehyde, epoxy, polyimide, organosilicon, etc. Sometimes inorganic binders are used, based on nonferrous metal silicates and phosphates. Fabrics made of various fibres are used as fillers—cotton, glass, asbestos, carbon, basalt, etc. Textolite production involves impregnating the fabric with resin and then pressing it under high temperature. Despite the fact that this technology was developed in the 1920s, the same principles are applied now. Currently, textolites are used to make not just slabs but also complex-shaped products such as bushings, rods, etc. Textolites have properties such as high strength, high impact resistance, and a broad operating temperature range (from −65 to 105 °C). Textolites are applied as electrical and heat insulation and heat-resistant materials. One of the first applications of textolites was to produce kitchen table surfaces. Due to their hardness, textolites are also used as a construction material.

2.3.3 Advantages of CF Composites

Comparing various fibre-matrix combinations (Table 2.17), it should be noted that coal plastics (CFs) have the best balance and offer the best physical and chemical properties—a high modulus of elasticity and strength.

Table 2.17 Comparative analysis of composites based on various fibres and matrices

Matrix type	Carbon fibre		Polymeric fibre		Ceramic fibre	
	Advantages	Disadvantages	Advantages	Disadvantages	Advantages	Disadvantages
Polymeric	High elasticity and strength properties under stretching	Average compression strength	High tensile strength	Low compression strength	High compression strength	Average tensile strength
Metallic	High modulus of elasticity value	Low resistance to metal melting, low corrosion resistance	–	Fibres unstable during composite production	High resistance to metal melting	–
Carbon	High elasticity and strength properties	Low resistance to thermal oxidation	–	Fibres unstable during composite production	High mechanical and thermal oxidation properties	–
Ceramic	Average mechanical properties	Low resistance to thermal oxidation	–	Fibres unstable during composite production	High mechanical and thermal oxidation properties	–

Source: HSE

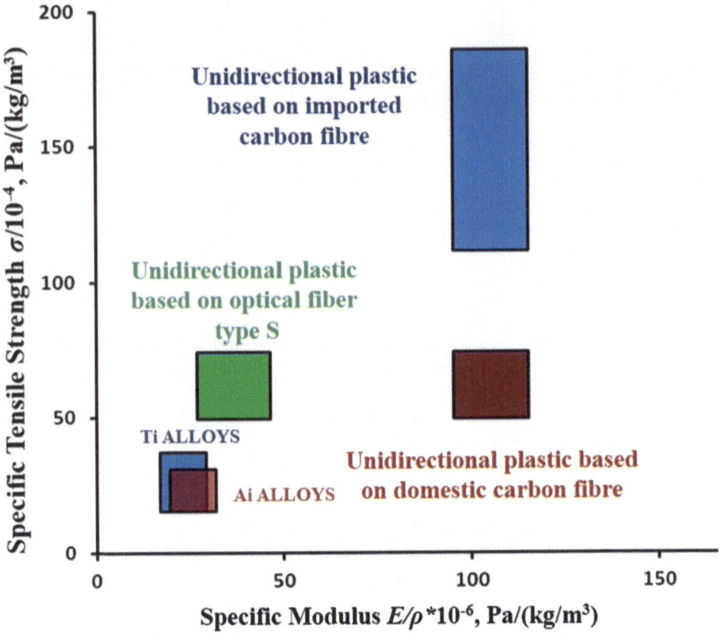

Fig. 2.1 Properties of coal plastics and alternative products. Source: HSE

Table 2.18 Main properties of coal plastics and alternative products

Properties	Coal plastics		Organic plastics	Glass-fibre plastics
	Ribbon	Bundle		
Breaking point, mPa	900	700	1600	600
Modulus of elasticity, hPa	150	130	80	50
Density, g/cm^3	1.5	1.53	1.35	2.2

Source: HSE

Figure 2.1 compares the characteristics of coal plastics and alternative materials: Russian- and foreign-made glass-fibre plastics, titanium, and aluminium alloys.

Thus coal plastics are superior to glass-fibre plastics in terms of modulus of elasticity, but organic plastics typically have higher tensile strength and lower density. Nevertheless, organoplastics' compression strength is 5–10 times lower than their tensile strength, and their production costs remain quite high (Table 2.18).

2.4 Demand for CF Products

Demand for CF materials has grown strongly across the world in the last few decades due to their increased application in special-purpose product manufacturing for nuclear power, aerospace, and other industries and for consumer goods production.

As to global CF production, its volume is expected to reach about 50,000 tons a year (Ponomarev 2012). Below, some of the key industrial markets for CF products are presented.

2.4.1 Markets for CFs

Most of CF demand is expected to come from seven large industries: aerospace, construction, energy, manufacturing, sports and recreation, medicine, and oil and gas production and transportation. Table 2.19 shows products to be demanded by each of these industries.

CF-based composite materials are increasingly applied in aircraft and products for which the inertia moment is crucial (such as centrifugal energy storage systems and high-speed centrifuges). CFs can also be usefully applied in deepwater drilling installations, for shelf development, and to ensure a presence in strategic areas such as the Arctic. It is becoming obvious that the application of CF in manufacturing should be extended to make equipment with advanced operational parameters in terms of their average price, strength, and elasticity. Figure 2.2 illustrates each of these parameters with future forecasts and their particular relevance to the aforementioned industries.

These parameters are keys for the widespread use of CF composite materials—e.g. in the automobile industry (in particular to reduce cars' weight) and in shipbuilding (primarily, to make hull skins). CFs can also be successfully applied in medical products (such as medical tissues, wheelchairs) and sports and recreation goods (Dislife 2008).

2.4.2 Market Pricing of CFs and Composites

The leading countries, primarily the USA, are actively working on reducing CF's costs. Several years ago, Russia also began to realise the importance of this issue, but low current demand from civilian industries does not contribute to dynamic development in this sphere. Only special application industries (such as space, nuclear energy, etc.) can afford to buy expensive CF. In addition to scaling, the following precursor innovations to reduce production costs are proposed: adopting textile PAN fibre as a precursor and using lignin as raw material to make CF fibre and polyolefins. Other technological innovations suggested include advanced stabilisation, plasma oxidation, MAP carbonization, and advanced surface treatment.

The current misbalance (demand for CF exceeding supply) contributes to the growth of prices. For example, the price of a kilo of 24K CF in 2010–2011 grew from 24 to 30 euros (ROSSTS 2011). Future international CF prices should be comparable with prices for the materials they replace (such as steel). Currently, the average world price for CF is about USD37.5 per kilo (other estimates say up to USD50 per kilo). In the future, these prices are expected to decrease. Based on

Table 2.19 Industrial demand for carbon fibre products

Aerospace industry	
Products	
• Airplane, helicopter, and rocket engine components • Airplane, helicopter, and glider structural components	• Stealth-type craft skin • Space-based antennas' bodies • Landing modules' skins
Demand drivers	Barriers
• Improved mass/dimension product characteristics • Improved competitiveness of the industry • Long-term agreements • Increased share of composite materials	• Low productivity (long production cycle, long periods required for R&D) • High costs and long time required for testing of materials • Hard-to-process raw materials • High costs of carbon fibre
Advantages of alternative products	
Glass-fibre plastics	Organoplastics
• Relatively low cost (300–500 roubles per kilo) • Transparent to radio waves	• Lower density • Can operate for a long time under increased temperatures
Silicon carbide fibre-based composites	Steel
• High shock strength, bending strength, and tensile strength • Durability	• Relatively low costs • Repairability • Perfected technologies
Titanium	Aluminium
• Good price/strength ratio • Highly corrosion-resistant	• Low costs • Perfected technologies
Trends	
Increased share of composite materials applied to improve products' operational characteristics (in Boeing and Airbus aircrafts over 50% of the mass is composites)	
Strategy	
• Implementation of the full cycle 'R&D—pilot production—mass production' in Russia • Procurement of cutting-edge technologies and equipment	
Construction	
Products	
• Bridge structures • Reinforced covers and elements for high-rise, seismic, and coastal construction • Light mobile shelter supports	• Cases and components of chemically stable equipment, pipelines, and fixtures • Concrete reinforcements • Elements for recovering concrete products
Demand drivers	Barriers
• Number and state of repair of bridges • Increased production of new materials for the industry • Reduced time and costs	• Lack of construction standards and rules • High costs of materials • Insufficient availability of information about materials' testing
Advantages of alternative products	
Glass-fibre plastics	Organoplastics
• Relatively low cost (300–500 roubles per kilo)	• Relatively low cost • Suitable for application under increased temperature and humidity

(continued)

Table 2.19 (continued)

Trends	
• Increased share of composite materials applied to improve products' operational characteristics • Poor state of bridges (in Russia over 20% of bridges have structural defects)	
Strategy	
• Implementation of the full cycle 'R&D—pilot production—mass production' in Russia • Development of relevant regulations (construction standards and rules, etc.)	
Energy	
Products	
• Gyroscopes as energy storage systems • Latticed power line girders • Carrier high-voltage cable cores	• High-speed rotors • Next-generation gas centrifuges • Wind power plants' blades
Demand drivers	Barriers
• Increased share of applied composite materials • Energy efficiency as a national priority • Next-generation gas centrifuges for uranium enrichment • Growth of global wind power market • Demand for open sea wind turbines, lightweight, water-resistant, and high-strength	• High CF costs • High risks associated with nuclear power generation • Vague prospects of wind power generation in Russia
Advantages of alternative products	
Glass-fibre plastics	Silicon carbide fibre-based composites
• Relatively low cost (300–500 roubles per kilo)	• High shock strength, bending strength, and tensile strength • High durability
Steel	Aluminium
• Relatively low cost • Repairability • Perfected technologies	• Low cost • Perfected technologies
Trends	
• Growth of the global wind energy market (CF in blades); bigger wind turbine blades • Growth of the nuclear energy market (CF in uranium enrichment centrifuges)	
Strategy	
• Implementation of the full cycle 'R&D—pilot production—mass production' in Russia • Promoting demand for CF from manufacturers of end products which may contain CF • Expansion of the global wind energy market	
Oil and gas production and transportation	
Products	
Demand drivers	Barriers
• Presence in strategic areas (the Arctic) • Increased share of applied composite materials • Long-term agreements	• High costs of CF • Investment risks • Limited number of consumers—companies using relevant technologies
Advantages of alternative products	
Basalt plastics	
• Relatively low cost	• Resistant to increased temperatures and humidity

(continued)

Table 2.19 (continued)

Trends	
• Increased depth drilling installations are required to reach (over 2.4 km) • Increased importance of being present in strategic regions (the Arctic shelf)	
Strategy	
• Active long-term government support due to the segment's moderate appeal to private investors • Development of relevant regulations (including national standards, specifications for end products containing CF)	
Medicine	
Products	
• Wheelchairs • Artificial prostheses	• Medical tissues • Dialysers
Demand drivers	Barriers
• Development of new materials sector in medicine • High share of applied innovations	• High costs of CF
Advantages of alternative products	
Titanium	
• Good price/strength ratio	• High corrosion resistance
Trends	
• Increased application of new materials (CF for prostheses, medical tissues, and wheelchairs)	
Strategy	
• Supporting Russian CF producers' participation in the final links of technological chains, jointly with leading companies and countries • Investing in highly market-ready products	

Source: HSE

production volume, relevant prices, and the actual growth rate, three price growth forecasts can be developed for the global CF market (Fig. 2.3).

According to forecasts drawn up using calculations, under the most optimistic scenario (discovery of a new raw materials source or development of new production technology), by 2020 the price of CF will be around USD15–20 per kg. It is widely believed that the industry has already reached the maximum decrease in production costs. A kilo of raw material (e.g. PAN fibre) currently costs at least USD6, and its price cannot be expected to drop further. However, hopes associated with scientific progress indicate that prices may fall to USD20 (or even 17–18) per kilo of CF.

Based on the above production volume and prices and on the actual growth rates, three forecasts can be developed for global CF production (see Fig. 2.4).

According to forecasts, a radically different optimistic scenario is possible. It may be the result of national policy (e.g. introducing limitations for harmful emissions) or global need (in particular, development of sea-shelf oil fields). It is difficult to forecast the effects of such factors at this stage.

The Russian CF market is relatively immature, which results in a higher growth rate and more diverse scenarios (Fig. 2.5). The actual volume of the Russian carbon fibre market in 2011 equalled to USD10–15 million (Gavriliuk 2011).

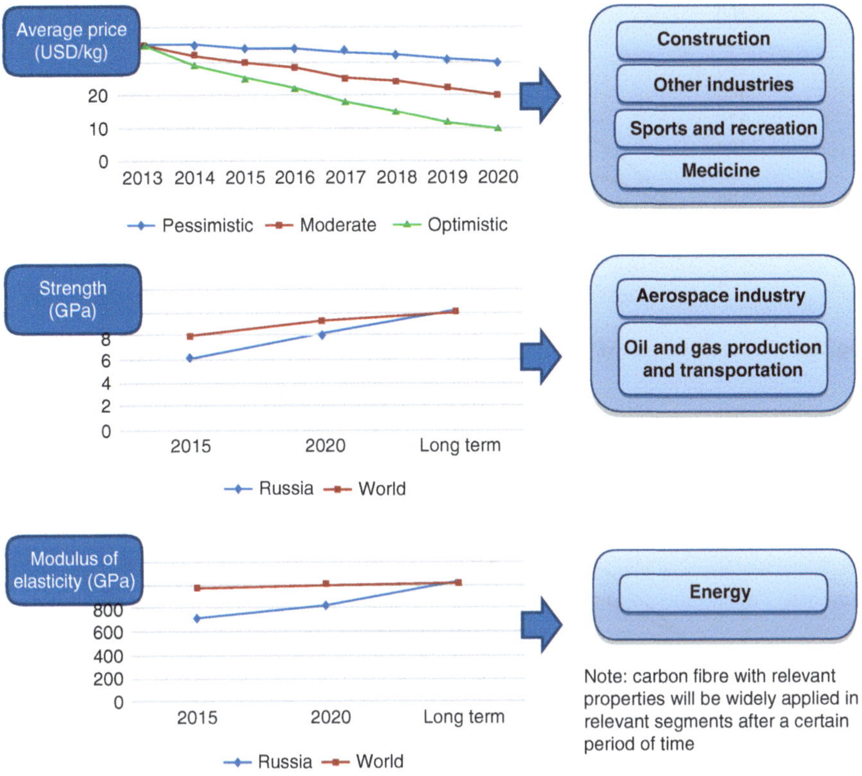

Fig. 2.2 Target CF properties for various segments. Source: HSE

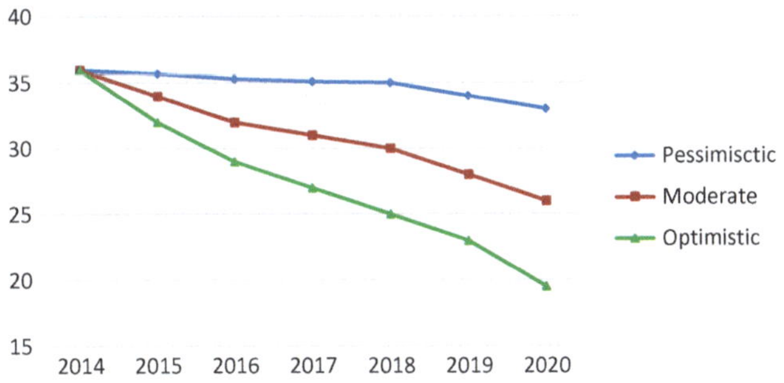

Fig. 2.3 Actual and expected CF prices (USD per kilo). Source: HSE

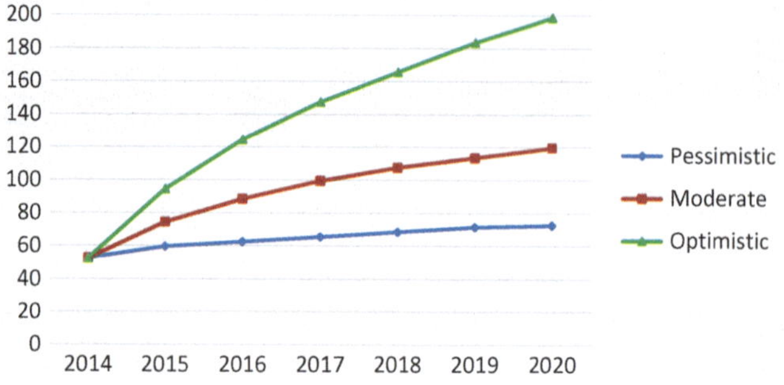

Fig. 2.4 Actual and expected volume of the global CF market (million USD). Source: HSE

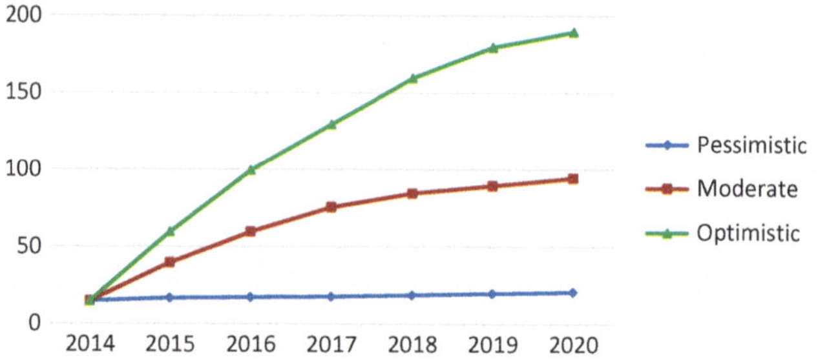

Fig. 2.5 Actual and expected volume of Russia's CF market (million USD). Source: HSE

2.5 Future Prospects and Scenarios for CF Markets

Currently, the production of CF and derived composites is limited to a small number of Japanese, American, South Korean, Taiwanese, and Chinese high-tech companies. Leading producers of technological chain include the full production cycle from PAN fibre to composite materials. Currently, many products used in almost all areas of human activities are made from—or contain—CF. Below, two sets of scenarios are presented based on the forecasts of the global and Russian markets. Each set considers optimistic, moderate, and pessimistic scenarios.

2.5.1 Global Market Scenarios

Three possible scenarios for the global market are presented in Fig. 2.6.

The **optimistic scenario** assumes that the global economic recession will manifest in a decline of the CF market's growth rate to 10% (or perhaps remaining at the current 15%). Subsequently, when global economic growth resumes, the market's actual growth rate will return to 20–25% a year. However, due to reduced prices, the

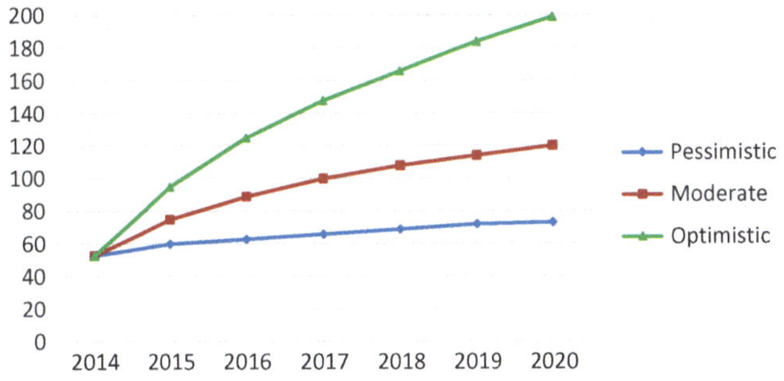

Fig. 2.6 Actual and expected volume of global CF production (thousand tons). Source: HSE

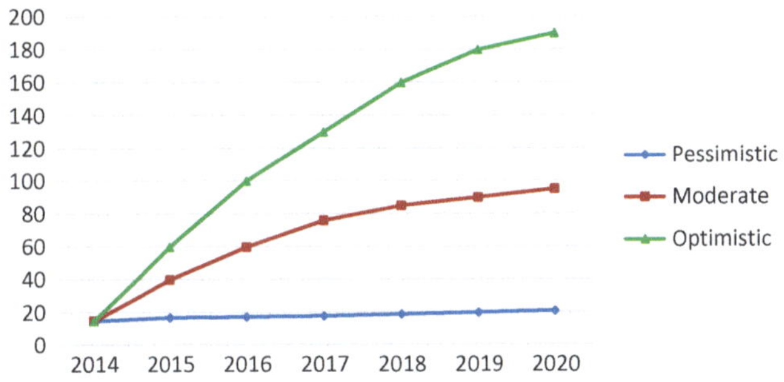

Fig. 2.7 Actual and expected volume of Russian CF production (thousand tons). Source: HSE

growth rate in monetary terms will be lower, 10–15%. The main growth drivers will be (1) increased demand for CF (among other things due to introduction of relevant regulations and standards), (2) development and improvement of technologies, and (3) reduced CF prices.

The **moderate scenario** indicated a decline in growth rates to 5%, followed by a fast growth. On average, the growth rate of 10–15% in 2013–2015 was reflected as a 5–10% in monetary terms (due to reduced prices). Declining rate of growth is expected towards 2020. The main growth drivers include the support of strategic industries, development of innovative industries, and the gradual growth of demand from the defence industry complex.

The **pessimistic scenario** is based on the premise that global economic problems will affect research-intensive industries first. In that case, in the next 2 years, market growth is expected at 2–3%; after that, it would be rather 'inflationary', i.e. 5–6% a year. However, experts estimate that the long-term probability of the occurrence of the pessimistic scenario is relatively low.

2.5.2 Market Scenarios in Russia

In addition to the global scenarios presented above, three scenarios are suggested regarding alternative development trajectories of the Russian market (Fig. 2.7).

The **optimistic scenario** assumes that the structure of CF demand in Russia would match the analogous international scenario, while production would move on to radically new qualitative, quantitative, and price levels. According to this scenario, annual growth in the next few years would be on average 48%. The explosive market growth would increase production volume to 3000 tons of average-quality material by the end of 2017, fully meeting the economy's demand. In 2013–2017, sufficiently moderate growth was predicted; after 2017, it is expected that the global average growth rate will reach about 20–25%, with prospective exports of about 20% of the total production volume. By 2020, Russia's carbon fibre production is expected to reach to 3–5% of the global output. Overall, experts believed this scenario was possible, although most of them still thought it to be overly optimistic. In 2019–2020, the growth rate may slow down due to a possible global economic recession which will affect the Russian economy about 1 year after other countries.

The **moderate scenario** assumes that Russian production would amount to 2.5% of global volume, with exports at about 5–10% of total output. Russia will start producing CFs with average characteristics for civilian applications, while also keeping its positions in high-modulus materials production. While increasing at a growth rate of 35% in 2013–2017, the production levels in the future is expected to remain at the average international level of about 15–25% a year.

The **pessimistic scenario** is based on the premise that Russia does not have resources to produce large volumes of high-quality products. Here, Russia's share of global production output may fall to 0.5% of global production output. Currently, the share of Russian carbon fibre is 0.25% of global output. Accordingly, the export prospects for CF-based innovative products remain vague. Internal demand for such products would be met primarily by imports.

Table 2.20 Development scenarios for Russian market

Optimistic	Moderate	Pessimistic
• 3000 tons of CF in the long term • Good export prospects • Production structure matches the global level • Annual growth rate of 45–50% by 2013 • After 2017, 10–20% growth in monetary terms • 5% of global CF production volume	• 600 tons of CF in the medium term • Good export prospects • Production structure matches the global level • Annual growth rate of 30–35% by 2013 • After 2017, 10–15% growth in monetary terms • 3% of global CF production volume	• Russia lacks the resources to produce large volumes of high-quality products • Slow production volume growth at 3–5% a year • 0.5% of global CF production volume

Source: HSE

The three scenarios for Russia are summarised in Table 2.20.

Thus, the Russian market for CF-based products shows the same trends as the global market, only with a certain lag. As in the world generally, aerospace industry and construction offer the highest market potential. By 2020, other industries of the Russian economy would also have significant demand for CF.

2.6 Technology Strategy for the Exploitation of CFs

The results of the aforementioned study indicate that to achieve a significant increase in CF production and improve its quality, a set of measures must be implemented to solve key technological problems. Efforts should be concentrated primarily on increasing CF's strength, reducing production costs, and improving the quality of CF-based composites. The technological objectives for CF-related production can be divided into three groups:

1. Development of technologies and equipment to produce high-strength CF
   - Development of PAN precursor to produce high-strength CF using wet moulding technique
   - Perfecting 'dry-wet' PAN production technology
   - Development of technologies to reduce PAN and CF' defects and impurities
   - Perfecting technological modes for thermal oxidation and carbonisation of PAN threads and bundles
   - Development of high-performance equipment to produce high-strength CFs in bundle form
2. Development of technologies and equipment to reduce CF's production costs
   - Development of high-performance equipment to produce technological PAN precursor in bundle form
   - Reducing the consumption rate of specific raw materials
   - Development of equipment to produce PAN bundles and CF based on textile PAN bundle
   - Development of technologies and equipment for efficient recycling and utilisation of waste, heat, and other emissions accompanying CF production
   - Development of new precursor compositions, switching to materials with high linear density
3. Development of technologies to improve the quality of CF-based composites
   - Improving coal plastics' structure to increase their strength
   - Development of technologies and launching production of advanced binder materials, including by adding nanoparticles
   - Development of technologies for surface treatment; improving composition of oilers used in CF production
   - Producing high-quality raw materials, primarily PAN (shown by our analyses to be the most important element of CF production's technological chain)

The main CF production technologies are defined by the *raw material* type. Accordingly, three groups of CF production technologies can be identified based on different materials:

- Polyacrylonitrile fibres
- Pitch fibres (oil and coal)
- Rayon (viscose) fibres

All CF production technologies involve in one way or another pyrolysis of raw material. No 'revolutionary' technologies for CF production are expected to emerge in the near future. Technologists' efforts in each of the above groups are concentrated on producing cheap and high-quality CF. Textile sorts of PAN—'thick' bundles—are becoming increasingly important in the production of PAN-based CF. Industrial production of relatively cheap CF based on polyfilament PAN bundles is expected to be launched in Russia in the next few years. Another important objective is to produce high-quality CF with high strength and modulus of elasticity values and relatively high rupture elongation.

The CF industry is currently changing from custom production (e.g. for the aerospace industry) to general mass market-oriented production. The most important aspect is believed to be increasing capacities of individual production lines. Currently Russian production lines' output is about 16–20 tons a year, while the average global figure is 500 tons. Russian producers should aim to achieve 100 tons output by 2020. Moreover, the product range is of crucial importance (threads, bundles). An adequate range would mean market demand can be met. Finally, low-cost CFs are becoming increasingly popular across the world.

However, besides their technological and market value, it should be borne in mind that new materials and nanotechnologies in general, and CFs in particular, have a number of broader impacts on society, economy, environment, politics, values, and culture (Loveridge and Saritas 2009, 2012; Cagnin et al. 2011). For instance, the process of manufacturing sheets of CF and plastic polymer resins is a wasteful process. Stakeholders from scientific and academic communities, policymakers, industry, and society should work together for a more responsible innovation in the domain and achieve increasing recycling and reuse of the material.

Acknowledgements The chapter was prepared within the framework of the Basic Research Program at the National Research University Higher School of Economics (HSE) and supported within the framework of the subsidy by the Russian Academic Excellence Project '5-100'.

References

AIMS – Academy of Industrial Market Situations (2008) Time to make glass-Fiber plastic pipes. http://www.vashdom.ru/articles/akpr_34.htm. Accessed 3 Mar 2017

Amanatidou E, Saritas O, Loveridge D (2016) Strategies for emerging research and innovation futures. Foresight 18(3):253–275

Aydogdu A, Burmaoglu S, Saritas O, Cakir S (2017) A nanotechnology roadmapping study for the Turkish defense industry. Foresight 19(4):354–375. https://doi.org/10.1108/FS-06-2017-0020

Battistella C, Pillon R (2016) Foresight for regional policy: technological and regional fit. Foresight 18(2):93–116

Burmaoglu S, Saritas O (2017) Changing characteristics of warfare and the future of military R&D. Technol Forecast Soc Chang 116:151–161

Cagnin C, Loveridge D, Saritas O (2011) FTA and equity: new approaches to governance. Futures 43(3):279–291

Dislife (2008) From racing bolides to wheelchairs, 20 September. http://www.dislife.ru/flow/theme/608. Accessed 2 Aug 2017

Erdmann L, Schirrmeister E (2016) Constructing transformative scenarios for research and innovation futures. Foresight 18(3):238–252

Gavriliuk A (2011) Development of techniques to produce next-generation CFs from coal isotropic pitch. Kulibin, 26 December. http://kulibin.org/projects/show/2915. Accessed 10 Sept 2012

Kim K-H, Shium W, Moon Y-H, Kwon O-J, Kim K-H, Son J-K (2014) The structure of bio-information-nano technology convergence from firms' perspective. Foresight 16(3):270–288

Loveridge D, Saritas O (2009) Reducing the democratic deficit in institutional foresight programmes: a case for critical systems thinking in nanotechnology. Technol Forecast Soc Chang 76(9):208–1221

Loveridge D, Saritas O (2012) Ignorance and uncertainty: influences on future-oriented technology analysis. Tech Anal Strat Manag 24(8):753–767

Miles I, Saritas O, Sokolov A (2016) Foresight for science, technology and innovation. Springer International, Heidelberg

OECD (2017) The next industrial revolution: implications for governments and business. OECD, Paris. https://doi.org/10.1787/9789264271036-en

OUSD (AT&L) Industrial Policy (2005) Polyacrylonitrile (PAN) CFs industrial capability assessment. DoD – U.S. Department of Defence

Ponomarev V (2012) Stake on composites. http://expert.ru/2012/05/25/stavka-na-kompozit/. Accessed 1 Dec 2016

ROSSTS (2011) Carbon Valley will replace Moskvitch factory, 27 July. http://rossts.ru/index.php/articles/6/4367/. Accessed 10 Sept 2015

Russian Corporation of Nanotechnologies (2010) Technology Roadmap "Nanotechnology Application in Manufacture of CFs and CF Products" Executive Summary. http://en.rusnano.com/upload/OldNews/Files/33822/current.pdf. Accessed 17 Jul 2017

Rutkovskaya MA, Koledayev OV (2010) Model of a spacecraft composite combined antenna. Reshetnev readings. In: Proceeding of the XIV International Scientific Conference, Siberian State University, Krasnoyarsk, 10–12 November 2010

Saritas O (2015) SPIEF2014: Technology, energy and Russia's future development agenda. Foresight 17(3):233–239

Saritas O, Burmaoglu S (2015) The evolution of the use of foresight methods: a scientometric analysis of global FTA research output. Scientometrics 05(1):497–508

Saritas O, Proskuryakova L (2017) Water resources – an analysis of trends, weak signals and wild cards with implications for Russia. Foresight 19(2):152–173

Saritas O, Taymaz E, Turner T (2007) Vision2023: Turkey's national technology foresight program: a contextualist analysis and discussion. Technol Forecast Soc Chang 74(8):1374–1393

Shmatko N (2016) Researchers' competencies in the coming decade: attitudes towards and expectations of the Russian innovation system. Foresight 8(3):340–354

Vaseashta A (2014) Advanced science convergence based methods for surveillance of emerging trends in science, technology, and intelligence. Foresight 16(1):17–36

Vishnevskiy K, Yaroslavtsev A (2017) Russian S&T Foresight: case of nanotechnologies and new materials. Foresight 19(2):198–217

Ozcan Saritas is Professor of Innovation and Strategy at the HSE and editor-in-chief of "Foresight"—the journal of future studies, strategic thinking and policy. He worked as Senior Research Fellow at the Manchester Institute of Innovation Research, The University of Manchester. His research focuses on innovation and policy research with particular emphasis on socio-economic and technological Foresight. With a PhD from the "Foresight and Prospective Studies Program," he introduced the "Systemic Foresight Methodology", and has produced a number of publications on the topic. Dr. Saritas has extensive work experience with the international organizations, including the United Nations, OECD, and the European Commission. He has been involved in large scale national, multinational and corporate research and consultancy projects on sectors including energy, climate change, agriculture, food, water, transportation, and ICT among others; published a number of articles in respected journals; and have delivered keynote speeches in more than 50 countries across the world. Besides his research and publication activities, he designs and delivers academic and executive education courses on Foresight and Strategic Planning. He has recently co-authored a book, entitled "Foresight for Science, Technology and Innovation" published by Springer, which has become one of the key readings in the field.

Alexander Sokolov is Deputy Director of HSE ISSEK and Director of HSE International Foresight Center. His main professional interests are related to Foresight, STI priorities, indicators and policies. Prof. Sokolov is tenure professor at HSE, he teaches Foresight for undergraduate and postgraduate students. He authored over 120 publications in Russia and internationally and managed many Foresight projects, including: Russian S&T Foresight: 2030; Foresight for Russian ICT sector (2012); Innovation Priorities for the Sector of Natural Resources (2008–2010); Russian S&T Delphi Study: 2025 (2007–2008); Russian Critical Technologies (2009) et al. Prof. Sokolov is member of several high-level working groups at OECD and other international organizations, serves for advisory boards at several international conferences and journals.

Konstantin Vishnevskiy is Head of the HSE ISSEK Department for Digital Economy Studies. He holds a PhD from Moscow State University, Faculty of Economics. Dr. Vishnevskiy has long-standing experience in the development of technology roadmaps, the elaboration of Foresight methodology and corporate innovation development programs, the integration Foresight into government policy as well as financial and econometric modeling. He participated in many high-level research projects on S&T Foresight on national and regional level both in Russia and abroad. Dr. Vishnevskiy authors about 70 scientific publications on long-term planning and Foresight, roadmapping, digital economy, macroeconomic regulation and government policy, innovation strategies for businesses. He presented about 100 reports on academic and professional conferences and workshops concerning Foresight, roadmapping and innovations.

Nanotechnology for High-Tech Industries: Light-Emitting Diodes

3

Vitaly Roud, Alexander Sokolov, and Dirk Meissner

3.1 Introduction

Enhancing energy efficiency has been one of the top policy goals in many countries in the last decade. Innovative lighting solutions, and light-emitting diodes (LEDs) in particular, are among the very much promising opportunities to increase energy efficiency. A modern LED is a multilayer thin-film structure with the thickness of layers in the range of nanometers. LED technologies are becoming dominant in a number of application segments. Demand for economic and energy security makes the development of the LED industry one of the national priorities in many countries including Canada, the USA, Japan, and China and European countries, among others. These innovative technologies have attracted the attention of major lighting industrial corporations such as General Electric, Philips, and Osram, among others. The key areas of the LED industry's development envisage designing materials with unprecedented characteristics and using nanoscale components. At the same time, the technology level of semiconductor industries as such to a large extent is determined by chip processing—the key stage of LED production chain.

The following chapter on LEDs is based on a foresight study undertaken by National Research University Higher School of Economics. The study involved multiple expert interviews and related foresight techniques involving technical experts and market experts in the field of lightening technologies.

V. Roud · A. Sokolov · D. Meissner (✉)
National Research University, Higher School of Economics, Moscow, Russia
e-mail: vroud@hse.ru; sokolov@hse.ru; dmeissner@hse.ru

D. Meissner et al. (eds.), *Emerging Technologies for Economic Development*, Science, Technology and Innovation Studies,
https://doi.org/10.1007/978-3-030-04370-4_3

3.1.1 LEDs

An LED is a semiconductor light source emitting incoherent light when voltage is applied to it. LEDs are multilayer thin-film structures where the thickness of layers may potentially be in the range of nanometers. Overall there are two types of LEDs, namely, inorganic and organic LEDs (OLEDs).

An *inorganic LED* is an LED with a structure consisting of inorganic compounds only; it was discovered in 1907 and its commercial use commenced in the mid-1960s. Its competitive advantages are ensured by a number of characteristics attractive for the customers, such as high energy performance in many applications, long lifetime, electrical safety due to low-voltage power supply, small sizes of devices, mechanical resistance, high switching rate, resistance to low temperatures, and absence of hazardous components (such as mercury). An *organic light-emitting diode* (*OLED*) is an LED whose layers include films of organic compounds. OLED was developed in 1950s; its commercial prospects were outlined in early 1990s. Meaningful commercial use of OLEDs began in the 2000s. Their competitive advantages are ensured by the expected decrease in production technology costs and a number of new opportunities for OLED use in a variety of technological processes, i.e., the construction of light-emitting panels, the use of polymer organic materials, development of a technological base for producing flexible light sources and displays, and its use as a component for hybrid and organic electronic devices.

Solid-state light source technologies are currently believed to be drivers for a number of industries. The attractive features of LED technologies include their compact size, high energy efficiency compared to alternative technological solutions, and the opportunity for quick control of light emission. Thanks to these characteristics, LEDs are used in lighting applications as energy-efficient light sources suitable for smart lighting schemes and for information display applications as individual indicators and display panels of both small and large sizes.

One of the most important socioeconomic impacts of the large-scale use of LED technologies is the prospect of a dramatic decrease in electricity consumption for lighting, which amounts, by various estimates, to 18–20% of all produced power consumption.

The future of electronic household appliance markets is also associated with the development of LED technologies, which are capable, in the long run, of surpassing a number of information display technologies. LEDs allow for the development of transparent and flexible display panels, as well as hybrid devices using organic electronic elements.

LEDs compete with other technological solutions such as incandescent light bulbs, low-pressure gas-discharge lamps, high-pressure and high-intensity gas-discharge lamps, as well as non-electric light sources each demonstrating advantages and disadvantages depending on the field of application (Table 3.1).

3.1.1.1 Incandescent Light Bulbs

This traditional lighting technology uses the heating of a wire filament to transform electric energy into light. The largest share of emission is not visible light, but the

Table 3.1 Lighting and illumination technologies

Advantages	Disadvantages	Prospects
Incandescent light bulbs		
Traditional incandescent light bulbs, halogen lamps, halogen lamps with infrared coating, etc.		
• Continuous light spectrum • Low cost • Small size • Simple light fittings • No toxic elements • DC (with any polarity) and AC operation • Manufacture of bulbs with a broad range of voltage ratings (from fractions of volt to hundreds of volts)	• Low luminous efficacy • Short lifetime • Luminous efficacy and lifetime depend on voltage • Color temperature is about 2300–2900 K which gives the light a yellowish tinge	• Engineering solutions allowing for increased energy efficiency of incandescent light bulbs by 40% and lifetime by 100% (IRC-halogen lamps) expected • Technology approaching its limit • Further advancement not anticipated
Low-pressure gas-discharge lamps		
Fluorescent and compact fluorescent lamps		
• Luminous efficacy about 40–60 lm/W may be higher for cold cathode lamps • Longer lifetime for commercial products about 5000–10,000 h • Low average luminance	• Contain mercury and require special recycling • Require time to be switched on and reach full capacity • More complicated dimming • Electromagnetic radiation • Side ultraviolet radiation • Mechanical fragility • Large sizes • Impossible to use at a low temperature • Recycling issues • Luminous flux ripple • Line light spectrum	• Expected luminous efficacy increase by 10–15% and cost reduction within 10 years • Challenges of white light quality and dimming addressed • Mercury content makes technology undesirable for most applications • Technological limit almost achieved
Plasma lamps, high-efficiency plasma (HEP) lamps, other low-pressure lamps		
• High energy efficiency • High luminous flux	• Large size • Require time to reach working mode • Specific light spectrum	• Increase in energy efficiency forecasted • Limited use of lamps possible due to light spectrum
High-pressure and high-intensity gas-discharge (HID) lamps		
Gas-discharge (sodium vapor and metal halide) high-pressure and high-discharge lamps, electric arc-based lamps		
• Highest energy efficiency • High luminous flux	• Low quality of light • Large size • require time to reach working mode • Require time after switch-off to "cool" before another switch-on • Uncertain switch-on at low temperatures	• Limited use due to low quality of light (narrow spectrum, unsuitable color temperature) • Improvement of light quality core of ongoing development work

(continued)

Table 3.1 (continued)

Advantages	Disadvantages	Prospects
	• Recycling issues • Require cooling when used for a long time	
Electroluminescent lamps with luminophors		
• Energy efficiency close to parameters of compact fluorescent lamps • Environmentally friendly • Possibility of dimming; instant switch-on	• Large size • Relatively short lifetime	• Limited spread and case record • Commercial prospects unclear
Electroluminescent wires		
• Flexibility • Low electricity consumption	• Relatively low output • Not suitable for lighting	• Good prospects (and current use) for decoration of buildings and vehicles, advertising
Non-electric light sources		
Fluorescent materials with prolonged afterglow		
• Don't require electricity for glowing	• Low brightness • Limited time of glowing	• Good prospects for road signs and information displays • Longer afterglow time expected
Combustion lamps (kerosene lamps, candles, wick-fed lamp, etc.)		
• Don't require connection to electric network • Technologically simple • Cheap to buy	• "Dirty" • Fire-prone • Service costs (fuel purchase)	• Technological advancement not envisaged
Inorganic light-emitting diodes		
Single LEDs and LED matrixes		
• High luminous efficacy • Long lifetime • Low source voltage • No hazardous elements • Possibility of luminance and color of spectrum control • Smart control • Possibility of operation at low temperatures (as low as tens of degrees K) • Wide range of color temperatures: 2500–10,000 K	• High current cost • Heat control issues • Efficiency drop with increased output and temperature	• Decrease of costs • Luminous efficacy increase • Higher quality of white light • Better integration into existing infrastructure
Organic light-emitting diodes		
Light-emitting OLED panels		
• In future—high luminous efficacy • Possibility to develop flexible and semitransparent light-emitting panels • Expected reduction of production costs • Low average luminance	• Technology under development • Insufficient lifetime (high degradation speed)	• Significant decrease of cost • Longer lifetime • Higher energy efficiency

Source: HSE

Table 3.2 Alterative lighting technologies: incandescent light bulbs

Advantages	Disadvantages	Prospects
Incandescent light bulbs		
Traditional incandescent light bulbs, halogen lamps, halogen lamps with infrared coating, etc.		
• Continuous light spectrum • Low cost • Small size • Simple light fittings • No toxic elements • Possible DC (with any polarity) and AC operation • Possibility to vary input voltage	• Low luminous efficacy • Short lifetime • Luminous efficacy and lifetime depend considerably on voltage • Limited color range 2300–2900 K (yellowish tinge)	• Increasing energy efficiency of incandescent light bulbs by 40% and lifetime by 100% (IRC-halogen lamps) • Technology close to limit • Further advancement not anticipated

Source: HSE

infrared band; thus, the bulb has an extremely low efficiency. For an average temperature of about 2700 °C, the lifetime of the lamp may be up to 1000 h. The key factor limiting the lifetime of an incandescent light bulb is the uneven evaporation of the filament material, resulting in its uneven thinning and subsequent break off. The key positive peculiarities of the technology are the continuous light spectrum, which is close to the solar one; the lack of glimmer that is hazardous for eyes; high switching rate; and allowing for the use of this technology in display and decorative applications (Table 3.2). Simple light fittings, the low cost of the device, and ease of recycling were the reasons for the widespread use of the incandescent light bulbs, which now is causing problems with the inefficient use of electricity on a large scale.

The advancement of this traditional technology resulted in the development of halogen bulbs. The increase in the lifetime and efficiency of the bulbs was achieved through the joint use of inert gas and halogen vapor for filling the bulb. The lifetime of such bulbs is extended up to 2000–3000 h, and the temperature of the wire filament—up to 3000 °C. The efficiency of halogen bulbs is up to 25–28 lm/W. Another area of technological development is the reduction of infrared light and to use it for the additional heating of the filament. Such devices were named IRC-halogen lamps (IRC—infrared coating). In the medium term, it is expected that new engineering solutions would allow increasing their energy efficiency by 40% and lifetime by 100%. However, further technological progress is not envisaged.

3.1.1.2 Low-Pressure Gas-Discharge Lamps (Fluorescent Lamps, Compact Fluorescent Lamps, Plasma Lamps, etc.)

Fluorescent lamps, falling into the category of low-pressure gas-discharge lamps, use the glow of gas-discharged mercury vapor modified into visible light by fluorescence. These sources compare favorably with incandescent light bulbs due to their lower power consumption and longer lifetime (5000–10,000 h) but at the same time are characterized by unsatisfactory quality of white light, which is only partially suitable for human eyes. Other negative factors include glimmer that causes higher eye strain and the complexity and large size of light fittings. The technology has been continuously improved since the 1930s, and in the 1990s a new generation of the devices has gone mainstream—compact fluorescent lamps, which are relatively energy-efficient, compact, and long-lasting light sources. However, a number of issues make this technology solution only temporary. The most important challenge is the presence of mercury vapor and other toxic substances, which causes short-term problems with regard to recycling the waste equipment. Experts note that the improvement of the characteristics of fluorescent lamps due to the advancement of technology in the next 20 years would not exceed 10–15% that, in addition to the aforementioned deficiencies, would urge the exploration of alternative lighting sources.

Apart from fluorescent lamps, this category includes plasma (common and high efficiency) and neon lamps. Being energy-efficient they feature specific spectral composition, which is extremely uncomfortable for human eyes, and extended switch-on time and time to reach working mode. Further research to enhance technology and increase energy efficiency is underway in the sector (Table 3.3).

3.1.1.3 High-Pressure and High-Intensity Gas-Discharge Lamps

Lamps within this category demonstrate high capacity (luminous flux) and energy efficiency, but have a very narrow spectrum; besides, they require time to reach working mode and to "cool" before another switch-on. Such equipment is used in large lighting systems without high requirements for emission parameters (e.g., in the illumination of large service areas, tall buildings, etc.). Technological advances are primarily aimed at improving the quality of light. Due to the wear and tear of such lamps when frequently switched on and off, they are efficient only when switched on for a long time (Table 3.4).

3.1.1.4 Electroluminescent Light Sources

Luminescence is known to be used in a number of ways. The most relevant of them in the context of this paper is LED lighting. Electroluminescent lamps with multi-color luminophors are used today for outdoor advertising and decorative illumination. Their distinct advantages are believed to be energy efficiency close to the parameters of compact fluorescent lamps, environmental friendliness, the possibility of dimming, and instant switch-on. Commercial prospects for this type of lamps are unclear.

Electroluminescent wires may also be used for decoration purposes. They consume very little power and their brightness and luminous flux are low (Table 3.5).

3.1.1.5 Non-electric Light Sources

The most widespread elements in this category are the traditional combustion lamps, generating light on the basis of fuel combustion. They play an important role when there is no access to power networks. However, they are fire-prone and harmful for

Table 3.3 Alterative lighting technologies: low-pressure gas-discharge lamps

Advantages	Disadvantages	Prospects
Low-pressure gas-discharge lamps		
Fluorescent and compact fluorescent lamps		
• Luminous efficacy of about 40–60 lm/W (may be higher for cold cathode lamps) • Longer lifetime for commercial products (about 5000–10,000 h) • Low average luminance	• Contain mercury • Require special recycling • Require time to be switched on and reach full capacity • More complicated dimming • Electromagnetic radiation • Side ultraviolet radiation • Fragility • Large sizes • Impossible use at low temperature • Luminous flux ripple, discrete light spectrum	• Luminous efficacy increase by 10–15% within the next decade • Cost expected to decrease • Challenges of white light quality • Possibilities of dimming addressed • Presence of mercury makes technology undesirable for majority of applications • Technological limit almost reached
Plasma lamps, high-efficiency plasma (HEP) lamps, other low-pressure lamps		
• High energy efficiency • High luminous flux	• Large size • Require time to reach working mode • Specific light spectrum	• Forecasted energy efficiency increase • Limited use of lamps due to light spectrum

Source: HSE

Table 3.4 Alterative lighting technologies: high-pressure and high-intensity gas-discharge lamps

Advantages	Disadvantages	Prospects
High-pressure and high-intensity gas-discharge (HID) lamps		
Gas-discharge (sodium vapor and metal halide) high-pressure and high-discharge lamps, electric arc-based lamps		
• Highest energy efficiency • High luminous flux	• Low light quality • Large size • Require time to reach working mode • Require time after switch-off to "cool" before another switch-on • Problematic switch-on at low temperature • Recycling issues • Require cooling when used for a long time	• Limited use due to low light quality (narrow spectrum, unsuitable color temperature) • Further development for light quality improvement

Source: HSE

Table 3.5 Alterative lighting technologies: electroluminescent light sources

Advantages	Disadvantages	Prospects
Electroluminescent light sources		
Electroluminescent lamps with luminophors		
• Energy efficiency close to parameters of compact fluorescent lamps • Environmentally friendly • Possibility of dimming • Instant switch-on	• Large size • Relatively short lifetime	• Limited spread and case record • Commercial prospects unclear
Electroluminescent wires		
• Flexibility • Low electricity consumption	• Relatively low output • Not suitable for lighting	• Good prospects for decoration of buildings, vehicles, in advertising

Source: HSE

Table 3.6 Alterative lighting technologies: non-electric light sources

Advantages	Disadvantages	Prospects
Non-electric light sources		
Fluorescent materials with prolonged afterglow		
• Don't require energy for glowing	• Low brightness • Limited time of glowing	• Good prospects for road signs and information displays • Development for longer afterglow time
Combustion lamps (kerosene lamps, candles, wick-fed lamp, etc.)		
• Don't require connection to electric network • Technologically simple • Cheap to manufacture/purchase	• "Dirty" • Fire-prone • Service costs (fuel purchases)	• Technological advances not envisaged

Source: HSE

the environment and require spending to purchase fuel. When choosing non-electric sources of light, it is preferable to use fluorescent materials, which accumulate energy of the sun (or other) light and, for an extended period of time, emit a pale glow suitable for household indication and marking (Table 3.6).

LEDs still occupy a rather small market share indicated by the Russian market figures (Fig. 3.1). Compact fluorescent lamps, incandescent, and linear fluorescent lamps remain the dominating light sources in the market with more than two-thirds of all lamps. However, LEDs have a significant potential for lightening applications taking account of the ongoing technological progress for LEDs as products and for the manufacturing process supporting the application diffusion and penetration at broader scale and speed.

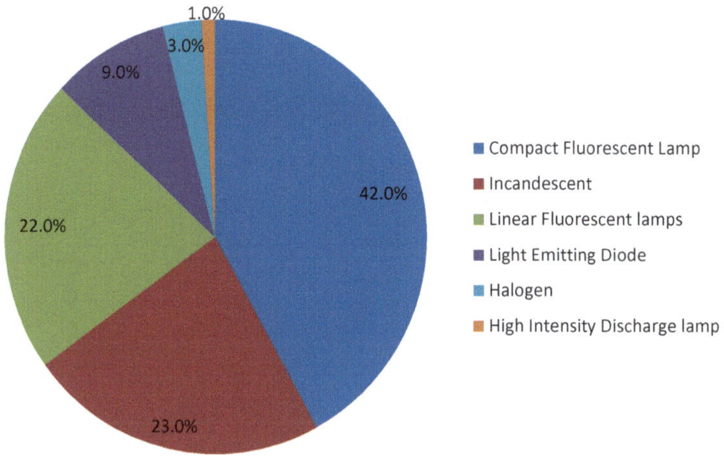

Fig. 3.1 Russian light sources market segmentation by types of lamps (%). Source: European Commission (2014)

3.2 Information Display Technologies

Another application field is the information display technologies, in which LEDs also compete with other solutions, namely, E-ink, EL (electroluminescent) panels, LCD (liquid crystal display) panels, plasma display panels, projection systems, and laser projection with each technology providing specific advantages and disadvantages (Table 3.7).

3.2.1 E-Ink

Technological advancements in this area are aimed at the production of displays with features similar to traditional paper in optical and mechanical parameters. Usually this refers to making static pictures or images with low update rate. The technology is based on microcapsules filled with transparent liquid and pigment particles of various colors and charge. The electric charge directs their movement and shapes an image. The main advantages of such displays are ultralow power consumption (exclusively at the moment of image change) and reduced eye strain for users; major disadvantages are the limited number of displayed colors and long response delay (200–500 ms). This technology is developing rapidly: research is aimed at shortening response delay and wider color range. Ultimately color displays are expected (Table 3.8).

Table 3.7 Information display technologies

Advantages	Disadvantages	Prospects
E-ink		
• Energy efficiency • Lower eye strain	• Limited color range • Limited size • Low frame rate (for existing samples)	• Technology under development • Wider use in mobile devices • Envisaged
EL panels		
• Possibility for developing thin and flexible displays • Potentially low cost	• Low contrast	• Commercial prospects unclear
LCD displays		
• Established production technology • Contrast ratio control (LED-backlit displays) • Wide dynamic range	• Low efficiency • Low quality of picture in open sunlight	• Pronounced tendency to shift to LED-backlit displays • Higher energy efficiency
Plasma display panels		
• Wider color gamut (as compared with LCD—CFFL) • High luminance • Suitable for outdoor use	• Large size • High granularity • High cost • Burnout of display elements • Degradation of luminophors • Leakage electromagnetic radiation • High power consumption	• Substantial technology improvements not expected short term • Application for outdoor information indication
OLED displays		
• Wide color gamut • Potentially high energy efficiency • Potentially low price • Low operating voltage	• Short lifetime • Limited display size • High current cost • Unbalanced degradation of colors	• Midterm (7–8 years) wider use possible provided cost and lifetime challenges solved • Rapid advancement of operating parameters
Matrix LED displays		
• Mass use of large images • Long lifetime • Energy efficiency • High luminance	• Almost no disadvantages • Low resolution	• Wide use of LED displays for alarm and advertising applications
Projection systems		
• Possibility making large-size images	• Low luminance • Limited lifetime • Low contrast ratio • Size	• Smaller sizes • Dissemination of LED technologies as light sources (for mobile applications)

(continued)

Table 3.7 (continued)

Advantages	Disadvantages	Prospects
Laser projectors		
• Wide color gamut • High contrast ratio	• High cost • Granularity of an image • Emission coherence (unsafe, capacity restrictions) • Low definition • Limited use	• Cost reduction • Development of powerful continuous emission sources in green zone of light

Source: HSE

Table 3.8 Alternative information display technologies: E-ink

Advantages	Disadvantages	Prospects
– Energy efficiency – Reduced eye strain	– Limited color gamut – Limited size – High response delay (for existing samples)	The technology is being developed. Wider use in mobile devices is envisaged. The development of color displays is expected

Source: HSE

Table 3.9 Alternative information display technologies: LCD displays

Advantages	Disadvantages	Prospects
– Established production technology – Contrast ratio control (LED-backlit displays) – Wide dynamic range	– Low efficiency – Low quality of picture in open sunlight	– Pronounced tendency to shift to LED-backlit displays – Higher energy efficiency

Source: HSE

3.2.2 Liquid Crystal Displays (LCDs)

This category includes a number of products based on various principles of image generation and backlit mechanisms. A number of basic versions of LCD displays appeared over the last 40 years, including passive- and active-matrix displays, fluorescent lamp-backlit displays, and modern LED-backlit displays. The latest solutions allowed, preserving traditional production technology, for reaching a new level of image quality through higher contrast ratios and considerably reduced power consumption by using high brightness, super-efficient LEDs. The commercial future of display technologies for the next decade is likely to be closely associated with the production of LED-backlit LCD displays (Table 3.9).

Table 3.10 Alternative information display technologies: plasma display panel

Advantages	Disadvantages	Prospects
– Wider color gamut (as compared with LCD—CFFL) – High luminance – Suitable for outdoor use	– Large size – High granularity – High cost – Burnout of display elements – Degradation of luminophors – Leakage electromagnetic radiation – High power consumption	– Substantial technology improvements are not expected in the short term – Key area of application is outdoor information indication

Source: HSE

3.2.3 Plasma Display Panel (Gas-Discharge Screen)

The technology is based on small cells combined in a matrix and filled with gases (typically—xenon or neon) with different glow spectrums, placed between glass plates. When a high voltage is applied across the cell, the resulting plasma emits ultraviolet (UV) light, which is subsequently transformed into visible light by phosphor.

The manufacturing of plasma displays is challenging because it is necessary to use large-size transparent front electrodes. Other barriers in using this product are strict limits on the sizes of the devices—the lower limit of the diagonal of existing panels is about 80 cm. Other disadvantages of the technology noted by experts are required high voltage, side electromagnetic radiation, and extremely high power consumption. It should be noted that plasma technology is one of the few that are suitable for use in large-scale outdoor displays. Experts believe that it is in this segment that plasma displays would be used in future (Table 3.10).

3.2.4 Electroluminescent Panel (EL-Panel or Electroluminescent Paper)

The electroluminescent panel developed by Ceelite (USA) looks like a thin glowing sheet of paper. Separate flexible and thin glowing panels (plastic light panels, glowing plastic) and similar devices are produced using the technology of a light-emitting capacitor (LEC). Polymer film components for stencil screen printing are used to manufacture Ceelite EL-panels.

The range of sizes of EL-panels is wide: from several centimeters up to a meter. These light and flexible devices ensure even glow (illumination); they can be included in any construction, be cut into pieces (it is possible to make glowing letters or shapes), or be used as an element of design for clothes, cars, or rooms. The considerable enhancement of this technology is expected in the future (Table 3.11).

Table 3.11 Alternative information display technologies: electroluminescent panel

Advantages	Disadvantages	Prospects
– Possibility for developing thin and flexible displays – Potentially low cost	– Low contrast ratio	– Considerable technology enhancement is expected

Source: HSE

Table 3.12 Alternative information display technologies: projection systems

Disadvantages	Prospects
– Low luminance – Limited lifetime – Low contrast ratio – Large size	– Smaller sizes – Dissemination of LED technologies as light sources (for mobile applications)

Source: HSE

3.2.5 Projection Systems

Projection systems form an important area of information display technology's development. A great variety of technologies for projecting images exists, e.g., those employing cathode-ray tubes, combinations of lamps and LCD matrixes, and LCD matrixes on silicon base, among others. The optic system focuses the resulting image on a screen, where the dimensions of the image are smaller than, for example, on traditional LCD displays. Major technological restrictions are attributed to use of powerful lamps that require timely replacement and efficient cooling. The current shift to using LED backlit would allow for the reduction of the size and power consumption of these devices as well as extending their lifetime (Table 3.12).

3.2.6 Laser Projectors

These systems produce images using a laser beam, which "draws" each pixel in pulse mode. The employment of lasers allows for simplifying the optic system, which should only control the beam's direction and shape high-contrast pictures at long distances. Disadvantages of these projectors include the high cost of equipment, granularity and the low definition of an image, as well as emission coherence. Prospects for the technology depend upon the development of enhanced lasers (especially in the green zone of light) and the lower cost of the devices (Table 3.13).

Table 3.13 Alternative information display technologies: laser projectors

Disadvantages	Prospects
– High cost – Granularity of an image – Emission coherence (safety hazards, capacity restrictions) – Low definition – Limited use	– Cost reduction – Development of powerful continuous emission sources in the green zone of light

Source: HSE

3.3 Lighting Applications and the World Market

3.3.1 Fields of Application

Lightening applications are demanded by the corporate sector (in particular assembly and tool engineering), municipalities, and households, as well as by companies with dedicated sophisticated and partially unique requirements like railways and aviation companies as well as corporate lighting. LEDs are commonly used in mobile devices, large-size displays, industrial and household electronic devices, signaling devices, vehicles, lighting, and outdoor decorative illumination. LED applications are offered by companies in different industries and markets (Table 3.14).

The segmentation of demand for LEDs by intended use (fields of application) requires special attention. An in-depth analysis of the first-level segments is required for the comparison of LED development potential with the potential of alternative products, with which they "compete" for the field of application, as well as for revealing leaders in LED production. The status of major fields of worldwide application of LEDs in absolute and relative terms is shown, respectively, in Figs. 3.2 and 3.3.

The current market share of lighting (roughly 30%) as of the main field of application for LEDs will change in the future, with the saturation of the market, although the scale of LEDs' use for lighting and in the automotive industry would be relatively stable (Lux Research 2015).

The world market is so far dominated by "standard" LEDs with a typical operating current of 20 mA. However, the segment of devices with power intake of 0.5 W and more has been growing since 2004 by some 50% per year. High-radiance LEDs (with operating current from 20 mA) have the following market shares: white LEDs, 48%; blue and green LEDs based on InGaN structures, 28%; yellow, red, and orange LEDs based on AlGaInP, 17%; and RGB LEDs, 7%. The rate of LED performance enhancement is higher than it was forecasted at the beginning of twenty-first century. Laboratory samples with a luminous flux of 100 lm/W were produced as early as in 2007.

As for UV band LEDs, more than 90% of the market is used in the medical industry and in the production of tools—this sphere requires A-type sources (with a

Table 3.14 LED application fields and suppliers

Application field	Suppliers
1. Mobile devices	– Organizations specializing in the design and development of devices: cell phones, mobile messengers, etc.
	– Organizations using small-size panels to improve the visual appearance of products, i.e., package producers
2. Large-size displays	– Manufacturers of outdoor advertising
	– Companies specializing in the design and development of tools
3. Industrial and household electronic devices	– Manufacturers of various indication devices for industrial and household purposes
	– Manufacturers of household and office equipment (printers, scanner, etc.)
4. Signaling devices, information signs	– Manufacturers of outdoor advertising
	– Highway police – Organizations ensuring traffic lights maintenance
5. Vehicles	– Organizations selling spare car parts
	– Car service enterprises, car dealers, and centers specializing in automotive tuning
	– Car manufacturers and vehicle fleet operators
6. Lighting	– Specialist companies for maintenance of street lighting systems
	– Construction and design organizations
	– Manufacturing enterprises—subdivisions of plants, factories, warehouses, and vehicle fleet operators, responsible for the maintenance of external and internal lighting on the territories of the enterprises
	– Housing and utility infrastructure services—subdivisions responsible for the lighting of yards and area surrounding the building, public, and utility service areas in houses
7. Outdoor decorative illumination	– Specialist companies for installation and maintenance of the municipal illumination system
	– Private companies using illumination of their buildings for brand image promotion
	– Owners of single-family detached homes (for landscape design)

Source: HSE

wavelength of 400–315 nm) and B-type sources (315–280 nm)—as well as for forgery detection. The remaining 10% were used for photocatalytic treatment of water and air, primarily by UV-A LEDs. Today it is one of the key products on the UV LED market (Fig. 3.4). This situation will not change in the short term at least.

Medical equipment is an important and dynamic segment of the UV LED market with a volume of \$120 mln and growth rate of about 10%. Here LEDs may compete with traditional mercury lamps. Beam output power of these LEDs has increased up to several W/cm^2. Many new players have joined this segment over the past few years.

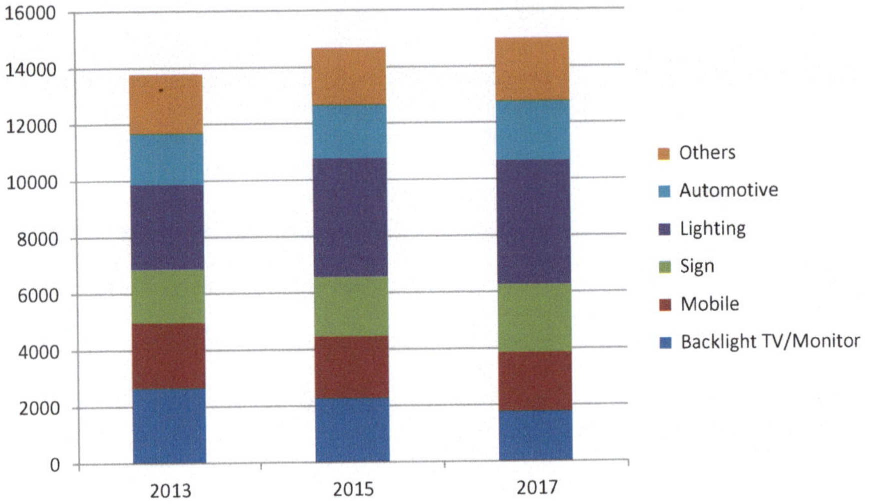

Fig. 3.2 Dynamics of LED use worldwide ($bln). Source: Strategies Unlimited (2013)

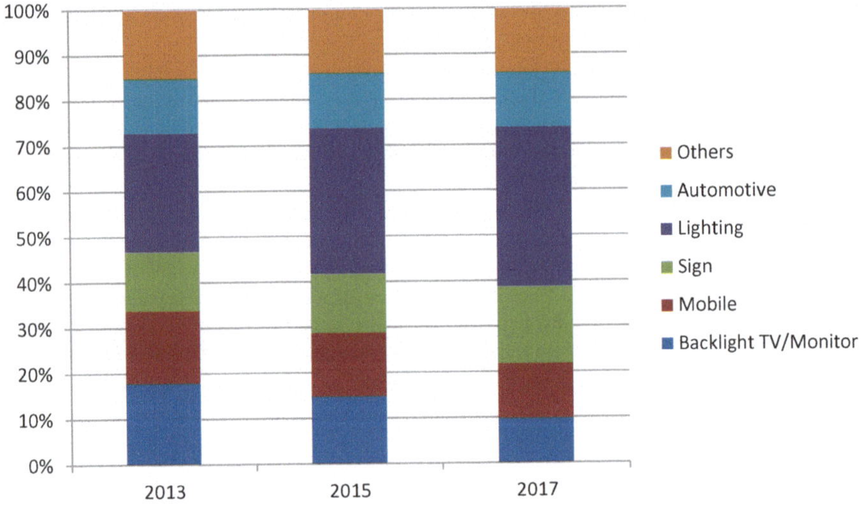

Fig. 3.3 Structure of LED use worldwide (%). Source: Strategies Unlimited (2013)

The introduction of UV-C LEDs (with a wavelength of 280–100 nm) in the cluster of cleaning and disinfection, which was expected by the experts several years ago, has not yet commenced due to a number of technical and economic problems: low output capacity and efficiency, short lifetime, and high costs. They have been primarily used in the sphere of the development of scientific analytical tools. Growth of UV-C LED market is first of all associated with the need for the mass production of AlN substrates that may be the basis for LEDs with a beam output power that

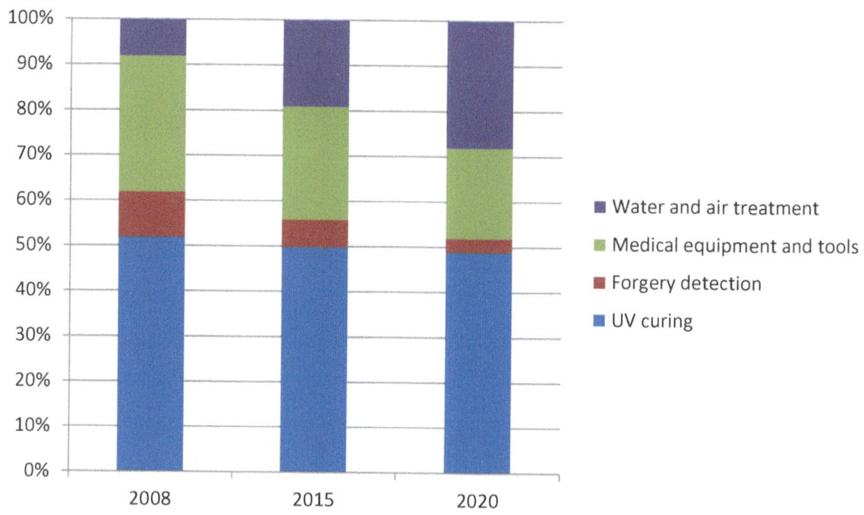

Fig. 3.4 Retrospective structure of UV LED market (%). Source: Yole Development data (2013)

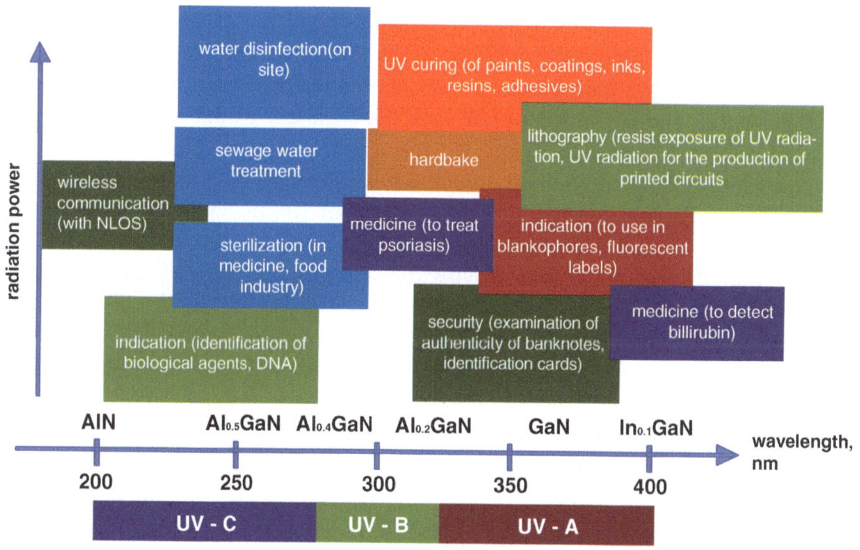

Fig. 3.5 Fields of application of LED sources. Source: HSE

exceeds the current level by 100 times. A certain trend exists for increasing the production of UV LEDs by AlN producers to get additional profit and wider presence in the supply chain. Fields of their application are shown in Fig. 3.5.

The production of diode (semiconductor) lasers is based on the technology and construction, which were invented by Russian scientist Zhores Alferov in 1963. Currently such emitters are used almost in every industry. The most promising area

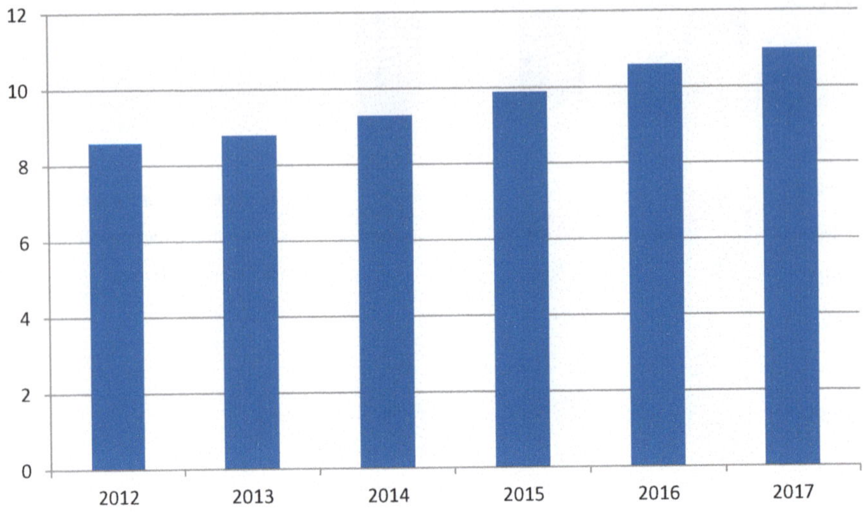

Fig. 3.6 Dynamics of the world laser market ($bln). Source: Strategies Unlimited (2014)

of use is in fiber-optic communications (Great Stone 2013). Visible lasers may be used as a central element of the projection system of a laser display (television).

The market of lasers for projection systems is on the rise, but in absolute terms, it is not that large and is limited by the lack of red lasers of the required capacity (Nogee 2014). Market growth is expected due to microprojectors and pico-projectors using laser LEDs. Intensive market growth is possible in the case of mass production of powerful RGB lasers for home cinema systems of 100-in. and over category and digital cinemas. Currently mass production is impeded by the lack of compact lasers of the required capacity. Vertical cavity surface-emitting lasers (VCSEL) satisfy these requirements (Figs. 3.6, 3.7, and 3.8).

In Russia, this type of laser suitable for commercial use is being developed by Spectralaser (St. Petersburg). A photonic crystal laser (PCL) designed by this company has narrow beam and large aperture. These parameters allow for using direct photon energy conversion of the PCL, increasing energy efficiency and achieving the dramatic simplification and reduction of costs as compared to existing samples developed by Novalux (USA). The fastest demand growth for VCSEL components is expected in HDTV and USB 4.0 segments. The primary advantage of VCSEL as compared to Fabry-Perot and DFB (distributed feedback) lasers is low power consumption. This point requires special attention taking into account the increased interest in data transmission at the bit rate of 40 and 100 Gbit/s using protocols defined in IEEE 802.3ba and the need to solve the complicated technological problem of reducing power dissipation for keeping transceivers' size small. The 10-fold rise of the bit rate (from 10 to 100 Gbit/s) while using widespread types of lasers causes a substantial increase in power consumption and dissipation, which is not consistent with the market requirements. Therefore, long-wavelength VCSELs have become an attractive alternative to conventional lasers in the area of high-speed

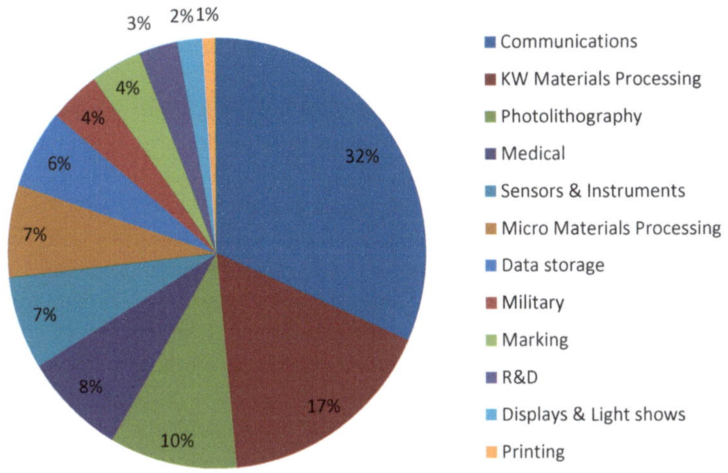

Fig. 3.7 Structure of use of the lasers (%). Source: Strategies Unlimited (2014)

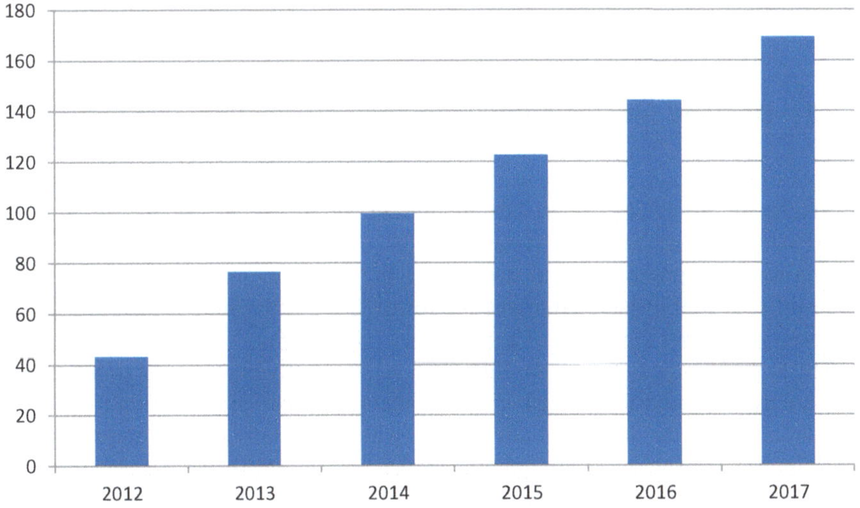

Fig. 3.8 Dynamics of green laser diodes market to be used in projection systems ($mln). Source: Strategies Unlimited (2014)

technologies. However, the degree of their reliability is unclear as producers have not yet revealed this information.

The process of manufacturing the VCSEL laser and making the body at the base level is well-proven and gives accurate and reproducible results. It is widely used for manufacturing SOI (silicon-on-insulator) microcircuits, MEMS devices, and gallium arsenide structures for the mass production of red LEDs (with a structure similar to VCSEL). VCSEL lasers, emitting in wavelength ranges of 1270–1330 nm and

1500–1600 nm, are of utmost interest as an alternative to lasers with emitting cut due to low power consumption and a potentially more efficient interface with single-mode optical fiber.

Built-in projection systems and LD flashlights, backlit for photo and video equipment, have the best growth potential in the mobile devices segment in the Russian market. The share of the display sub-segment will also increase in the medium term, but in the long term, a decline is expected.

The share of LED market in electronic home appliances, including LCD displays, will not change in the long-term perspective due to the growth of production of these devices by about 20% per year. In the large-size display segment, experts forecast a larger share of video screens. The share of "scrolling text" panels would go down.

LED applications in the external light market (parking lights, turn signals, etc.), as well as of parking systems, are expected to demonstrate significant growth which might turn similar as in the fields of decorative elements and internal car illumination.

It is presumed that LEDs have a rather high degree of interchangeability within a field of application. So, market potential for each type of LEDs was assessed for a situation when a specific LED type is outgrowing all the others. This may happen, for example, when projects are selectively funded.

3.3.2 Demand for LED Applications and Competing Technologies

Factors slowing down extensive use of LEDs have technological and market aspects. Enhancement of LED characteristics involves in-depth studies, design and development works, and institutional measures (e.g., implementation of procedures and criteria for LED products certification, studying implications of narrowband light on health), as well as targeted marketing researches to determine demand for LED characteristics in all key fields of application (Table 3.15).

The key advantage of light-emitting diodes is critically higher level of luminous efficacy as compared to alternative light sources. Their implementation may have substantial economic and social advantages, with the most important one being the lower power consumption for lighting, which, by various estimates, amounts to 18–20% of all produced electricity. Today LEDs significantly outweigh other light sources by their efficiency and lifespan (Table 3.16).

Analysis of LEDs' luminous flux growth and cost reduction rate in the last 45 years suggests that every 10 years there is a 20-fold increase of LEDs' luminous efficacy and 10-fold decrease of cost of produced light (Fig. 3.9).

Currently LEDs are about 20 times more expensive than bright white lamps and 2.6 times more expensive than compact fluorescent lamps; however, their cost has gone down considerably in the last years and this trend should preserve. Experts believe that sharp cost decline of LEDs is expected in a few years due to technology advancement and this would ensure their competitiveness. Comparative analysis of costs for various light sources with luminous flux of about 500 lm, used for local light, shows that in spite of high initial cost, XL7090 and XR7090 LEDs with

Table 3.15 Market potential development for key LED groups in Russia

	Market capacity, \$mln	
	2008	2020
Phosphide- and arsenide-based LEDs (infrared, red, yellow, and orange)		
Phosphide-based color LEDs with enhanced luminous efficacy ($\eta > 100$ lm/W)	27	190
Phosphide-based color LEDs: infrared, red, yellow, orange, and green ($\eta = 80$ lm/W)	27	190
Phosphide-based color LEDs with substantially enhanced luminous efficacy ($\eta > 140$ lm/W)	27	190
Nitride-based LEDs		
Color nitride-based LEDs: blue, green, and UV	27	190
Color nitride-based LEDs with plasmonic effect	27	190
Color nitride-based LEDs with photonic crystals	27	190
Nitride-based warm white glow LEDs with one or two luminophors ($\eta = 80$ lm/W, $P <$ \$25/kilolumen, Lt = 50,000)	42	350
White light generation		
Nitride-based warm white glow LEDs with one or two luminophors ($\eta = 100$ lm/W, $P =$ \$10/kilolumen)	47	460
Nitride-based warm white glow LEDs with one or two luminophors ($\eta = 126$ lm/W, $P =$ \$10/kilolumen)	47	460
Nitride-based warm white glow LEDs with one or two luminophors ($P <$ \$2/kilolumen)	47	460
Other inorganic LEDs		
Nitride-based UV LEDs with luminophors	33	300
White LEDs with color mixing	33	300
Blue LEDs with ZnO	27	190
OLED: $\eta = 25$ lm/W, $P <$ \$100/kilolumen, Lt = 5000 ч	42	350
OLED: $\eta > 45$ lm/W	42	350
OLED: $P <$ \$30/kilolumen	42	350
Organic LEDs in lighting applications		
OLED: $\eta > 100$ lm/W	47	460

Source: HSE
η luminous efficacy, P price of light, Lt lifetime

luminous efficacy of 55 lm/W are more cost-effective than incandescent light bulbs and halogen lamps and only slightly behind the energy-saving lamps. In general light systems, where a high luminous flux is required, LEDs are at this point much less economically sound than fluorescent, energy-saving, or metal-and-halogen lamps. However, if electricity is produced by more expensive sources (petrol or diesel generators, accumulator batteries, etc.) or tariffs for electricity are commercial ones, then economic efficiency of light-emitting diodes improves considerably Karasev et al. (2014).

Apart from economic benefits, LEDs have a number of promising features ensuring their advantages over alternative technologies. First of all, LED technologies provide for high-quality and electricity- and fire-safe lighting in houses,

Table 3.16 Comparative analysis of various light sources

Type of the light source	Efficiency of the light source (luminous efficacy) (lm/W)	Efficiency of a product with this source of light (lm/W)	Lifetime (h)
Incandescent light bulbs	8–13	6–10	1000
Halogen lamps	16–37	12–20	50–6000
Compact fluorescent lamps	50–70	35–50	6000–15,000
Metal-and-halogen lamps	60–100	<40	6000–10,000
Fluorescent lamps	60–100	55–70	15,000–32,000
Semiconductor LEDs (Cree XR-E)	100–110	90–100	>50,000
High-pressure sodium vapor lamps	90–130	<50	15,000–32,000

Source: HSE

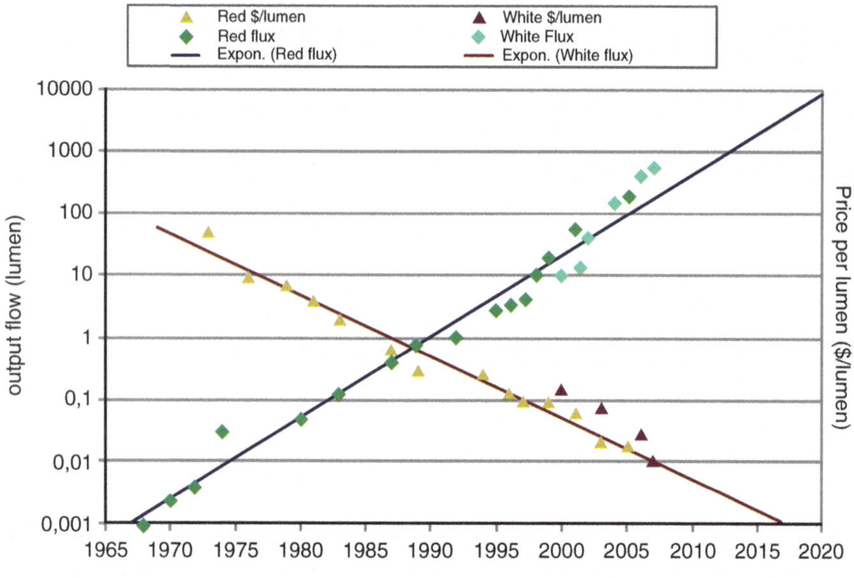

Fig. 3.9 Luminous flux growth and cost reduction rate of LEDs. Both curves have the same numerical scale (with different units). Source: HSE

public and industrial premises, on the roads and large public places, resulting in higher quality of housing and social and utility infrastructure, better health, and safety conditions. Besides, LED usage would reduce the cost of recycling of outdated light sources with hazardous components, thus supporting higher level of environmental safety and better ecology. Advantages of LED lighting used in the preservation system for especially high-value items are evident. In particular, the low level of UV band emission (for a combination of "blue" diode plus one luminophor) is a critical factor for preserving illuminated pictures, photographs, and objects of virtu. Small sizes of LED lamps allow installing them in places that are hard to reach for other light sources (except for miniature incandescent light bulbs), making them less noticeable and energy-efficient. Besides, long lifespan and high vibration resistance of these sources of light are preferential qualities for manufacturing of vandal-proof equipment.

Light-emitting diodes are most suitable for usage in extreme conditions. High light directionality (small beam angle) is a quite satisfactory replacement for complicated systems of directed beaming. Apart from that LED-based lamps have lower heat radiation than other light sources of the same size.

3.4 Conclusions: LED Development Perspectives

LED are established light sources with multiple application fields. It shows that the different applications provide significant opportunities for LEDs in the short term and also long term (Russian Electronics 2013). It appears that LED technology although reasonably developed in scientific terms is lacking full application. It has been frequently observed that there remains a significant time interval between development of a laboratory and an industrial sample which might be shortened if equipment procurement and development of a production base would go simultaneously. In addition, there appears a reasonable need for fundamental researches in this, which needs to be complementary to ongoing groundwork performed in related sectors of organic chemistry. It may be used for capacity development in the field. Table 3.17 summarizes the main perspectives for LED and OLEDs in the respective application fields.

OLED research and development is particularly aimed at display technologies advancement. Top-priority technical challenges in this area fundamentally differ from those in general lighting. Thus, the key point in assessing display quality is division of pixels, while lifetime is not that crucial. Within the frameworks of forecasts for the various segments of LED market would develop unevenly. The organic LEDs market environment requires special consideration. Scientific research and development in the sphere of organic LEDs are still at the initial stage and are led mostly by separate departments of major companies. They so far consider OLED technology development works as risky projects and insecure investments. In particular, materials research is a very time-consuming and expensive line of research for small organizations.

Table 3.17 LED technologies perspectives by segments

Inorganic LEDs	Organic LEDs
1. Mobile electronic devices	
Application	
• Color status indicators • LD flashlights • Backlit for photo and video equipment • LCD displays	• Displays
Competing technologies	
• No direct competitors in the field of indication • Share of single indicators drops as the number of integral display panels increases • Competitor: xenon flashlights, alternative display technologies	• LCD displays • E-paper • EL displays (including flexible displays)
LED technologies perspectives	
• Absolute dominance of LEDs • Small cheap LEDs in demand • Trend for higher demand for powerful LEDs (annual growth of demand for LEDs with power intake >1 W 20%) • Share of separate indicators drops with transition to new display technologies	• OLED video screens for mobile devices would dominate in short- and midterm perspective • In long-term perspective are display technology for hybrid devices (with organic elements)
2. Large-size displays	
Application	
• LCD displays backlit • Large-size LED displays	• Displays
Competing technologies	
• Halogen lamps displays • Alternative LCD displays backlit projection technologies • Plasma panels • OLED displays	• LCD displays • Plasma panels • Laser and projection technologies
LED technologies perspectives	
• LED displays hold firm position in the niche of large-size video screens • Use of LED backlit expanding in the niche of LCD displays backlit • Promotes demand for bright LEDs	• Short-term and midterm development of OLED technologies for TV and computer displays
3. Industrial and household electronic devices	
Application	
• Color and monochromic indication displays • Used in telecommunication and electronics (fiber-optic communication lines, fiber lines for digital television, etc.) • Metering equipment • LCD displays backlit • Other (UV disinfection of water and premises, banknotes authentic control units, plant growth stimulation tools, etc.)	• Displays

(continued)

Table 3.17 (continued)

Inorganic LEDs	Organic LEDs
Competing technologies	
• Display technologies in indication application • Laser technologies in the fields of telecommunication and electronics, metering equipment • Variety of competitors in niche applications	• Other display technologies
LED technologies perspectives	
• Dominance in specific niches	• Dominance of OLED displays in small-size devices
4. Signaling devices	
Application	
• Traffic lights, railway semaphore signals • Traffic situation displays • Safety marking and inscriptions • Control systems, presence sensors, smoke detectors, including in security systems	• Light-emitting panels of predetermined form
Competing technologies	
• Halogen and fluorescent lamps • Electroluminescent lamps and wires • Neon lamps	• Halogen and fluorescent lamps • Electroluminescent lamps and wires • Neon lamps
LED technologies perspectives	
• Promising segment • Lifespan, high energy efficiency, emission spectrum control, and instantaneous switching make inorganic LEDs dominant technology in short-term and long-term perspectives	• Currently application is limited by insufficient lifetime and low environmental resistance
5. Transport	
Application	
• Parking and position lights, turn signals, stop signals with color LEDs • Headlights • Internal car illumination • Car dashboards • Built-in electronic systems • Decorative elements	• Internal car illumination • Indication on windscreen
Competing technologies	
• Incandescent bulbs (halogen lamps) • Fluorescent and compact fluorescent lamps • Electroluminescent lamps	• Incandescent bulbs (halogen lamps) • Fluorescent and compact fluorescent • Electroluminescent lamps • Electroluminescent displays
LED technologies perspectives	
• LEDs steadily replace alternative technologies in most applications	• Projects for development of transparent windscreen built-in displays are underway • Broad-scale implementation of these technologies in future

(continued)

Table 3.17 (continued)

Inorganic LEDs	Organic LEDs
6. Outdoor architecture and decorative illumination	
Application	
• Advertising and decorative billboards • Architecture • Decorative and landscape lighting	• Light-emitting panels including flexible and semitransparent
Competing technologies	
• Incandescent bulbs • Fluorescent lamps • Electroluminescent lamps and wires • High-pressure gas-discharge lamps	• Electroluminescent displays • Electroluminescent wires
LED technologies perspectives	
• Durability, lifespan, controlled spectrum, directionality, small size, high switching speed, high energy efficiency, and smart control ensure LED leadership in decorative applications • Use of smart control systems provides for more capabilities in architecture illumination, advertising applications	• OLEDs promising for indoor decoration • Short-term perspective manufacturing of commercial equipment is not expected
7. Lighting	
Application	
• LED lamps and light sources	• Light-emitting panels
Competing technologies	
• Incandescent bulbs • Fluorescent lamps and compact fluorescent lamps • Other types of lamps	• Incandescent bulbs • Fluorescent lamps and compact fluorescent lamps • Other types of lamps
LED technologies perspectives	
• Universal use of LED lamps and light sources supported by state programs in numerous countries	• New type of light sources—light-emitting panels (sectional at first, then one-piece large-size panels) forthcoming • Trend for reducing average luminance

Source: HSE

Acknowledgments The book chapter was prepared within the framework of the Basic Research Program at the National Research University Higher School of Economics and supported within the framework of the subsidy by the Russian Academic Excellence Project '5-100'.

References

European Commission (2014) Update on the status of LED market. http://iet.jrc.ec.europa.eu/energyefficiency/sites/energyefficiency/files/reqno_jrc92971_jrc92971_online.pdf.pdf

Great Stone (2013) Production of LEDs and LED lighting devices. Great Stone. http://investinbelarus.by/docs/production_of_leds_and_led_lighting_devices.pdf

Karasev O, Sokolov A, Roud V, Doroshenko M, Vishnevskiy K, Veselitskaya N, Afanasiev A, Hahanov Y, Pashkov V (2014) In: Gokhberg L, Karasev A, Malyshev M (eds) LED Industry: innovation technologies, products and markets. A Roadmap. Rusnano, HSE, Moscow

Lux Research, Inc. (2015) Scaling up carbon fiber: roadmap to automotive adoption. http://speautomotive.com/SPEA_CD/SPEA2015/pdf/KEY/KEY1.pdf

Nogee A (2014) The worldwide market for lasers market review and forecast 2014. Strategies Unlimited, 2014. http://cdn-content.opticsinfobase.org/WW%20Laser%202014%20Final.pdf

Russian Electronics (2013) Moscow international LED forum 2013: jeto «konec starogo sveta». http://www.russianelectronics.ru/engineer-r/review/doc/65459/

Strategies Unlimited (2013) Worldwide LED component market grows 9% with lighting ranking first among all application segments, according to strategies unlimited. http://www.strategies-u.com/articles/2013/02/worldwide-led-component-market-grew-9%2D%2Dto%2D%2D13-7-billion-with-lig.html

Yole Dévelopement (2013) LED applications. http://www.yole.fr/iso_upload/News/2013/PR_BulkGaN%20_YOLEDEVELOPPEMENT_Nov.2013.pdf

Vitaliy Roud is Senior Researcher at the HSE ISSEK Laboratory for Economics of Innovation. Vitaliy has participated in a number of research and policy advice projects initiated by public agencies, international organization and companies including several national-level foresight initiatives. His academic interests include empirical studies of innovation, evidence-based innovation policy, methodology of innovation surveys, STI statistical indicators, STI policy design and evaluation. Mr. Roud holds a lecturer position in the Master's Program 'Governance of STI' at HSE.

Alexander Sokolov is Deputy Director of HSE ISSEK and Director of HSE International Foresight Center. His main professional interests are related to Foresight, STI priorities, indicators and policies. Prof. Sokolov is tenure professor at HSE, he teaches Foresight for undergraduate and postgraduate students. He authored over 120 publications in Russia and internationally and managed many Foresight projects, including: Russian S&T Foresight: 2030; Foresight for Russian ICT sector (2012); Innovation Priorities for the Sector of Natural Resources (2008–2010); Russian S&T Delphi Study: 2025 (2007–2008); Russian Critical Technologies (2009) et al. Prof. Sokolov is member of several high-level working groups at OECD and other international organizations, serves for advisory boards at several international conferences and journals.

Dirk Meissner is Deputy Head of the Laboratory for Economics of Innovation at HSE ISSEK and Academic Director of the Master Program "Governance for STI". Dr. Meissner has 20 years experience in research and teaching technology and innovation management and policy. He has strong background in policy making and industrial management for STI with special focus on Foresight and roadmapping, funding of research and priority setting. Prior to joining HSE Dr. Meissner was responsible for technology and innovation policy at the presidential office of the Swiss Science and Technology Council. Previously he was management consultant for technology and innovation management with Arthur D. Little. He is member of OECD Working Party on Technology and Innovation Policy. He is Associate Editor of Technological Forecasting and Social Change, IEEE Transactions on Engineering Management, Journal of the Knowledge Economy, member of Editorial Review Board at Small Business Economics and Journal of Knowledge Management. He guest edited Special Issues in Industry and Innovation journal, Journal of Engineering and Technology Management, Technological Analysis and Strategic Management among others.

Technology Use in Traditional Industries: Catalysts for Petroleum Refining

4

Dirk Meissner and Pavel Rudnik

4.1 Global Technological Development Trends in the Oil Refining Sphere

National oil refining markets are all integrated into the global petroleum production and consumption system, and that it is affected by global trends, their dynamics and consequences, both within and outside industry-specific processes—in the framework of international paradigms. Therefore, the strategic objectives set in the oil refining sphere can only be accomplished if the sphere's development is approached in a systemic way—i.e. domestic demand and available S&T potential are viewed in the context of global technological trends and development vectors on the global catalyst market.

4.1.1 Oil Refining Structure

The technological processes that are applied to extract products from crude oil required in practically all spheres of modern life demonstrate a complex structure: the current scheme is based on double processing and includes over 20 technological processes performed at various installations, using different kinds of catalysts. During primary processing, crude oil is distilled under atmospheric pressure and separated into fractions with different boiling temperature. The heavy fraction (boiling temperature >350 °C) is often subjected to additional vacuum distillation. Catalysts are actively applied in secondary oil refining processes, including catalytic cracking, reforming, hydrofining, isomerisation, alkylation, hydrocracking,

D. Meissner (✉) · P. Rudnik
National Research University, Higher School of Economics, Moscow, Russia
e-mail: dmeissner@hse.ru; prudnik@hse.ru

D. Meissner et al. (eds.), *Emerging Technologies for Economic Development*, Science, Technology and Innovation Studies,
https://doi.org/10.1007/978-3-030-04370-4_4

hydrodesulphurisation, hydrodewaxing, deasphalting, and selective treatment of lubricants.

Various processes' specific roles vary depending on the fluctuations of prices of and demand for oil and oil products and are affected by the emergence of new requirements for oil products' properties, new technologies, and catalysts. For example, in regions where only basic oil refining is performed and consumption of fuel oil is high, the predominant technologies include catalytic reforming, petrol isomerisation, hydrofining of diesel and jet fuels, and hydrofining of fuel oil for the production of commercial-grade fuel. In regions where oil is processed more deeply and consumption of fuel oil is low, the predominant technologies include destructive refining (cracking and hydrocracking) and upgrading of oil products (via reforming and hydrofining) and fuel oil. In addition to oil refineries, some of these regions also have petrochemical synthesis facilities. In that case, pyrolysis and catalytic reforming processes are applied to maximise lower olefins and aromatic hydrocarbons output (Krylov 2004). Among these processes the development prospects for secondary oil processing nanotechnologies involve upgrading six technological processes, which are believed to have the biggest catalyst markets:

- Hydrofining
- Catalytic cracking
- Isomerisation of light gasoline fractions
- Catalytic reforming
- Alkylation
- Hydrocracking

Key oil refining development trends and potential for applying nanotechnologies to produce catalysts are analysed below in the framework of this technological processes' structure.

4.1.2 Oil Refining Trends and Development of Global Catalyst Market

Recently, developed countries were showing a trend towards abandoning the construction of new crude oil refining facilities and switching to upgrading production processes by introducing new installations and applying catalytic processes to upgrade and improve the quality of intermediate primary refining products. Having more secondary oil refining installations allows for an increase in the output of high-quality, expensive light oil products, with lower output of cheap fuel oil. Table 4.1 shows that leading global producers of oil products (Western European countries, the USA, Russia, and Japan) set the technological structure of global oil processing industry, which comprises five key production processes: catalytic cracking, reforming, hydrofining, alkylation, and isomerisation.

Thus, light oil products are becoming increasingly important on the global oil refining market, and catalysts are actively applied in their production. Catalytic

Table 4.1 Current technological processes applied for oil refining in Russia and other countries (% of total crude oil refining)

Main secondary processes	Russia	USA	Western Europe	Japan
Catalytic cracking	7.1	35.8	15.8	19.8
Hydrocracking	2.6	9.1	7.5	4.0
Thermal cracking and viscosity breaking	4.0	0.2	12.2	0
Coking	1.0	16.2	2.5	23.8
Hydrofining of distillates	26.0	41.3	35.3	52.5
Catalytic reforming	10	18.3	12.7	13.9
Alkylation	0.3	5.6	1.4	0.8
Isomerisation	0.9	3.0	2.7	0.3

Source: HSE

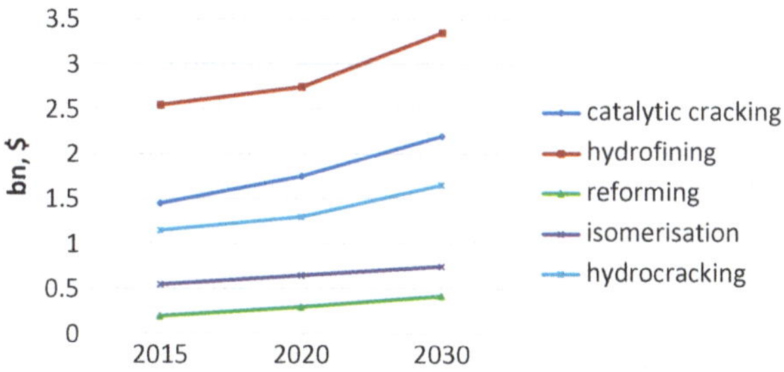

Fig. 4.1 Forecasted growth rates for global catalyst market. Source: HSE

processes' capacity is rapidly growing. This trend is expected to continue and will largely determine the industry's development vectors. Another global trend, particularly evident in developed countries—importers of oil products—is the increasingly stringent environmental laws aimed primarily at reducing harmful emissions from burning fuel and constantly growing requirements for oil products' quality, which among other things resulted in the EU member countries' switching to Euro-5 fuel with extremely low sulphur content. Also, the demand trends of recent years show that consumption of distilled diesel fuels and high-quality petrol in the EU countries is rapidly growing. Petrol consumption in the USA and Asia-Pacific region is also growing. Demand for jet fuel grows less rapidly, while demand for fuel oil is gradually falling. All these trends contribute to the growth of global catalyst market, which is estimated at 15–20 billion USD annually, and the range of available industrial catalysts is annually renewed by 15–20%. Currently the global catalyst market has the following structure: hydrofining catalysts amount to about 40%; cracking catalysts, 30%; hydrocracking, 7%; reforming, 5%; and other catalysts, 20%. Figure 4.1 shows the forecasted growth rates for main oil refining technologies' markets (in value terms).

Overall the global oil refining market is determined by a few key trends noted in the framework of the five selected technological processes.

Catalytic Cracking

- Installations' total capacity will remain at the current level until 2030.
- A slight growth of cracking catalysts consumption is expected.
- In value terms, the global market is expected to significantly grow by 2030 due to considerable growth of catalyst prices.

Hydrofining

- Hydrofining installations' total capacity by 2030 will grow by 25%.
- Consumption of sulphide hydrofining catalysts is expected to grow both in absolute and value terms.
- Catalyst prices are generally expected to remain at the current level.

Isomerisation of Light Gasoline Fractions

- Global isomerisation catalysts market is expected to grow by 40–50% by 2030.
- Low-temperature isomerisation is the most advanced and promising process.

Catalytic Reforming

- Reforming installations' capacity is expected to grow by 20% by 2030.
- Reforming in moving catalytic layer is considered to be the most advanced and promising way to apply this process.
- Annual consumption of catalysts is expected to quickly grow, both in value and absolute terms, due to increased production capacity.

Hydrocracking

- One of the most promising technologies; by 2030 installations' capacity is expected to double.
- Annual consumption of catalysts is also expected to double.

4.1.3 Russian Oil Refining Industry and Catalyst Market

Major changes were noted in Russian oil refining industry in recent years. The output grows, and the quality of produced motor fuels is gradually improving. Several Russian oil refineries are building new facilities for deep oil processing; some of them are already operating. The backbone of the Russian oil refining industry is 27 large refineries with total designed capacity of 260 million tons of crude oil a year (which amounts to 95–98% of all processed oil). Lately Russian oil companies have been significantly increasing their investments in oil processing, which resulted in growing oil refining volume and gradual improvement of motor fuel quality—due to discontinuing production of leaded petrol, there is a higher share of high-octane

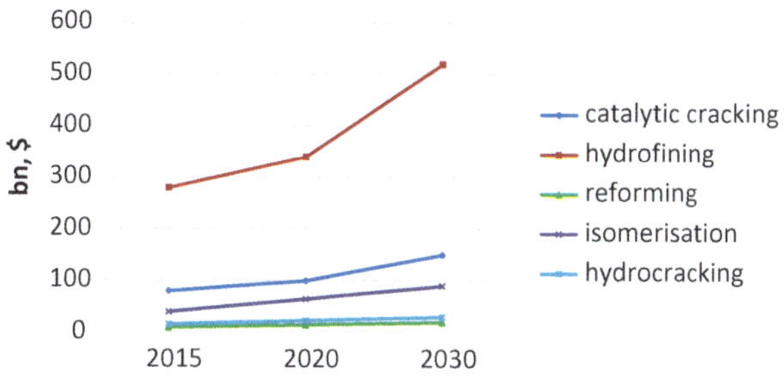

Fig. 4.2 Forecasted trends for Russian catalyst market. Source: HSE

Table 4.2 Market development trends

Hydrofining
• Capacity growth will mostly occur due to wider application of nickel-molybdenum aluminium oxide-based catalysts
• By 2030 possible industrial application of oxidation technology—a radically new catalyst production technology is expected
• New carrier types are expected to be developed by 2030, e.g. nanostructured titanium dioxide
Catalytic cracking
• Cracking installations' total capacity will be growing
• Bead catalyst-based installations are expected to be decommissioned by 2030
• Capacity will grow mostly due to wider application of microspheric silica-alumina zeolite-containing catalysts with average particle diameters between 10 and 150 microns
Isomerisation of light gasoline fractions
• The number of isomerisation installations in Russia is expected to double by 2030, reaching 30
• Consumption of catalysts will significantly grow, both in absolute and value terms
• Catalyst prices are expected to moderately grow, due to growing prices of precious metals
Catalytic reforming
• Total installations' capacity is expected to grow by 50% by 2030
• Capacity growth will be due to increased number of oxide- and zeolite-based catalyst installations
• Catalyst prices are expected to significantly grow due to growing prices of precious metals
Hydrocracking
• More stringent requirements for oil products, primarily in terms of sulphur content, will prompt the development of more efficient catalysts, integration with hydrofining and de-metallisation processes, removal of heavy multiring hydrocarbons, hardware improvement, development of more efficient technological schemes (Kapustin 2007)
• Installations' total capacity will increase
• By 2030 the number of such installations is expected to reach seven in Russia
• Catalyst prices are not expected to significantly change until 2030

Source: HSE

petrol and environmentally clean diesel fuel. Forecasts for major types of oil refining catalysts' markets (in value terms) are presented in Fig. 4.2.

The development of the Russian market during the time horizon of the study will be affected by the following key trends (Table 4.2).

Thus an analysis of global and Russian trends revealed vectors affecting the development of technological processes suitable for the application of catalysts with the required properties. Improvement of these processes will allow for the production of competitive oil refinery products for the Russian and international markets in the next 15 years.

4.2 Catalytic Refining Technologies' Characteristics

Oil refining catalysts (further on, catalysts) are nanostructured substances which, in the course of a cycle of intermediate interactions, speed up oil refining chemical reactions without being spent in the process. Russian oil refining nanotechnologies are developed to improve catalysts for cracking heavy crude oil to refine it into light products, for gasoline fractions' reforming, for deep hydrofining of diesel fuel to remove sulphur, to process accompanying oil gases, etc. Analysing the structure of demand for catalysts, the dynamics of key technical and economic properties of major installation types and the catalysts they use, allows one to identify opportunities for and barriers hindering technological development of oil refining industry. The development of oil refining technologies went through several key stages connected with the introduction of specific catalytic processes. The 1960s was the time of major application of catalytic reforming of gasoline fractions (aromatisation, isomerisation, and de-hydrolysation of original hydrocarbon compounds), which allowed for the significant improvement of the efficiency of internal combustion engines with the spark ignition of air-fuel mixture—carburettor ones, and currently engines based on fuel injection into cylinders. The increased efficiency of such internal combustion engines is thermodynamically linked to the increased compression ratio of the air-fuel mixture, which is limited by petrol's antiknock quality. The acceptable compression ratio increases with the reduction of the engine cylinders' diameter, while the volume power is boosted by increasing the number of crankshaft revolutions—which complicates adjusting the composition of the working mixture under varying speeds and loads.

As of 2002, leading automobile manufacturers started the mass production of petrol internal combustion engines with forced injection of fuel into cylinders and electronic adjustment of spark ignition's advance. This solved the problem of preparing the working mixture and providing optimal conditions for combusting it in high-speed engines, but made requirements for petrol's antiknock quality and the range of its components' boiling temperatures even more stringent. Traditional oil refining processes needed novel technological solutions not just abroad but in Russia as well.

Next, the main characteristics of processes, installations, and catalysts to a greater or lesser degree represented and demanded on Russian and global markets are described.

4.2.1 Catalytic Cracking

Catalytic cracking is a major technology for producing a high-quality petrol component with octane number of 91–93 (RM). It also produces fractions which can be used as diesel or jet fuel components. Catalytic cracking produces a significant amount of gas rich with propane-propylene and butane-butylene fractions (raw materials for making high-octane MTBE ether, alkyl benzene, and other valuable motor fuel components). Catalytic cracking installations also supply materials for producing high-quality "needle" coke. Catalytic cracking gas-oils are used to make carbon-black materials and naphthalene. During the long period of development (starting in the 1930s), catalytic cracking was significantly improved, both in terms of contact between raw material and catalyst (in the stationary layer, bead catalyst moving layer, microspheric catalyst "boiling" layer) and the applied catalysts (tableted catalysts based on natural clays, bead synthetic aluminium silicates including nanostructured zeolite-containing ones). These improvements resulted in radical changes in the technological process as a whole, which allowed for increasing end product output (petrol component) from 30–40% to 50–55% mass (Melik-Akhnazarov 1980). The catalytic cracking process is largely determined by the properties of the applied catalyst. To process heavy oil residue and secondary products, modern cracking catalysts must provide a high output of high-octane petrol (over 50% of raw material mass), with minimum required coke content in the catalyst. A typical modern cracking catalyst is a very complex nanostructured composite, whose all-important component is 0.2–0.6 micron nanocrystals of ultra-stable sheet-morphology zeolite (Parmon 2008). Cracking catalysts are heterogeneous dispersed multicomponent porous systems, with the following overall functional structure:

- Active phase (3–20% mass)—zeolite Y in various cation-exchange forms
- Binder—aluminium silicates, oxides Al, Si
- Filler—clays of kaolin or montmorillonite type (Gerzeliyev 2008)

To synthesise a cracking catalyst with specific properties, the components (zeolite, filler, binder) must be combined in the optimal way, taking into account their physical–chemical interaction during production of the catalyst. Cracking catalyst's activity is primarily determined by the quality of the nanostructured zeolite used, which in turn is adjusted both at the synthesis and processing stages. X and Y zeolites in rare-earth or ultra-stable forms are highly active and highly selective in oil fractions' cracking; they are stable under high-temperature impact in air and water vapour environments and have adequate entrance window sizes in structural cavity; accordingly, they are frequently used to synthesise cracking catalysts (Gerzeliyev 2008).

Cracking catalyst's matrix must ensure the catalyst has the necessary basic properties: high mechanical strength, primary cracking of the raw material's heavy component (the matrix's adjustable activity to process vacuum gas-oil with high-end boiling point (up to 600 °C)), efficient heat removal from zeolite crystals during the catalyst's regeneration, and high heat capacity to provide heat for endothermic cracking reaction; nanostructured zeolite's surface must be accessible to reacting molecules (i.e. it must be sufficiently porous); and the catalyst must have acceptable volume weight so it could be retained in the system. Currently two kinds of cracking catalysts are produced:

- Granulated catalyst in the form of beads or granules (average particle size from 2 to 5 mm), for application in installations with moving catalytic layer
- Microspheric catalyst (maximum particles diameter up to 120–150 microns), for application in installations with aerated catalyst layer

The demand for granulated catalyst is estimated at 6–8 thousand tons a year, depending on installations' utilisation rate. As to microspheric catalyst, demand for it, again depending on installations' utilisation rate, amounts to about 8–9 thousand tons a year. These catalyst types are used in both traditional and advanced "process-catalyst" complexes (the catalytic process's scheme and the applied catalyst type):

1. In lift reactor installations, microspheric (powdered, with average particle diameter of 10–70 microns) silica-alumina zeolite-containing catalysts are used.
2. In double-regeneration cracking installations, microspheric (powdered, with average particle diameter of 10–70 microns) silica-alumina zeolite-containing catalysts are used.
3. In X-design technology, microspheric (powdered, with average particle diameter of 10–70 microns) silica-alumina zeolite-containing catalysts are used.
4. In millisecond cracking installations, microspheric (powdered, with average particle diameter of 10–70 microns) silica-alumina zeolite-containing catalysts with optimal rare-earth elements content are used.
5. In moving catalytic layer installations, bead silica-alumina zeolite-containing catalysts are used.
6. In aerated catalytic layer installations, microspheric (powdered, with average particle diameter of 50–150 microns) silica-alumina zeolite-containing catalysts are used.

Lift reactor installations are currently considered the mainline type; they offer several advantages such as reduced contact time (the catalyst does not have time to coke which allows for the use of heavier raw materials); increased temperature (in reactor, 530 °C; in regenerator, about 700 °C) and petrol output (52–56%); the octane number is over 90 and catalyst circulation ratio is 6–7 (the amount of catalysts per ton of raw material); and high productivity. However, it also shows certain drawbacks, such as complex operation and high energy consumption. Installations of this type are operating in several Russian cities; their optimal capacity is up to 2 million tons of raw material a year.

Double-regeneration cracking installations are an analogue of lift reactors. There are no such installations in Russia, but in foreign countries they amount to 7–10% of the total number of lift reactor installations. Such installations have several significant advantages including the possibility of using heavy raw materials; finer catalyst cleaning (resulting in higher activity); opportunity to change circulation ratio; more flexible temperature controls; and more efficient regeneration due to an independent air supply to each regenerator. There are also certain drawbacks including significant emissions, loss of catalysts, high energy consumption, and problems with retaining the installation's thermal balance.

X-design technology is mentioned among the prospective ones due to several advanced properties, e.g. low energy consumption during regeneration, opportunity to prolong service life, and more efficient use of catalyst. A drawback is the high energy requirements for transporting the catalyst.

A millisecond cracking installation is operating in Belarus, but there are no plans to build one in Russia so far—though the costs are quite comparable with others. Among its strengths, the experts noted a compact reactor, convenient operation, low energy consumption, and opportunity to use very heavy raw materials. Weaknesses mentioned by the experts included ultrashort (in terms of time) contact; cracking of heavy raw materials; and the opportunity to obtain better characteristics using vacuum gas-oil and fuel oil mix.

Installations with a moving catalytic layer belonging to an already obsolete type are no longer built because their weaknesses are well known. These include small contact surface, low petrol output and low octane number, and quick coking of catalyst which occurs due to long contact of raw material with catalyst. Other obvious advantages of this type are the use of air as a transportation agent to move the catalyst in the installation and the separation of reactor and regenerator.

In the late 1990s, such installations were widely used in Russia, amounting to about 30% of all installations of the 43–102 type. Currently, 11 such installations remain in Russia. Each processes about 600 thousand tons a year. In 20 years' time, they are expected to become completely obsolete and will be replaced with new ones.

Aerated catalytic layer installations are also obsolete now. They are large, difficult to operate, and have high energy requirements. However, they also have certain advantages such as higher output and octane number (the main advantage) and possibility of maintaining higher temperatures (which provides a deeper cracking producing more light fractions). Three such installations currently remain in operation in Russia. In other countries they are no longer in use. This type of installation uses a microspheric catalyst with a pore diameter between 50–70 and 150 microns. Total contact time is between 2 and 6 min; contact time in aerated layer, under 2 min; reactor temperature, about 500 °C; regenerator temperature, 650–670 °C; catalyst circulation ratio, about 4–5; and petrol output, 48–50%.

4.2.2 Catalytic Reforming

Catalytic reforming is a complex chemical process which involves reactions resulting in formation of aromatic hydrocarbons. It is applied in the petrochemical industry to perform some of the most important, large-scale, and common processes: the production of highly aromatic distillates from low-octane straight-run gasoline fractions, i.e. high-octane petrol components, and separating from them individual aromatic hydrocarbons such as benzene, toluene, and xylenes (Kapustin 2007).

Aromatic compounds and benzene are now officially recognised as harmful cancerogenic substances, and a decision was made to radically reduce their content in petrol. However, reducing the reforming petrol's share in the total pool of petrol firstly would be quite expensive for the society, and secondly, reforming is one of the main sources of cheap hydrogenous gas at oil refineries, which is necessary for hydrogenation processes (hydrofining, hydrocracking).

Reforming's share in oil processing at global oil refineries is on average 14%. Commercial petrol produced in the USA on average contains 20–25% of reformate; in Western Europe the relevant figure is about 30–40% and in Russia 45–50%. Therefore, completely abandoning reforming is hardly going to happen in the near future (Krylov 2004).

Straight-run gasoline fractions (their content in oil is usually 15–20% mass) due to their chemical composition have low antiknock value (MON-50 and RON-55). Gasoline fractions of most oil types contain 60–70% of paraffin hydrocarbons, about 10% of aromatic, and about 30% naphthenic. However, the process's product—reformate—has high antiknock value (MON = 80–90 and RON = 90–100), due to results of various chemical reactions (the main one being cyclodehydrogenation of hydrocarbons) leading to changes in its chemical composition (Borodacheva and Levinbuk 2008).

The process's evolution during the last 60 years amounted to an increased depth of raw material processing, the selective aromatisation of paraffin hydrocarbons, and an increased stability of catalysts. Thus output of aromatic hydrocarbons and hydrogen increased by more than 1.5 times, while catalysts' inter-regeneration cycle grew by 4 times. The process's technological progress amounted to a reduced working pressure, from 3.0 to 0.35 MPa, due to the development of new highly stable catalysts and the application of a modified technology with continuous catalyst regeneration.

Catalysts used in reforming perform two major functions: dehydrating-hydrating and acidizing. Catalyst's dehydrating-hydrating function is normally performed by metals—platinum, palladium, and nickel. Platinum component has the highest dehydrating properties. Its function is to speed up dehydration and hydration reactions, which stimulates formation of aromatic hydrocarbons, continuous hydration, and the partial removal of intermediate reaction products which lead to coking. Platinum content in the catalyst is usually at 0.3–0.6% (mass). Lower platinum content reduces the catalyst's resistance to poisons, while a higher concentration reveals a tendency towards stronger demethylation reactions. Another factor limiting the catalyst's platinum content is high cost. Platinum-conditioned activity of the catalyst can be adjusted by changing platinum concentration and the degree of its dispersion in the carrier, up to values measured in whole nanometres.

Acidizing function is performed by the catalyst carrier—aluminium oxide. The catalyst's acidic properties determine its cracking and isomerising activity. A halogen is added to the catalyst to increase its acidizing function. Recently, chlorine was frequently used for this purpose; a while ago it was fluorine (and occasionally still is), which stabilises high dispersion of platinum by forming complexes with it and with aluminium oxide. Chlorine's advantage is that it promotes cracking reactions to a lesser extent (this is particularly important under a strict regimen). Chlorine content is 0.4–2.0% (mass). Loss of chlorine in the catalyst during its oxidising regeneration is replenished by periodic or ongoing chlorine feed (the amount is 1–5 mg/kg of raw material). The catalyst's acidity is very important for achieving a certain depth of raw material's processing, for making a product with the required octane number during the specified period for staying in the reaction zone, and under the specified temperature. Catalysts combining both these functions (dehydrating and acidizing) are called bifunctional. They are also used for cyclodehydrogenation reactions which are particularly relevant for processing raw material with high paraffin hydrocarbon content.

The main catalyst evaluation criteria are the raw material's feed rate; stable reformate (catalysate) output; product's octane number or output of aromatic hydrocarbons; light fractions' content in the reformate; output and composition of gas; and catalyst's service life. An efficient way to increase catalysts' activity, selectivity, and stability is to introduce special elements into them—promoters, which ensure desirable effects. Russian refineries initially used AP-56 aluminium-platinum catalysts, based on fluorinated aluminium oxide and containing 0.56% (mass) of platinum. The catalyst was applied without preliminary hydrofining of raw material, to produce catalysate with MON 75. Under such conditions, the main reaction leading to formation of aromatic hydrocarbons was naphthene dehydrogenation. The second stage of the Russian variant of reforming was connected with development and application of AP-64 chloride-containing aluminium-platinum catalyst and application of such technologies as raw material hydrofining and dehydration, chlorination of the catalyst during the reaction cycle, and reducing the process's pressure from 4.0 to 3.0 MPa. Such intensification of the process allowed the production of up to 40% (mass) of aromatic hydrocarbons in the reformate through cyclodehydrogenation of paraffin hydrocarbons.

However, each of the applied catalytic reforming types has its own advantages and drawbacks.

1. The process in the stationary catalytic layer with intermediate heating between reaction zones (several reactors are positioned in a row one after another, with reaction's products being heated in between them, on their way from the previous to the next reactor). At the same time hydrogen is circulating through the system, which requires high-pressure compressors. The catalyst is platinum (including other metal additives, i.e. bimetallic) overlaid on aluminium oxide. This catalytic complex has a number of advantages and currently meets the main consumer

requirements; also this technology is believed to be the best-developed one. The drawbacks are a low volume rate (13 inv.h.); high pressure (35 atm.) (which increases the costs); high sensitivity to harmful contaminants; and limited regeneration of catalyst, which reduces its activity.

2. Process in the stationary catalytic layer with intermediate heating between reaction zones—zeolite platinum-containing (including with other metal additives, i.e. bimetallic) catalyst. This catalyst's advantages (compared with platinum on aluminium oxide) include lower pressure (20 atm.), high volume rate (2–2.5 inv. h. and potentially 3–3.5), higher octane number of the reformate, and increased hydrogen output under the same process regimen. This catalyst shares its drawbacks with platinum on aluminium oxide, though it has fewer conditions for continuous regeneration (which results in reduced activity), and is sensitive to harmful contaminants. Another major drawback of the process is high pressure—higher than in moving layer.
3. Process in moving catalytic layer—platinum (including other metal additives, i.e. bimetallic) overlaid on aluminium oxide. The attractive properties including lower sensitivity to harmful contaminants, higher octane number of reformat, and higher hydrogen output, plus the opportunity to process heavier raw materials. On the other hand, higher strength is required to improve product quality.
4. Process in moving catalytic layer—zeolite platinum-containing (including with other metal additives, i.e. bimetallic).

Significant strengths of this technology are the highest octane number of the reformate, lowest pressure (9 atm.), possibility of processing heavier raw materials, and higher hydrogen output. But at the same time production costs are high, including high catalyst prices. It's expected that catalysts' development moves towards increasing octane number and reformate output and reducing the content of aromatics and benzene in it. Overcoming this contradiction requires strengthening catalysts' isomerisation function—and the zeolite component does exactly that.

4.2.3 Isomerisation of Light Gasoline Fractions

The isomerisation process is applied to increase C5–C6 oil fractions' octane number by transforming the normal structure paraffins into isomers with a higher octane number. Isomerisates' most important consumer property (unlike reformates) is a very small difference between their motor and research octane numbers. Accordingly, a high-octane isomerisate can be considered an adequate component for mixing with reformate, because of the following important reasons:

- To increase light gasoline fractions' octane characteristics (overpoint fraction—100 °C)
- To reduce the difference between MON and RON in commercial petrol and increase octane index
- To reduce overall aromatic hydrocarbons content, including benzenes

- To level petrol's octane numbers for the whole mass of evaporant fuel

In the last decade isomerisation process became one of the most profitable technologies for the production of high-octane, environmentally clean petrol components, widely applied in many countries to increase octane numbers of C5–100 °C gasoline fraction by regrouping molecular structure of regular paraffins into their isomers with higher octane number.

Isomerisate of C5–C6 light gasoline fraction has a low content of aromatic and olefinic hydrocarbons with research octane characteristic between 68 and 83 points, depending on the processed fraction's quality, isomerisation type, and process scheme.

It should be stressed that combining isomerisation and reforming processes to meet the current and prospective requirements for motor petrol is a high-priority and economically sensible option. If reforming removes heavy low-octane petrols and thermal processes' petrols from the overall petrol pool and transforms them into aromatic high-octane components, isomerisation extracts from the petrol pool low-octane light straight-run and secondary components, replacing them with high-octane components—isomers which do not contain benzene and other aromatic hydrocarbons. The isomerisation process as such has a number of positive technological and environmental features, such as low costs, 100% product output, and low hydrogen consumption.

And the end product—isomerisate—is the most valuable component of commercial petrol, because it does not contain benzenes, aromatic hydrocarbons, sulphurous compounds, and olefinic hydrocarbons and has high RON and MON values. The purpose of applying catalytic isomerisation in modern oil refining is to produce high-octane isocomponents of motor petrol or materials for the petrochemical industry, primarily isopentane, to synthesise isoprene rubber.

There is high-temperature, medium-temperature, and low-temperature isomerisation. Currently low-temperature and medium-temperature processes are applied most commonly.

Low-temperature isomerisation catalysts are divided into three types:

(a) Chlorinated metal oxides, primarily chlorinated ή- and ά aluminium oxide
(b) Wide-pore zeolites of the faujasite type
(c) Zirconium oxide promoted with sulphate, molybdate, or tungstate ions

To prolong this type of catalyst's stable action, 0.3–0.5% mass of platinum or palladium is added to it. The working temperature range is between 100 and 200 °C, the process pressure is 10–30 atm., and raw material load is 0.5–4 h^{-1}. Isomerisation reaction is conducted under presence of hydrogen, with molar ratio to raw material of 0.02–4. Due to the above thermodynamic characteristics of isomerisation process, this class of catalysts seems to be the most promising one for processing light paraffin fractions from butane to NK-70 fraction. A serious weakness of the (a) and (b) catalysts is their sensitivity to presence of contaminants such as water, sulphur, and nitrogen in the raw material, which under low temperatures are efficiently absorbed on the catalyst's surface, negatively affecting its acidizing and hydrating functions. For chlorine-promoted aluminium oxide, the acceptable content of water, sulphur and nitrogen is about 2–5 ppm. For zeolites and promoted zirconium oxide, acceptable content of contaminates in raw material is much higher than for aluminium oxide.

Medium-temperature isomerisation catalysts are represented by zeolite-based catalysts of the mordenite type, with a sodium content of about 2–3 ppm and modified with 0.4–0.5% mass of platinum. Their working temperature range is 260–300 °C, process pressure is 10–30 atm., and raw material load is 0.5–1.5 h^{-1}. Hydrogen/raw material ratio is 0.5–1.5. Mordenite-based isomerisation catalysts are much more tolerant to raw material contaminants—up to hundreds of ppm. This type of catalysts is mainly used for isomerisation of paraffin fractions; they are not suitable for the isomerisation of *n*-butane.

High-temperature isomerisation catalysts include catalysts based on fluorinated aluminium oxide and medium-pore zeolites such as ZSM-5 type, etc.; their working temperature range is 320–380 °C, process pressure is 10–30 atm., raw material load is 0.5–1.5 h^{-1}, and hydrogen/raw material ratio is 0.5–1.5.

High-temperature isomerisation catalysts can accept contaminant content in raw material of up to hundreds of ppm. They are applied for the isomerisation of light, medium, and heavy paraffin fractions.

4.2.4 Hydrofining

The development of advanced processes for the hydrofining of petrol, kerosene, and diesel fractions is mainly aimed at reducing concentration of sulphurous, olefinic, and partially nitrogenic oxygen-containing compounds. This is due to the growing share of sulphurous oil in the overall oil production, coupled with stricter requirements for sulphur content in fuel—due to the corrosion of fuel storage equipment and engines caused by sulphurous compounds and atmospheric pollution by sulphur oxides in exhaust gases (Kapustin 2007).

Several types of catalysts are currently applied, namely:

- Molybdena-alumina catalyst is synthesised by applying ammonium paramolybdate on γ-Al_2O_3 particles' surface from water solution of a particular concentration.
- Cobalt-alumina catalyst is produced by saturating aluminium oxide tablets γ-Al_2O_3 with $Co(MO_3)_3$ water solution at room temperature.
- Cobalt-alumina molybdenum hydrofining catalyst is produced by saturating aluminium hydroxide with Co and Mo salts solution.
- Nickel-alumina-molybdenum catalyst is applied for hydrofining of diesel fuel and other oil fractions (e.g. vacuum gas-oil in catalytic cracking installations).
- Zeolite catalyst is based on gamma-aluminium oxide and PdY and $AlPO_4$ zeolite types, using as source material NY, γ-Al_2O_3, $PdCl_2$, zeolites,

ethylenediaminetetraacetic acid, NH_4OH (water-based), NaOH, and sulphuric acid.

Most promising "process-catalysts" complexes are:

1. Process with low (100–50 ppm) residual sulphur content (for standard material with initial sulphur content of about 1%)—aluminium oxide (carrier), cobalt-molybdenum, or nickel-molybdenum. The process is regularly performed under the pressure of 35–38 atm.; the higher the pressure, the purer the fuel. Many believe that pressure of 45–50 atm. may be applied. However, the higher the pressure that is used, the more metal would be needed—for thicker walls and use of different pumping equipment, because equipment suitable for pressure range 35–38 atm. could not be used under 45–48 atm. Switching to high-pressure equipment would increase energy consumption accordingly.
2. Process with ultralow (50–10 ppm) residual sulphur content—Nebula-type catalyst. After the first regeneration, its properties drop to the coated catalysts' level.
3. Process with ultralow (50–10 ppm) residual sulphur content (with low volume rate)—aluminium oxide (carrier) and cobalt-molybdenum (active component). According to the experts, this catalyst is applied at nine out of every ten installations in the world (and only one of ten uses Nebula-type catalysts).
4. Process NZSD (under 10 ppm)—Nebula type.
5. Process NZSD (under 10 ppm)—aluminium oxide (carrier) and nanomodified cobalt-molybdenum (active component) is thought to have very high potential.

4.2.5 Hydrocracking

Hydrocracking is among the most quickly advancing oil refining processes. Suffice it to say that in the last 20 years, hydrocracking installations' capacity grew by four times globally: they process almost 250 million m^3 of raw materials a year. The process allows, through application of appropriate catalysts and operating parameters, to achieve a high output of a wide range of high-quality components of mainline oil products from practically any hydrocarbon raw material—such as liquefied gases, jet and diesel fuel, lubricant components, etc. The difference between hydrocracking and hydrofining is that the former includes processes where over 10% of the material is subjected to destruction, reducing molecular size. Depending on the conversion rate, hydrocracking processes are divided into light (soft) hydrocracking (LHC) and deep (hard) one. Accordingly, in the first case the conversion rate is between 10% and 50% and in the second over 50%.

The first group of processes is designed for both raw material treatment (for subsequent processing) and for increasing "light" oil products' output. The second group is applied exclusively to increase "light" oil products' output. Hydrocracking of oil distillates and residue to produce "light" oil products is a relatively new destructive oil refining process and is widely applied in the USA. Hydrocracking's main strength is the opportunity to process both distillate and residue materials, producing high-quality products such as liquefied gases, high-octane petrols, waxy diesel, and jet fuel. Hydrocracking is the only secondary oil refining process which allows jet fuel resources to be significantly increased. Most of hydrocracking processes are designed to process distillate material (heavy atmospheric and vacuum gas-oils, cracking and coking gas-oils, deasphaltisates). Hydrocracking is a highly selective and flexible technology; slight modification of the process's conditions significantly affects end products' characteristics. The following types of industrial hydrocracking processes are currently applied in oil refining industry:

- Light hydrocracking of vacuum gas-oils, to upgrade catalytic cracking materials while at the same time producing diesel fractions
- Hydrocracking of vacuum distillates under pressure, to produce motor fuels and high-index lubricant bases
- Hydrocracking of oil residue, to produce motor fuels, low-sulphur fuel oil, and materials for catalytic cracking

Three "process-catalyst" complexes can be identified at this stage:

- Single-stage hydrocracking—catalyst based on amorphous aluminium silicates containing NiWS-phase sulphide nanoparticles
- Single-stage hydrocracking—catalyst based on crystal aluminium silicates (zeolites) containing NiWS-phase sulphide nanoparticles
- Two-stage hydrocracking—catalyst based on crystal aluminium silicates (zeolites) containing platinum nanoparticles

Single-stage hydrocracking is currently considered to have the second highest potential for practical application in Russia; it offers a number of advantages including comparatively low capital investments in installations' construction, low running costs, and a very high quality of kerosene (sootless flame height more than 28 mm). But it also has a low conversion rate into target products and low quality compared with the two-stage process. Single-stage cracking is based on two catalysts:

1. Catalyst based on amorphous aluminium silicates containing NiWS-phase sulphide nanoparticles provides high selectivity for producing motor fuels and low selectivity of gas formation (few side products). On the other hand, the catalyst is unstable—it quickly deactivates so its service life is short (1 year), while the zeolite-based catalyst serves for 2–3 years.
2. Catalyst based on crystal aluminium silicates (zeolites) containing NiWS-phase sulphide nanoparticles is believed to be the best for the single-stage scheme. It has higher stability compared with amorphous aluminium silicate.

Selectivity, as a highly relevant consumer characteristic, is currently a factor contributing to the development of this catalyst's production the world over. It is expected to significantly improve by 2030. Service life without regeneration is also expected to increase. The latter parameter deserves special attention in production of catalysts for moving systems; its importance is quite comparable to selectivity. The experts also expected significant improvement of such catalyst parameters as activity and strength.

4.2.6 Alkylation

4.2.6.1 Alkylation of Isobutane with Butylenes

In the process of butylene's and isobutane's amalgamation in the reactor, butylene has a tendency to amalgamate with itself—which reduces productivity. This problem is solved by increasing the amount of isobutane which envelopes butylene and does not allow other butylene to join it, while the catalyst allows to reduce the number of isobutanes required to isolate 1 butylene (beta, 10 to 1; mcm22, 6 in total). To achieve maximum efficiency of the process, a catalyst is required which would create conditions for forming alkylbenzene from 1 isobutane and 1 butylene. In that case separation becomes unnecessary, and heat release is greatly reduced. There are two "process-catalyst" complexes available (catalytic process scheme and relevant catalyst):

1. Standard process—hydrogen fluoride- or sulphuric acid-type catalyst. It's a zeolite catalyst which is not applied at industrial-grade installations.
2. Standard process—Y (faujasite)-type catalyst was developed after fluoride- or sulphuric acid-type catalysts. Its advantage over the latter is that it is non-toxic, non-corrosive, and does not generate acidic drainage or acidic gases. Catalysts of this type, due to their high activity, perform the process at a relatively low temperature and in a liquid-phase reactor. Currently this complex is believed to have the highest potential.

4.2.6.2 Alkylation of Benzene

Alkylation with Ethylene

In the process of ethylene's and benzene's amalgamation in a reactor, ethylene has a tendency to amalgamate with itself and with ethylbenzene—which reduces productivity. This problem is solved by increasing the amount of benzene which envelopes ethylene and does not allow other ethylene to join it, while the catalyst allows for the reduction of the number of benzenes required to isolate 1 ethylene (pentasil, 40 to 1; mcm22, 6 in total). To achieve maximum efficiency of the process, a catalyst is required which would create conditions for forming ethylbenzene from 1 benzene and 1 ethylene. In that case separation becomes unnecessary, and heat release is greatly reduced.

Y-type catalyst—due to their high activity—can perform the process at a relatively low temperature and in a liquid-phase reactor. zsm5-type catalyst (pentasil) is the second generation of catalysts (after Y). Its advantage over the Y type is that the pore size is the same as benzene (raw material). The zsm5 catalyst can work under higher temperatures and lower pressure than the Y type (which reduces the amount of steel, etc. needed to build installations). Since zsm5 type can operate under higher temperatures, its activity is an order of magnitude higher than that of the Y type, so it is beginning to supplant the latter.

On the other hand, zsm5 catalysts have the lower selectivity than the Y type, and also they can only continuously operate for half a year and require oxidative regeneration twice a year—which is hardly profitable since the facility must operate two reactors and switch production between them—thus increasing energy consumption.

Zeolite beta-type catalyst belongs to the next (third after zsm5 catalysts') generation. Its activity is similar to that of zsm5, but it has higher selectivity than zsm5. This type of catalyst also can work in a liquid-phase reactor under low temperatures—which according to the experts is an advantage because you only need one reactor and do not have to switch production from one to another. Also, zeolite beta-type catalyst has a longer "mileage" without regeneration than zsm5.

The mcm22-type catalyst is the latest-generation catalyst which can work in liquid-phase reactor under lower temperatures than zeolite beta type. The mcm22 type also has higher productivity and selectivity.

Alkylation with Isopropylene

In the process of propylene's and benzene's amalgamation in the reactor, propylene has a tendency to amalgamate with itself and with isopropylbenzene which reduces productivity. This problem is solved by increasing the amount of benzene which envelopes propylene and does not allow other propylene to join it, while the catalyst reduces the number of benzenes required to isolate 1 propylene (beta, 10 to 1; mcm22, 6 in total). To achieve the maximum efficiency of the process, a catalyst is required which would create conditions for forming isopropylbenzene from 1 benzene and 1 propylene. In that case separation becomes unnecessary, and heat release is greatly reduced. It turns out that the following "process-catalyst" complexes (catalytic process scheme and relevant catalyst) are most promising:

- Y (faujasite)-type catalyst is developed after aluminium chloride or phosphoric acid carrier-type catalysts. Its advantage over the aluminium chloride or phosphoric acid carrier-type catalysts is that it is non-toxic, non-corrosive, and does not generate acidic drainage or acidic gases. Catalysts of this type, due to their high activity, can perform the process at a relatively low temperature and in a liquid-phase reactor.
- Zeolite beta-type catalyst belongs to the third after Y catalysts' generation. Its activity is similar to that of the Y type, but it has higher selectivity than the Y type. Also, zeolite beta-type catalyst has a longer "mileage" without regeneration than the Y type.
- mcm22-type catalyst is the latest-generation catalyst which can work in liquid-phase reactor under lower temperatures than zeolite beta type. The mcm22 type also has higher productivity and selectivity.

An analysis of traditional oil refining processes' technological characteristics revealed the most efficient "process-catalyst" complexes with the best short- and long-term development prospects. These prospects are connected with the application of nanotechnologies for the production of next-generation catalysts.

4.3 Future Outlook for Nanotechnologies in Refining Technologies

Oil refining catalysts (further on, catalysts) are nanostructured substances which, in the course of a cycle of intermediate interactions, speed up oil refining chemical reactions without being spent in the process. The development of nanotechnologies is thought to provide the main solutions to improve catalysts for cracking heavy crude oil to refine it into light products, for gasoline fractions reforming, for deep hydrofining of diesel fuel to remove sulphur, to process accompanying oil gases, etc., is the main way to solve numerous oil refining and petrochemical problems.

4.3.1 Catalytic Cracking

Solid and plausible estimates assume that Russian producers of cracking catalysts by 2020 may control up to 80% of the domestic market and achieve the level of product quality on a par with world leaders' products. To accomplish strategic objectives, the country must by 2030 build a powerful production base capable of meeting a significant proportion of domestic demand. Also Russia should become an international technology development hub, offering conditions for creating and disseminating far beyond its borders the most advanced industrial and research technologies, attracting international intellectual capital and high-tech companies, and—most importantly—achieving competitive positions to secure advantages in a specific segment of the technology chain, to successfully integrate into the international division of labour. In the context of the set strategic objectives, four "process-catalyst" complexes appear as having the highest potential for Russian producers: lift reactor, double-regeneration, millisecond, and X-design cracking installations.

Each catalyst's production stage may be implemented using one or more alternative technologies—each of the six stages of producing beads of microspheric catalysts can involve various technological solutions (2–3). For example, forming and drying can be conducted in liquid or gaseous environments while burning and thermo-steam stabilisation in fire or electric furnaces.

It was established that the lowest catalyst production costs can be achieved through a particular combination of various bead catalyst production techniques. Given further technological development, production costs are expected to drop during the next 10 years.

The lowest production costs for microspheric catalysts also can be achieved through a particular sequence of production processes. Given further technological development, production costs are expected to drop during the next 20 years. The highest quality of catalysts also can be achieved by combining the following granulated catalysts' production processes. A high level of quality is expected to be achieved by 2020.

Top quality of microspheric catalysts is achieved by combining production techniques, e.g. obtaining high-quality raw materials and zeolites; ensuring sufficient number of ion exchanges; using gas drying in the optimal way; and removing as much moisture and residual compounds as possible, preferably completely (recrystallization quality). Dynamic development of microspheric catalysts' production technology is likely to be completed until 2030. Such parameters as capital intensity, labour intensity, required skill level of personnel, and energy consumption are expected to rapidly grow. An additional stage of the technology is expected to emerge—the treatment or purification of raw materials, which would allow the improvement of the catalyst's properties such as channel size and configuration, and the number of active centres. This will lead to increased selectivity and activity of the catalyst. Relevant technological solutions are expected to be implemented by 2020.

4.3.2 Isomerisation of Light Gasoline Fractions

By 2020 the top Russian producers of isomerisation catalysts may control 50% of the domestic market, exceeding world leaders in terms of product quality. Especially low-temperature isomerisation catalysts are the most efficient which enable the production of products with 88–89 octane number in a raw material's single pass through the reactor. Currently the low-temperature catalyst-based complex is not present on the Russian market, but such complexes can be implemented quite quickly—which would enable the use of the catalyst in the most efficient way for the isomerisation of *n*-butane, pentane-hexane, or pentane-heptane fractions, at the lowest production costs. Among new unconventional "process-catalyst" complexes designed recently is the BIMT process—single-stage catalytic technology which allows for the production of a wide fraction of light oil hydrocarbons (overpoint to 380 °C) or gas condensates without preliminary distillation of the fraction into three commercial products, e.g. liquefied propane-butane fraction, Euro-4 petrol, and Euro-3 and Euro-4 diesel fuel. Compared with existing classic industrial technologies for production of these fuel types, BIMT technology allows for the reduction of capital investment and running costs by 6–8 times and energy consumption by 3–4 times. Another advantage is that it can be applied in a commercially viable way starting on a very low scale (20–50 thousand tons a year)—i.e. it can be implemented in hard-to-reach Far North areas.

The catalyst for this technology is currently produced at the rate of 150 tons a year (and there is potential to increase the output by six times). Currently the catalyst's production technology involves six sequential stages. Each stage can be implemented in several alternative ways, i.e. there are various (5–6) available technological solutions. The lowest production costs of low-temperature catalysts can be achieved by combining various production technologies for various stages of the process. If the technology is improved further, production costs are expected to drop further. Meanwhile it should be noted that the catalysts' quality is not affected by specific combination of production processes and their application. Any of the above combinations, and any way of applying them, result in the same quality of catalyst. For isomerisation of wider gasoline fractions, it would make more sense to use medium- or high-temperature isomerisation catalysts (at least until 2020).

4.3.3 Hydrofining

The top Russian diesel fuel hydrofining catalyst producers by 2020 may control 50% of the domestic market and by 2030 60% and reach the level of product quality on a par with world leaders'. Taking into account the practical aspects of super-low sulphur content diesel fuel production, the following conclusions can be made.

1. Changes in quality requirements for diesel fuel aim to reduce sulphur content. Changing octane number, density, boiling temperature for the 95% fraction, etc. can be a matter of the not so distant future.
2. The potential for restructuring existing installations for the production of super-low sulphur content diesel fuel (SLSDF) will largely depend on their design and operational parameters. Many refineries can discover that the major reconstruction of their existing single-stage installations would not be the most practical solution for SLSDF production. For many of them, reconstruction coupled with the integration of the second production stage, or the construction of a new installation for the second production stage (additional purification), would make more sense economically.
3. A more demanding operating regime required for SLSDF production can negatively affect output and lead to increased energy consumption. Also, SLSDF-producing oil refineries will in effect have to produce transport fuel identical to chemically pure products, so operating the installation is going to be much harder.

All of the above will significantly affect oil refineries' operations and distribution infrastructure. They will need additional reservoirs, extra facilities for waste disposal, loading/unloading of catalyst, new distribution techniques, etc. Within the next decade, aluminium oxide in these complexes will be replaced with a new carrier, e.g. nanostructured titanium dioxide coated with cobalt-molybdenum or platinum (if under 10 ppm), to reduce sulphur content (purification rate of 1–5 ppm) and nitrogen, since platinum hydrates nitrogen compounds quite well. From an environmental point of view, the nitrogen removal problem is not yet very relevant, but people are beginning to talk about it in Europe. The new purification process will include five stages: carrier preparation, impregnating solution preparation, impregnation, drying, and burning. In turn, more thorough purification will allow for meeting the demand of one or several consumer groups more fully. Relevant technological solutions may be applied by 2030. New ways of implementing the technological stages of existing catalyst production processes are new *drying* technologies, such as microwave drying which can reduce capital investments and increase productivity and contact drying, i.e. a more uniform distribution of active components in granules which improves quality, increases activity, and reduces energy consumption (by 15–20%).

4.3.4 Hydrocracking

Hydrocracking is one of the processes with the highest potential for destructive processing of heavy distillate and residual materials. In recent years the work on improving residual hydrocracking technologies has been significantly stepped up. The wide application of hydrocracking is hindered by high capital investments and high running costs, due to the need to apply the process under high pressure, and with high hydrogen consumption. Accordingly, light hydrocracking processes (LHC) are widely applied, which provide a sufficiently high output of medium distillates and a significant amount of high-quality FCC materials under moderate pressure (less than 10 MPa). Modern hydrocracking catalysts allow for applying this process at ordinary vacuum gas-oil hydrofining installations, after some minor reconstruction.

A major direction of distillate materials' hydrocracking development is designing highly efficient, stable, and easily regenerated catalysts. Currently, along with advanced amorphous catalysts, nanostructured zeolite catalysts are widely applied. Zeolite-containing catalysts provide the highest output of medium distillates, are highly flexible, and allow the application of the production process under a softer regimen. Another important area of modernising the hydrocracking process is finding better ways to remove heavy multiring aromatics (HMA, i.e. containing 11 or more rings). Experts on the operation of installations agreed that the problem of removing emerging HMA cannot be solved exclusively by selecting appropriate catalysts: it requires targeted reconstruction to remove contaminations emerging in the course of hydrocracking.

In recent years a targeted modular PNA management process was developed, specifically for selective adsorptive removal of HMA from recycling hydrocracking flow. Currently this system is successfully applied at a number of industrial installations in various regions of the world. It is different from single-stage hydrocracking in that it has an adsorber. Adding an adsorptive HMA removal module as part of a hydrocracking installation's reconstruction pays off in less than a year. Removal of heavy multiring aromatics can also be useful in the case of upgrading recycling products in adjoined systems, e.g. hydrocracking-FCC, which allows for increasing the overall output of light products (petrol from catalytic cracking and diesel fuel from hydrocracking), and reducing microspheric catalyst consumption during FCC.

Modernisation of internal reactor devices recently became an equally important process development area, in order to, among other things, increase the output of valuable liquid products by reducing the formation of gases. Generally, in recent years, hydrocracking technology became significantly different from previous hydro-destructive process models. Taking into account these changes, and new opportunities to produce cheap hydrogen, hydrocracking can be definitely expected to play a leading role in solving major oil refinery problems during the next 20 years, in particular production of non-sulphurous de-aromatised and waxy diesel fuels. A modern modification of hydrocracking process is catalytic isocracking, which in addition to removing hetero-compounds and aromatic hydrocarbons performs selective hydroisomerisation of *n*-paraffins. Due to stricter requirements for oil products, first of all concerning sulphur content, hydrocracking will be developing towards the application of more efficient catalysts; the combination with hydrofining and de-metallizing; the removal of HMA; the improvement of hardware solutions; and the development of more efficient technological schemes.

For the next two decades, the complex based on single-stage scheme and catalysts based on crystal aluminium silicates (zeolites) containing NiWS-phase sulphide nanoparticles is seen as a technology with the lowest production costs compared with other complexes and catalysts in this market segment—since it has a longer operational life without reloading and allows for the processing of cheaper raw materials. Two-stage hydrocracking with a catalyst based on crystal aluminium silicates (zeolites) containing platinum nanoparticles is seen as a priority complex in terms of application by Russian producers; it offers several advantages, the main ones being deep processing and the opportunity to meet the demand of one or more consumer groups as fully as possible. On the other hand, the experts noted high capital intensity and production costs. If productivity, labour intensity, and demand for skilled personnel remain unchanged, capital intensity is expected to grow by 2030, while energy consumption should decline.

4.3.5 Alkylation

According to certain forecasts, Russian producers of isobutane-butylene alkylation catalysts by 2020 may control between 80% and 100% of the domestic market if the first installation is built in Russia, and up to 20% if the first installation is built abroad, and achieve the level of product quality on a par with world leaders. Also, Russian producers of benzene-ethylene-isopropylene alkylation catalysts by 2020 may control 50% and by 2030 80% of the domestic market and achieve the level of product quality on a par with world leaders. As of 2020 they are expected to hold and strengthen these leading positions. During 2015–2020 labour intensity, capital intensity, energy consumption, and the share of faulty products are all expected to drop. At the same time demand for highly skilled personnel will grow.

Production technology for zeolite-based catalysts for alkylation of isobutane with butylenes (alkyl benzene production) and alkylation of benzene with ethylene (ethylbenzene production) and with propylene (isopropylbenzene production) allows for production of catalysts with a more ordered crystal structure, which increases their activity and selectivity, and definitely has development priority for the period until 2020, leading to reduced labour intensity and increased productivity. Also, the most promising combinations of processes for each stage of these complexes' production technologies have been identified in recent years, allowing for a reduction in production costs compared with other complexes and catalysts in this market segment. It is expected that in the future new ways to implement these technological stages will be developed; in particular, raw materials will be treated by ultrasound crushing (alkylation); solutions will be prepared for activating the material by ultrasound or magnetic radiation. Continuous crystallisation and ion exchange techniques will also be applied, as well as granulation with a reduced share of binder and contactless drying. By 2020 catalyst production technology may emerge based on the introduction of nanosized precursor in the soaking system with the subsequent extraction of precursor from reaction products and its recycling. This technology is based on a trial way of producing platinum salt and offers a distinct cost advantage: it will allow to reduce production costs of petrol, etc.

4.3.6 Catalytic Reforming

Significant progress has been made in recent years in developing catalysts and technologies for reforming gasoline fractions. In the late 1990s to early 2000s, a breakthrough in the process's qualitative characteristics was achieved. The depth of raw material aromatisation came close to the thermodynamic equilibrium value, while the length of technological cycle in the RON 96–98 reformate production mode has reached 2 or more years. This progress was largely achieved by applying latest-generation catalysts. The main directions of the reforming process development include the following:

1. The development of dual-purpose process: producing high-octane petrol component and aromatic hydrocarbons, which increases the process's flexibility.
2. The development of new reforming catalysts. In the case of aromatic hydrocarbons, it is the catalysts that increase benzene and xylene output. In case of high-octane petrols, it is the catalysts where high octane number is achieved by increasing the share of isoparaffins and reducing the share of aromatic hydrocarbons.
3. The reconstruction of operating installations with stationary catalytic layer, by upgrading the continuous regeneration stage (dual forming, octanising) or by upgrading equipment.
4. Combining the reforming process with various others (first of all isomerisation), to improve product quality.

Currently there is an emerging trend to modernise the existing catalytic reforming installations (their combined capacity is about 20 million tons a year) to apply isomerisation (up to 5 million tons a year) and pentaforming (up to 15 million tons a year) processes, to minimise the need for new construction. The essence of pentaforming is, using technological scheme of traditional catalytic reforming installation as a basis, to radically change the process's chemistry—to make isomerisation and C_5-cyclodehydrogenation the main direction of paraffin hydrocarbons' transformations—which increases the octane number of the non-aromatic part of petrol reforming up to 70. Calculations show that in this case the total content of aromatic hydrocarbons can be reduced to 40–45% while retaining reforming petrol's octane number at the level of 95 (Pashigreva et al. 2007). The implementation of this approach requires developing technology for low-temperature isomerisation of light gasoline fractions and producing solid catalysts based on nanostructured zeolite materials. Meanwhile the technology allowing for meeting the demand of the majority of consumers, which provides reformate output with octane number of up to 105, is frequently considered possessing the highest potential for the next 20 years: the process in moving catalytic layer with zeolite-based platinum-containing catalyst, including with other metal additives, i.e. bimetallic. During the next decade, new technological solutions are expected to emerge, in particular for producing highly porous carrier with competitive advantages—wider fraction range of raw material. Production technology for "platinum on aluminium oxide" catalyst will be supplemented with a way of evenly applying platinum throughout the zeolite and carrier volume, which would allow for increasing reformate's octane number and reducing the requirements for contaminants' content in raw materials. Relevant technological solutions may be implemented by 2020.

4.4 Measures for Spurring Technology and Economic Development

Obviously, producing competitive oil refinery products (as opposed to supplying raw materials) requires the significant growth of consumption of advanced high-quality catalysts—which would be able to produce higher-quality end products. As shown above, supporting the industry primarily involves promoting demand for catalysts. It needs to be noted that the catalyst is not an end product—which means that demand for catalysts depends on demand for oil refining products, which in turn is determined, among other things, by global and local trends. Therefore, several demand factors may be named. Firstly, the price includes catalyst prices and engineering services prices (i.e. application of catalysts and equipment maintenance). That means the actual price of the catalyst is not the crucial consumer decision factor. Also selectivity, i.e. desired products' output, is noted among the most important demand-affecting characteristics. In addition, the stability of characteristics declared at the application stage is very important: varying quality of end products is unacceptable. The catalyst's service life is also important, since replacing it is a complex and highly expensive procedure.

Environmental characteristics become increasingly important, since a major factor affecting demand for end products, e.g. petrol, is their compliance with environmental safety standards. Thus demand drivers can be notionally divided into market-related and partially market-related ones. We say "notionally" because even environmental aspects are up to a point market-related: the consumer will not buy petrol which does not meet environmental standards, though obviously this factor is primarily ecological and social in nature.

For the further development of Russian oil refining industry, it requires the adoption of legislation introducing stricter requirements for oil products' quality and a different taxation policy for oil refineries. Accelerated restructuring of the industry and promoting development and implementation of competitive Russian technologies involve the restructuring of the design services market, first of all by creating a public research and engineering oil refining and petrochemical centre. In order to create an advanced catalyst industry in Russia that is competitive both in Russia and internationally, a whole set of various steps must be taken, including the following:

1. The development of advanced domestic technologies
2. Building advanced catalyst factories, possibly through public-private partnerships
3. Providing support to R&D centres, including guaranteed public funding at the initial stages of research, development, and experimental application
4. Providing economic incentives for major aspects of the industry's activities, in particular:
   - Deep and integrated oil processing
   - Deep processing of natural and associated gases
   - The production of high value-added products, to export products meeting current international standards and those forecasted for the medium term
   - Steps to promote internal demand for high-quality oil refining and petrochemical products, in particular through accelerated development of Russian automobile industry and other industries

Creating economic conditions need to motivate the industry to export high value-added, high-quality oil refining and petrochemical products, instead of crude oil. Following the introduction of Euro-4 and Euro-5 motor fuel standards, new (higher-quality) diesel fuel and vacuum gas-oil hydrofining catalysts became the most relevant. Further on, to increase oil processing depth, advanced cracking and hydrocracking catalysts' role will also increase.

Naturally, Russia, as a leading oil producer, is a major consumer of oil refining catalysts. Therefore, all international companies—catalyst producers, both the leaders and the "rank-and-file" ones—are quite active in Russia. Foreign companies see the Russian market as constantly growing, with a high potential. At the same time, it should be stressed that Russia possesses technologies on a par with foreign ones or even exceeding them, but their large-scale application requires much more development effort and major investments. Among other things, Russia has serious research results in the field of new catalysts, not yet mass-produced anywhere in the world. It is important to realise that Russia has practically all conditions for implementing the full oil refining cycle: cheap raw materials, technological capacity, and R&D potential. The only inadequate things are the main development factors: government support of the industry and investments in production.

Acknowledgements The book chapter was prepared within the framework of the Basic Research Program at the National Research University Higher School of Economics and supported within the framework of the subsidy by the Russian Academic Excellence Project '5-100'.

References

Borodacheva AV, Levinbuk MI (2008) Oil refining industry's development trends and specific features of Russian oil refining. Russ Chem J 6:37–43
Gerzeliyev IM (2008) Review – cracking catalysts
Kapustin VM (2007) Oil processing technology. Part 2. Destructive processes. Kolos, Moscow
Krylov OV (2004) Heterogeneous catalysis. Academkniga, Moscow
Melik-Akhnazarov NK (1980) Catalytic processes for deep oil processing
Parmon VN (2008) Catalysis and nanotechnology: from basic research to major industry in modern Russia. In: Catalysis in industry. Moscow, p 6–17
Pashigreva AV, Bukhtiyarova GA, Klimov OV, Noskov AS, Polunkin YM (2007) Deep hydrofining of primary and secondary oil distillates with new-generation catalysts. Oil Refin Petrochem 10:19–23

Dirk Meissner is Deputy Head of the Laboratory for Economics of Innovation at HSE ISSEK and Academic Director of the Master Program "Governance for STI". Dr. Meissner has 20 years experience in research and teaching technology and innovation management and policy. He has strong background in policy making and industrial management for STI with special focus on Foresight and roadmapping, funding of research and priority setting. Prior to joining HSE Dr. Meissner was responsible for technology and innovation policy at the presidential office of the Swiss Science and Technology Council. Previously he was management consultant for technology and innovation management with Arthur D. Little. He is member of OECD Working Party on Technology and Innovation Policy. He is Associate Editor of Technological Forecasting and Social Change, IEEE Transactions on Engineering Management, Journal of the Knowledge Economy, member of Editorial Review Board at Small Business Economics and Journal of Knowledge Management. He guest edited Special Issues in Industry and Innovation journal, Journal of Engineering and Technology Management, Technological Analysis and Strategic Management among others.

Pavel Rudnik is Director of the HSE ISSEK Centre for Strategies and Programs. He is the author of academic papers on STI development published in leading Russian and international journals, and co-author of monographs on economics of innovation and S&T development. In 2012–2015 Mr. Rudnik was Deputy Director of the Innovation Development Department at the Ministry of Economic Development of the Russian Federation. His main areas of responsibility included methodological and organisational support for technology platforms and innovative territorial clusters, and designing and implementing innovation development programes of state-owned companies.

Part II

Energy and Transport

Renewable Energy Technological Potential Assessment for Evidence-Based Policy-Making

5

Boris Ermolenko, Georgy Ermolenko, and Liliana Proskuryakova

5.1 Introduction

Global energy markets are constantly on the move. Over the centuries we have seen various energy sources dominating the world economy. Waves of innovations in energy technologies led to the paradigm shifts in the industrial and societal development. The ever-increasing competition among nations over fossil fuels has constantly led to military and political conflicts. While some countries earn national incomes on extracting and exporting hydrocarbons, others have applied efforts to secure the uninterrupted import of oil, gas, and coal. The centuries-long dominant position of fossil fuels as economic drivers has led to the establishment of major players—multinational companies that would like to preserve their business as long as possible.

However, over the past decade, the situation has been changing: global investments in renewables surpassed investments in power generation from traditional sources (IRENA 2016). Annual investments in Europe alone amount to \$0.9–1.6 trn (REN21 2017). In 2014 new capacities based on renewables accounted to more than one half of all capacities installed worldwide (OECD/IEA 2015a). One hundred seventy-six countries have recently announced various targets on advancing the renewables, and 154 countries provide government support in attaining these targets (REN21 2017). Among the developed countries, Germany has the highest share of renewables in the total electricity generation (25.2%). Three sources proved especially dynamic: wind, solar PV, and hydropower. Contemporary wind power, hydropower, geothermal power, and biomass power facilities prove to be competi-

B. Ermolenko · G. Ermolenko · L. Proskuryakova (✉)
National Research University, Higher School of Economics, Moscow, Russia
e-mail: germolenko@hse.ru; lproskuryakova@hse.ru

D. Meissner et al. (eds.), *Emerging Technologies for Economic Development*, Science, Technology and Innovation Studies,
https://doi.org/10.1007/978-3-030-04370-4_5

tive with gas-based generation (IRENA 2016). The share of renewables in today's energy consumption is over 19%, including over 10% of the "new" renewables (REN21 2015). In 2000–2014 the number of renewable power plants in Germany has grown 50 times and reached 1.5 mln units. Germany follows an ambitious goal to increase the share of renewables in final electricity consumption from today's 28% to 80% by the year 2050. Italy intends to level off the shares of natural gas and renewables in electricity production. Kuwait and Oman plan to increase the share of renewables to 10% by the year 2020, Saudi Arabia to 50% (without hydropower) by the year 2040. Through scaling up the renewables, the United States could cut greenhouse gas emissions from power production by up to 78% below 1990 levels by 2030 while meeting increased demand (MacDonald et al. 2016).

The main benefits of renewables are common for all countries including zero cost of the energy resources and in some cases lower capital costs for construction of generating facilities (as compared to those working on the fossil fuel). Due to the increasing vehicle-to-population ratio, producers of alternative motor fuels (i.e., biofuels) and producers of vehicles with new engine types (i.e., hybrid, fuel cells) will strengthen their market positions.

At the same time, there are additional specific positive effects of renewables. For a country like Russia, these involve a growing investment attractiveness of end users' energy saving projects, modernization of the energy sector through the introduction of advanced and energy efficient technologies, and a growing demand for decentralized efficient energy generation. The import substitution policy in the energy sector that was launched due to sanctions may open new opportunities to the producers of new energy products and services (including those related to renewables), technology transfer from Asia (China, Japan, South Korea), and localization of the full cycle of renewable energy equipment production.

In order to realize the transition to green growth and mitigate the impact of climate change, IEA suggests to increasing the investments in renewable-based power production by 1.5 times. The most important policy measure to stimulate the development of renewables is the stable demand for the produced energy at a fixed price, assured by long-term energy supply agreements that allow investors and producers calculating payback period of capital investments. This sectoral policy instrument has the largest structural, social, and economic effects. However, a guaranteed connection to the grid is another important condition.

Scholars often analyze the deployment of renewables in different countries and use various methods to assess their impacts on the society, environment, and the economy (Bhattacharya et al. 2016; Hua et al. 2016; OECD/IEA 2015b; Kim et al. 2015; Kumar and Agarwala 2016; Atalay et al. 2016; Mondal et al. 2016). However, the prospects and cumulative effects of renewable energy technology application in Russia are little researched as compared to some other countries. Among the few contributions of Russia are studies on the prospects of bioenergy in Russia (Pristupa and Mol 2015), application of solar energy solutions and the use of other renewables in rural areas (Shepovalova 2015), analysis of green energy production in the North-West Russia for further export to the European Union (Boute and Willems 2012), and production and consumption of wooden pellets (Proskurina et al. 2015).

Studies of technical potential of renewable energy are usually devoted to certain aspects, such as effects of renewable energy use in power supply systems and CO_2 emission reduction (Kumar and Madlener 2016; Boubaker et al. 2016). There are also attempts to make comprehensive national assessments of one or more renewable energy sources (Ghafoor et al. 2016; Abanda 2012; Shepovalova 2015). One of the most detailed studies was performed by Osmani et al. (2013) who presented the renewable energy potential for electricity generation of the United States, taking into consideration multiple criteria (including cost, environmental, and social impacts). This study analyzes the average annual wind speed for onshore power plants, but the height over the ground for wind turbines and technical features of wind turbines under consideration (significant factors that impact the output) is not specified. Moreover, the authors don't assess the quantities of saved fossil fuels and avoided greenhouse gas emissions.

A long-term (30 years) wind observations for three locations in Hong Kong based on the Weibull density function (WDF) was introduced by Lun and Lam (2000). Weibull PDF is the most commonly used function to describe wind distribution (Arslan et al. 2014). Another study by Andresen et al. (2015) presents a detailed 32-year-long hourly model of wind power time series for Denmark for the period of 1980–2035 based on a global renewable energy atlas—RE atlas (the US National Renewable Energy Laboratory computer program). This study aims at illustrating how current differences in model wind may result in significant differences in technical and economic model predictions.

Previous studies on solar and wind potential are based on data from the National Center for Environmental Prediction and the National Center of Atmospheric Research Re-analysis 1 (NCEP/NCAR) (Gunderson et al. 2015) and International Energy Agency (Ramli et al. 2017) and local measurement stations (Bilir et al. 2015; Islam et al. 2011; Azad et al. 2015). Other authors use the NASA Surface Meteorology and Solar Energy (SSE) as a source of reliable data for their analysis (Fathoni et al. 2014). However, most studies provide calculations only for selected locations.

Having reviewed the studies on solar and wind technical potential, we conclude that most studies don't provide a detailed methodology for calculating this potential that makes it impossible to validate and further use the outcomes (Khare et al. 2013; Aslani and Wong 2014; Ahmad and Tahar 2014; Alemán-Nava et al. 2014). Moreover, none of the research papers reviewed above attempts to systematically assess avoided greenhouse gas emissions or fossil fuel savings.

The chapter consists of seven sections. First the methodology section features terms, methods, and information (data) sources followed by a description of the most recent renewable policy and economic developments in Russia. Then a methodology for the assessment of the wind energy and solar PV technical potential and operating conditions for solar PV- and wind power-based generation is introduced. Thereafter, the authors provide actual calculations of various types of renewables technical potential in the Russian Federation. Finally, the conclusions sections discusses policy and social and research implications of the findings.

The approach used in the chapter involves several steps and methods. First, we describe the current position of renewables in the Russian energy sector in terms of capacities, the place of renewables in the energy mix, and the latest developments in the sector. Second, we identify social and economic preconditions and effects (gains) stemming from a faster development and deployment of renewable energy in Russia, including reliability of power supply and the cost of heat and power in different regions of the country, energy supply as a contribution to the quality of life, as well as negative impacts on environment and people's health and conservation of fossil fuels. Third, we offer the latest calculations of solar PV and wind exploitable technical potential for Russia.

This is done through assessment of the following potentials:

- *Fuel potential*—renewables potential in tons of oil equivalent.
- *Heat energy potential*—amount of heat energy that may be produced through renewable energy conversion into heat and the subsequent economy of thermal energy, released by the combustion of organic fuels.
- *Electrical energy potential*—quantity of electricity in kWh, obtained through the use renewable energy sources and the subsequent decrease of electricity produced using traditional energy sources.
- *Resource saving potential*—volume of saved organic fuels (in physical and monetary terms), which would otherwise be used in the amount that corresponds to the energy potential of renewables.
- *Environmental potential* (impact prevention potential)—volume of avoided pollutant emissions in the atmosphere that are typical for combustion of each of the organic fuel types necessary to obtain energy that corresponds to the energy potential of renewables. Assessment of the environmental potential may be done by type of pollutants (emissions) in physical terms, as well as in tons of CO- and CO_2-equivalents. We also assess the prevented environmental-economic damage emanating from the local and global air pollution.

The sources of information for assessing the electrical energy, heat energy, environmental potential, and resource saving potential of various types of renewables are information on the fuel potential of these sources with disaggregation by region of the country. The assessment of the potential of various types of renewables is based on the interrelated indicators:

- Consumption of fuel (in oil equivalent) versus electricity production
- Consumption of fuel (in oil equivalent) versus heat production
- Amount of oil equivalent versus equivalent volumes of natural gas, coal, fuel oil, and diesel
- Amount of combusted natural gas, coal, fuel oil, and diesel versus amounts of discharged air pollutants
- Amounts of certain air pollutants discharged versus their cumulative reduced mass in CO- and CO_2-equivalents
- Reduced mass of air pollutants discharged versus magnitude of environmental-economic damage (losses) due to negative impact on the environment

This approach allows making the up-to-date assessments of the exploitable technical potential of renewables by type and region of Russia with their fuel, power, heat, resource saving, and environmental effects for the country and its 85 regions based on the data from Kiseleva et al. (2015). The proposed classification by type of potential is linked with several goals of renewable use, including reducing the fossil fuel consumption for heat and power generation and the subsequent reduction of air pollution from fuel combustion. In doing so the approach overcomes methodological limitations of previous country-level assessments, such as incomplete or unreliable data sources and absence of hourly, daily, and monthly assessments among others. The present assessments were done with the use of the NASA SSE database, the network of Russian state weather bureau and aerological stations data, official statistical data (by region). Comparison with previous calculations of renewable energy potentials in Russia, namely, Bezroukikh et al. (2007) and Nikolaev (2011), allowed identifying and overcoming some of the previous methodological limitations and to cross-check and verify the outcomes. The main outcomes of the study are a further elaborated methodological approach for calculating the technical potential of renewables compared with previous studies, as well as latest calculations of this potential for Russia. However, the study doesn't review the economic potential and possible variations under different future energy scenarios, i.e., only one "ideal" development option is considered.

5.2 The Development of Renewables in Russia

Russia has been predominantly dependent on fossil fuels, both for domestic use and as a source of country's export revenues (Proskuryakova and Filippov 2015). Mining and quarrying of fossil fuels represents nearly 37% of the gross value added in the Russian economy (Russian Federal State Statistics Service 2015). Advancing renewable energy technologies is a challenge for Russia, oriented toward the use of traditional sources. Natural gas, oil, and coal constitute over 91% of country's total primary energy supply (51.8%, 21.8%, and 17.3%, correspondingly), with nuclear at 6.2%, small hydro at 1.9%, biofuels and waste at 1%, and geothermal at 0.1% (OECD/IEA 2013). The country's difficulties associated with relying on the traditional energy mix include the exhaustion of traditional hydrocarbons, negative impacts on environment and climate, barriers to international energy technology cooperation, hardly predictable fossil fuel prices, and increasing competition at traditional energy markets.

In 2009 renewable-based electricity generation in Russia was found as not more than 1% of the total electricity production (UNDP 2010). The total capacity of renewable-based electricity generation facilities and power plants a few years ago was estimated at 2249 MW, 57% of which was produced at bioenergy-based heat and power plants, mostly located near wood, pulp, and paper processing plants, followed by small hydropower plants (37%), geothermal power plants at Kamchatka region (5%), solar power plants (0.5–0.8%), and wind power plants (0.1%) (UNESCO 2014).

Russia's abundant resource base of renewable energy sources is acknowledged by the International Renewable Energy Agency (IRENA), REN21, International Energy Agency (IEA), and the International Finance Corporation (IFC). Legislation in force and industry support mechanisms would allow Russia achieving its initial 4.5% target of installed renewable electricity generation by 2020 (IFC 2016). One of the notable barriers for investments and more-precise monitoring measures in Russia is the lack or insufficiency of statistical data on energy production and use (REN21 2016). Over the last decade, renewable energy development in Russia and worldwide was spurred by the continuous growth of energy consumption and increasing competitiveness of renewable energy technologies. Development of renewables in Russia has sped up after the year 2010 (Government of the Russian Federation 2015). The government has set targets for wind and solar energy at the wholesale market until 2024 and the respective conditions for the development of all renewables at retail power market and in the off-grid areas. Legislative and regulatory documents already helped launch the solar power projects with 400 MW of total installed capacity (and prospects for increasing up to 1.5 GW by 2024) and wind power projects of 780 MW (with prospects to increase this figure up to 3.6 GW by 2024) (Russian Ministry of Energy 2015). Moreover, decentralized power generation from renewables is rapidly growing, primarily among manufacturing and agricultural enterprises.

The Russian Statistical Service reports that of all renewables (excluding large hydro) the highest volume of installed capacity belongs to publicly owned solar power plants (227 MW), which is more than double the capacity of wind power plants (Table 5.1). Comparable volume of electricity is produced by privately owned off-grid solar power plants.

The main directions of state policy in the sphere of Renewable Energy Sources (RES) identify the following targets of renewable energy development in Russia:

- Increasing power sector energy efficiency on the basis of renewable energy sources, necessary for a reliable, sustainable, and long-term energy supply for economic development of the Russian Federation
- Involving innovative science-intensive technologies and equipment in the energy sector and the development of local production of RES-based high-tech generating and auxiliary equipment
- Reduction of greenhouse gas emissions as one of the most important activities related to the implementation of Russia's international commitments

The local production development of high-tech generating and auxiliary equipment for renewable energy sector is the goal of government measures to support renewable energy industry.

Energy policy in Russia has recently introduced several tools to support renewable-based electricity production:

– 2024 target indicators for the development of renewables by type.

Table 5.1 Technical and economic indicators of public-owned power plants (without nuclear plants) (2014)

			Electricity produced MW/h			Thermal output by power station, including boiler house, total, gigacal					Share of electricity consumed for own power station consumption	
							Of which					
	Installed capacity, kW	No. of hours of installed capacity use	Total	Including heat energy (in cogeneration mode)	Net output electricity, MW/h	Total	By turbine units	By peak water-heating boiler	By desuperheating and pressure reducing units	By boiler house	For electricity production, %	For heat output, kWh/gigacal
Russian Federation	204,281,376	4058	814,465,278	187,620,864	762,609,972	519,760,953	405,256,474	35,088,879	29,794,880	34,047,328	4.2	34.0
Heat and power plants, public-owned	153,228,592	4249	639,022,727	187,620,864	587,945,092	519,760,953	405,256,474	35,088,879	29,794,880	34,047,328	5.2	34.0
Hydropower plants, public-owned	49,435,295	3532	172,869,476	x	172,145,538	x	x	x	x	x	0.4	x
Hydropower-pumped storage plants	1,215,900	1541	1,873,897	x	1,851,140	x	x	x	x	x	1.2	x
Geothermal power plants	74,000	6035	446,589	–	416,330	–	–	–	–	–	6.8	x
Wind power plants, public-owned	100,308	1235	92,600	x	91,883	x	x	x	x	x	0.8	x
Solar power plants, public-owned	227,281	–	159,989	–	159,989	–	–	–	–	–	–	x

Source: Russian Federal State Statistics Service (2016). Approximately 19% of Russia's electricity is generated by large hydropower plants. However the small hydropower plants' installed capacity is small—around 250 MW, slightly above solar plants. Tidal energy is used only at one experimental Kyslogubskaya power plant (RusHydro 2016). None of publicly owned renewable-based power plants produce heat energy in cogeneration mode

x denotes not applicable; – denotes zero

- Target indicators for the degree of localized production of the main and auxiliary power generation equipment based on renewables.
- Trade of installed capacity of renewable-based power generation facilities at the wholesale market within the target indicators for renewables (by type) and capital cost limits for the construction of 1 MW of facility's installed capacity by type of renewables.
- Return on invested capital and currency fluctuations in Russia are included in the marginal capital costs.
- Utilities are obliged to buy electricity (that amount to 5% of power grid losses) at retail market from renewable-based facilities.
- Up to 50% of the cost for connection to the power grid is compensated to the renewable-based facilities.

These measures may bear fruit in the mid- to long-term perspective should the general economic conditions be favorable to the establishment and growth of small and medium energy companies, international renewable technology transfer, and the diversification of business for large energy companies.

The energy policy priorities toward decentralization of power and heat generation and diversification of energy resources in the supply-demand balance may contribute to a considerable structural shift not just in the energy sector but in the economy overall. A consistent policy of decentralized energy generation will allow many companies, which generate power for own needs, to be more flexible in planning their operations including the choice of production sites and even to cut costs.

Despite some Russian regions offer better conditions for the deployment of renewables, it is important to assess their prospects across the country at all accessible territories.

The deployment of renewables in many of Russia's regions appears economically, environmentally, and socially justifiable (Ermolenko et al. 2013a, b), primarily, for the following territories:

- Regions with decentralized power supply, which account for approximately 70% of the country's territory inhabited by 10–20 mln persons. Power production in these regions is made predominantly with the use of low-power petrol and diesel generators, which require expensive organic fuel transported from other regions. The prime cost and commercial tariffs for electricity in such regions could go as high as RUB 250 ($4.4) per kWh.
- Regions with centralized power supply and high capacity deficit, extremely high costs and extensive technical difficulties associated with connection of consumers to power grid and heating system, and substantial financial losses due to frequent disconnection of final users from the grid.
- Settlements with problematic environmental situation, including high air pollution from industrial and urban facilities that use fossil fuels.
- Regions with high renewables potential.
- Regions with obsolete and depreciated traditional generating equipment.
- Cities and places of public entertainment and medical treatment.

- Regions that face difficulties in power supply of private houses, private farms, places of seasonal employment, and private garden plots.

5.3 Renewables Technical Potential Assessment

5.3.1 Wind Energy

In order to define the wind energy potential for investment purposes, we suggest introducing a conceptual framework with differentiation by the following features:

- Type of potential (resource)
- Potential estimation method
- Territory for which the potential is assessed
- Validity period of particular wind speeds frequency when estimating potential
- Time period for which the energy potential is assessed
- Method for assessing the wind vector
- Wind speed characteristics, determining range of application of wind turbines
- Specific indicator type of the wind energy potential

The proposed classification by type of potential is linked with renewable energy development objectives, such as decreasing the consumption of heat and electrical energy, produced by combustion of fossil fuels, and, consequently, decreasing fossil fuels consumption and volume of air pollutants.

The available technical wind energy potential of a given territory is understood as part of long-time average annual total wind flow energy (gross natural wind energy resource) that can be converted into electrical energy by real wind turbines during the reference time interval on territories available for installation of wind turbines at the time of assessment, based on the efficiency of its use.

Factors for smart selection of the Russian territories for assessing the *available technical wind energy potential* are based on the deployment of wind turbines at distances ensuring minimal power losses associated with the mutual influence of wind turbines and the potential to assuring a minimum negative impact on the environment associated with the creation of acoustic discomfort zones and withdrawal of lands from the economic turnover.

The technical potential at selected available lands is defined as the quantity of power that can be generated by real wind turbines without any additional restrictions.

In this chapter we consider the technical potential in the following territories: federal districts (districts of the Russian Federation) and national (Russian Federation). The suggested classification of wind energy potential allows us to develop a set of indicators that may be used for a wide array of regional and local tasks associated with analysis of wind potential, technical, and economic possibilities of its use, potential resource (including fossil fuels) saving, prevention of local and global environment pollution, as well as for the development of regional wind energy programs and decision-making on the investments into centralized and decentralized energy supply systems partly based on renewables.

Assessment of the wind energy technical potential was made in the frame of the "Atlas of renewable energy resources in Russia" reference book (Kiseleva et al. 2015) and is substantially different in methodological terms from approaches used earlier in similar studies, i.e., for the development of a "Guide on renewable energy resources in Russia and local fuels" (Bezroukikh et al. 2007) and the studies performed by Nikolaev (2011). The analysis performed in 2015 and presented in this chapter is much more oriented toward practical projections of renewable (wind and solar, in particular) power plants both for centralized power supply and distributed hybrid systems that supply power to small consumers. The methodological differences between the present study and previous works on assessing renewable potential include:

The use of base information on the wind speed PDF at the 50 m height over the ground level on coordinate grid with 64,800 1° × 1° cells from the database of the US National Aeronautics and Space Administration "NASA Space Environments and Effects Program" (NASA SSE) (NASA 2015), instead of information on wind speed PDF obtained through statistical processing of the data at the very few and unevenly distributed through Russia, technically imperfect state meteorological (10–18 m height) and aerologic (100, 200, 300 m heights, and more) stations. Given the large territory of Russia, most of which is high-latitude, verification of the data is of particular importance. This was done for locations for which national meteorological stations data is available. Those national stations that are representative for typical Russia's regions with high wind potential were chosen for performing a correlation analysis. Statistical processing of data showed a high degree of correlation $K_{\mathrm{corr}} = 0.818$ between ground-based measurement data and NASA SSE data. The methodology was validated also through a comparative analysis of the data obtained in the course of the study with the data of wind monitoring campaign carried out by a German engineering company CUBE Engineering GmbH in Krasnodar region for 18 months using 70 m masts with wind sensors located at three levels in line with international standards. This data verification proves the validity of the study.

- Wind energy potential was calculated for a year, each of its months, average daily value for each month, and every 3 h of a day, while the other sources provide information only about annual potential.
- Mathematical model was developed and tested for identification of parameters for the Weibull wind speed PDF at any given height over the ground level on the basis of data on wind speed repeatability at a certain basic distance. This allowed assessing wind energy potential for various heights for the whole territory of Russia, resolving the problem of identifying wind speed profile for the full range of heights of rotor axis of existing wind turbines, and, consequently, addressing the issue of their optimal choice.

- Potential assessment was made both for each cell of the coordinate grid (map) and for each region of Russia and the whole country, while in previous calculations the assessments did not go into territories smaller than the country's regions.
- For this study the wind potential was estimated with a view of actual deployment of wind turbines on the whole or available part of any given territory, for its space unit and one wind facility.

One of the key parameters to evaluate the technical potential of wind energy of a Russia's region is the potential availability of the territory, at which the wind turbines may be located given the technical and environmental limitations. In line with the best international practice on assessing the lands for wind energy use and the existing Russian classification of land stock, the authors consider only those lands that are suitable and available for wind energy generation purposes, including certain categories of agricultural lands and pastures.

5.3.2 Solar Energy

For solar energy we introduce the conceptual framework of energy potential and resources with differentiation along the following features:

- Type of potential (resource)
- Potential estimation method
- Territory for which the potential is assessed
- Time period for which the energy potential is assessed
- Method for assessing position of the sun for assessing the potential
- Availability of unavailability the system to track the position of the sun
- Specific indicator type of the solar energy potential

Except features 5 and 6, we use the same features in the wind energy potential assessment.

Total solar radiation directed toward the solar receiver shows highest efficiency. However, as solar receivers are located at a certain angle to the earth's surface and cannot be placed immediately adjacent without shading one another, we used annual sums of total solar radiation on horizontal surface to calculate the technical potential. The source of data for calculating these figures, as well as yearly solar radiation, is the NASA SSE database (NASA 2015).

Analysis of NASA data reliability on solar radiation monthly amounts was carried out with the use of actinometrical data of Moscow State University Meteorological Observatory (Kiseleva et al. 2015) for the two land observation periods: 1961–1990 and 1983–1993. No major discrepancies in the assessments of monthly solar radiation distribution by NASA and Moscow State University were detected. For the majority of points, the difference does not exceed 5% that is considered satisfactorily for PV plants technical calculations. Differences exceed 10% at only a few points: Yuzhno-Sakhalinsk (46.9° N.L.), Yeniseisk (58.5° N.L.), Vanavary

(60.3° N.L.), Tura (64.3° N.L.), and Turuhanskaya (65.8° N.L.). Most likely, these significant deviations in solar radiation data are the result of random error or of inaccurate accounting in mathematical models of NASA local climate features in these regions.

In calculating the solar radiation technical potential (besides the amounts of total solar radiation), one should take into consideration factors that limit efficient conversion of solar energy into power and heat:

1. Territory coefficient, which reflects the share (size) of each category (and subcategory) of land that may be used for deployment of solar power and heat facilities
2. Technical characteristics of solar energy converters reflected by their efficiency coefficient: efficiency factor for conversion of solar energy into electricity

5.4 Operating Conditions for Solar PV and Wind-Based Power Generation

The main difficulty associated with operation of solar PV and wind power plants is that their performance capacity fully depends on the meteorological conditions. Therefore, the ability to forecast the distribution of wind and solar radiation in time is of special interest for any system (grid) operator. Daily wind power output curves for a typical month in certain regions by 3 MW wind turbines at the 100 m height (typical height of MW-sized power plants, connected to the grid) are shown in Fig. 5.1. The curves have some similarities, albeit there are considerable shifts by time of the day and values depending on the time of the year and region.

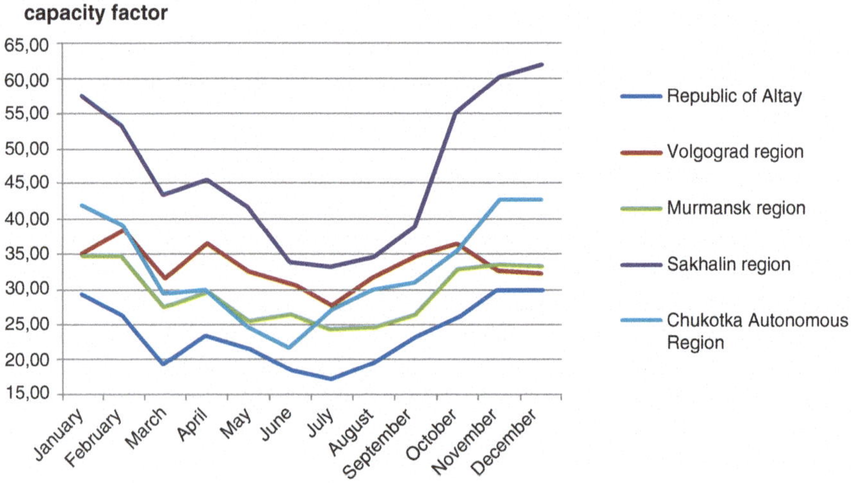

Fig. 5.1 Monthly power output curve (unit capacity factor) of a 3 MW wind facility at the 100 m height in some Russia's regions (Source: HSE)

For instance, in Murmansk and Sakhalin regions during winter months, we see sustainable energy production, while in all regions, the lowest output is observed during summer months with a highly visible minimum during morning hours. Besides daily fluctuations of power output, monthly uneven distribution of wind energy should also be taken into account (Fig. 5.1). The latter is as high as 50% in some regions of the country.

In spring solar power plants in all Russia's regions can operate for at least 9 h/day, and in 38 regions they can operate for between 12 and 15 h/day. In summer in 48 out of 85 Russia's regions, solar power plants can work 12–13 h/day and in 58 regions over 15 h/day. In autumn, solar power plants in 7 of Russia's regions cannot work at all, and in 20 regions they can work for no more than 3–4 h/day, in 32 regions 6 h/day, and in 24 regions 9 h/day.

Analysis of hourly changes in solar radiation on the ground surface shows that in winter in 48 of Russia's regions, solar power plants cannot work at all, in 36 regions they can work for no more than 3 h/day, and in 5 regions they can work for around 6 h/day.

The solar energy technical potential, which may be used both for conversion into electrical and heat energy, is substantial in the Russian Federation. The solar energy potential is highest in the South-West (Northern Caucasus, Black and Caspian Sea regions), Southern Siberia, and the Far East of Russia. Good prospects for solar energy use were found in the following regions: Kalmykia, Stavropol region, Rostov region, Krasnodar region, Volgograd region, Astrakhan region, and other regions in the South-West and Altai, Primorsky region, Chita region, Buryatia Republic, and other regions in South-East. Unexpectedly certain northern regions have higher technical potential than those situated in the South. This is linked to limited farmland territories and reserve lands in the northern regions, as well as with more territories theoretically available for solar receivers. Therefore, the use of solar energy for electricity and power generation by energy companies depends on the technical-economic justification for each power plant.

Among the areas with the highest wind energy technical potential at a given territory are Krasnoyarsk region, Republic of Sakha, Yamalo-Nenets, Chukotka, Nenets, Krasnodar, Altai, Saratov, Rostov, Volgograd, and Orenburg regions. Meanwhile, Kamchatka, Khabarovsk, Sakhalin, and Magadan regions (despite seemingly obvious potential) have minimal values due to lack of required categories of land. In other words, although the technical potential of wind is high in the Far East and some other Russia's regions, the land available for deployment of wind turbines is very limited. The efficiency of wind turbine operations also differs substantially depending on height: at 30 m wind activity drops so that wind turbines can utilize only 5–10% of their capacity; for higher capacity we recommend elevating the turbines to 100–140 m.

Maximum cumulative biomass energy potential was found in Irkutsk region, Krasnoyarsk region (South), Vologda region, Arkhangelsk region, Kirov region, Perm region, Leningrad region, Komi Republic, and Sverdlovsk region, while minimum potentials were found in Sakha (Yakutia) Republic (North), Nenets autonomous region, Chukotka autonomous region, Magadan region, and Murmansk region. Maximum total energy hydropower potential of small rivers was found at Siberian, Far Eastern, North-Western, Southern, Ural, and North Caucasus federal districts.

Active thermal waters are found in Buryatia Republic, Chukotka, Yakutia and Western Siberia, Krasnodar region, and Northern Caucasus.

Tables 5.2 and 5.3 feature the outcomes of the assessment of oil fuel, coal, and natural gas savings with the replacement of fossil fuels with various types of renewables, prevention of air pollution calculated as CO- and CO_2-equivalents.

The assessment of emissions and greenhouse gas prevention in tons of CO- and CO_2-equivalents through the use of technical potential of renewables instead of heat and power plants and facilities based on fossil fuel combustion revealed that replacement of oil products allows preventing the emission of 122,194 mln tons of CO-equivalent per year and 293,286 mln tons of CO_2-equivalent per year: coal, 3,836,579 mln tons of CO-equivalent per year and 523,189 mln tones of CO_2-equivalent per year, and natural gas, 24,125 mln tons of CO-equivalent per year and 231,961 mln tons of CO_2-equivalent per year.

Tables 5.4 and 5.5 demonstrate the regional differences of wind energy: the oil, coal, and natural gas savings if fossil fuels are replaced by wind energy technical potential and emissions of air pollutants calculated as CO- and CO_2-equivalents are prevented.

Based on the analysis of commercially available wind turbines and information about wind farm projects (those that are realized, under construction, and planned), the wind turbine model *ENERCON E-101 3050 kW* has been selected to assess technical potential.

Tables 5.6 and 5.7 demonstrate the regional divergences of solar PV: the assessed savings of oil, coal, and natural gas if fossil fuels are replaced with the technical potential of solar energy and emissions of air pollutants calculated as CO- and CO_2-equivalents are averted.

5.5 Conclusions

The chapter offers an overview of the major existing and future renewable energy potentials in Russia that may be explored for achieving the multi-sectoral gains. A methodology for physical and cost assessment of energy, fuel, resource, and environmental potential of various types of renewable energy sources, as well as a comprehensive multilevel system of wind and solar energy potential assessment, is suggested. High social and economic as well as environmental potentials of all major renewable energy sources in Russia are substantiated. This means that transition from traditional heat and power generation to renewable low-carbon energy resources will significantly reduce the rate of fossil fuel depletion and energy-related greenhouse gas emissions, as well as local and global negative environmental impact. The switch to renewables will also make it possible to preserve raw materials for manufacturing and for future generations.

Table 5.2 Fuel and power: oil fuel potential of renewables

Technical potential						
	Fuel and power potential			Oil fuel potential		
	Power	Heat	Fuel	Resource saving	Environmental	
Type of renewables	mln kWh/year	mln gcal/year	mln toe/year	mln tons/year	CO-equiv., mln tons/g	CO_2-equiv., mln tons/g
Solar energy	87,747,704.0	202,293.0	30,060.7	21,942.1	27,425.7	65,826.4
Wind energy	17,100,856.7	39,645.0	5891.2	4300.2	5374.8	12,900.5
Biomass energy	2,896,910.9	5783.6	859.4	627.3	784.1	1882.0
Geothermal energy	246,592,891.6	571,677.3	84,951.3	62,008.2	77,504.7	186,024.6
Low-grade heat	34,753,230.8	80,568.6	11,972.5	8739.0	10,923.0	26,217.1
Small hydro	584,534.9	1355.1	198.7	145.1	181.3	435.2

Source: HSE

Table 5.3 Coal and gas potential of renewables

Technical potential						
	Coal potential			Gas potential		
	Resource saving	Environmental		Resource saving	Environmental	
Type of renewables	mln tons/year	CO-equiv., mln tons/g	CO_2-equiv., mln tons/g	bln. m^3/year	CO-equiv., mln tons/g	CO_2-equiv., mln tons/g
Solar energy	39,141.6	861,099.3	117,424.8	25,969.0	5401.8	51,938.0
Wind energy	7670.9	168,756.6	23,012.7	5105.1	1061.9	10,210.1
Biomass energy	1119.1	24,619.1	3357.2	744.8	154.9	1489.5
Geothermal energy	110,613.6	2433,455.1	331,840.8	73,614.6	15,312.5	147,229.2
Low-grade heat	15,589.2	342,955.7	46,767.5	10,374.8	2158.0	20,749.5
Small hydro	258.8	5693.0	776.3	172.2	35.8	344.4

Source: HSE

Table 5.4 Technical potential of wind energy at 100 m height

Russia and federal districts (FD)	Fuel and energy potential			Oil fuel potential		
	Electric energy	Heat energy	Fuel	Resource saving	Environmental	
	mln kWh/year	mln Gcal/ year	mln toe/year	mln tons/ year	CO-equiv, mln tons/ year	CO_2-equiv., mln tons/year
Russia	17,100,857	39,645	5891	4300	5375	12,900
Central FD	1,491,619	3458.0	514	375	469	1125
North-West FD	1,481,854	3435.4	511	373	466	1118
South FD	1,980,809	4592	682	498	623	1494
Privolzhsky FD	2,664,236	6177	918	670	837	2010
Ural FD	2,098,280	4865	723	528	660	1583
North Caucasus FD	847,660	1965	292	213	266	640
Far East FD	2,708,869	6280	933	681	851	2044
Siberian FD	3,776,910	8756	1301	950	1187	2849

Source: HSE

Table 5.5 Technical potential of wind energy at 100 m height

Russia and federal districts (FD)	Coal potential			Natural gas potential		
	Resource saving	Environmental		Resource saving	Environmental	
	mln tons/ year	CO-equiv., mln tons/ year	CO_2-equiv., mln tons/ year	bln m^3/ year	CO-equiv., mln tons/ year	CO_2-equiv., mln tons/ year
Russia	7671	168,757	23,013	5105	1062	10,210
Central FD	669	14,720	2007	445	93	891
North-West FD	665	14,624	1994	442	92	885
South FD	889	19,547	2666	591	123	1183
Privolzhsky FD	1195	26,292	3585	795	165	1591
Ural FD	941	20,707	2824	626	130	1253
North Caucasus FD	380	8365	1141	253	53	506
Far East FD	1215	26,732	3645	809	168	1617
Siberian FD	1694	37,272	5083	1128	235	2255

Source: HSE

Table 5.6 Technical potential of solar energy

Russia and federal districts (FD)	Fuel and energy potential			Fuel oil		
	Electrical energy	Heat energy	Fuel	Resources saving	Environmental	
	Mln kWh/year	mln Gcal/year	mln toe/year	mln tons/year	CO-equiv., mln tons/year	CO_2-equiv., mln tons/year
Russia	87,747,704	202,293	30,060.7	21,942.1	27,425.7	65,826.4
Central FD	4,744,733.0	10,999.7	1634.6	1193.1	1491.3	3579.3
North-West FD	13,037,743.3	30,225.5	4491.5	3278.5	4097.8	9835.4
South FD	3,826,115.8	7737.2	1149.8	839.2	1049.0	2517.7
Privolzhsky FD	5,503,378.9	12,758.5	1895.9	1383.9	1729.7	4151.6
Ural FD	16,278,705.4	37,739.0	5608.0	4093.4	5116.4	12,280.3
North Caucasus FD	1,495,762.2	3467.6	515.3	376.1	470.1	1128.4
Far East FD	23,327,681.5	54,080.7	8036.4	5866.0	7331.9	17,597.9
Siberian FD	18,736,228.6	43,436.3	6454.6	4711.4	5888.8	14,134.2

Source: HSE

Table 5.7 Technical potential of solar energy

	Coal potential			Natural gas potential		
	Resource saving	Environmental		Resource saving	Environmental	
Russia and federal districts (FD)	mln tons/year	CO-equiv., mln tons/year	CO_2-equiv., mln tons/year	bln m^3/year	CO-equiv., mln tons/year	CO_2-equiv., mln tons/year
Russia	39,141.6	861,099.3	117,424.8	25,969	5401.8	51,938
Central FD	2128.3	46,822.6	6385.0	1412.1	293.7	2824.1
North-West FD	5848.3	128,660.7	17,545.0	3880.1	807.1	7760.3
South FD	1497.1	32,935.1	4491.2	993.3	206.6	1986.5
Privolzhsky FD	2468.6	54,309.1	7405.9	1637.9	340.7	3275.7
Ural FD	7302.1	160,643.5	21,906.3	4844.7	1007.7	9689.4
North Caucasus FD	671.0	14,760.7	2012.9	445.2	92.6	890.3
Far East FD	10,464.1	230,205.1	31,392.2	6942.5	1444.1	13,885.0
Siberian FD	8404.5	184,895.2	25,213.4	5576.1	1159.9	11,152.1

Source: HSE

Moreover, the assessment of wind (at 50 and 100 m) and solar energy technical potential in all regions of Russia undertaken with the proposed methodology showed that these clean energy resources may fully substitute fossil fuels for energy generation. Wind and solar energy potential surpass the natural gas-based (currently the main energy source in Russia) power generation several times to order of magnitude (depending on the region of Russia). The best-suited regions for deployment of renewables are identified, and recommendation to overcome the existing challenges and for effective development of renewable energy in Russia may be of interest to decision-makers, investors, and companies that work in renewables.

Future implications for research may include the development of mathematical models, information and software support to assess the public and private investments in renewables, and design in individual and hybrid power systems using different types of renewables. The successful development and deployment of solar and wind power capacities in Russia reply on application of innovative technologies, energy storage systems, and smart grind elements, and these interconnections may further be addressed.

The study's social implications are presented in the chapter devoted to effects (gains) from renewable energy capacity deployment and include better quality of life, reduction of emissions and corresponding improvement in health conditions of population, and reduction of electricity and heat costs for population.

The study outcomes can be used by institutional, private, and foreign investors for wind and solar power plant investment designing in centralized and distributed power systems.

Acknowledgments The book chapter combines two earlier published papers on methodology of renewable potential assessment and the first calculations of this potential for Russia:

Ermolenko B.V., Ermolenko G.V., Fetisova Yu., Proskuryakova L.N. (2017) *Wind and solar PV technical potentials: measurement methodology and assessments for Russia.* Energy, Vol. 137, pp 1001–1012.

Ermolenko G., Proskuryakova L., Ermolenko B. (2017) Switching to renewables: what will Russia gain? Foresight, Vol. 19, No. 5, pp. 528–540.

Contributions in this publication were supported within the framework of the Basic Research Program at the National Research University HSE and were funded within the framework of the subsidy by the Russian Academic Excellence Project '5-100'.

References

Abanda FH (2012) Renewable energy sources in Cameroon: potentials, benefits and enabling environment. Renew Sust Energ Rev 16(7):4557–4562

Ahmad S, Tahar RM (2014) Selection of renewable energy sources for sustainable development of electricity generation system using analytic hierarchy process: a case of Malaysia. Renew Energy 63:458–466

Alemán-Nava GS, Casiano-Flores VH, Cárdenas-Chávez DL, Díaz-Chavez R, Scarlat N, Mahlknecht J, Dallemand J-F, Parra R (2014) Renewable energy research progress in Mexico: a review. Renew Sust Energ Rev 32:140–153

Andresen GB, Søndergaard AA, Greiner M (2015) Validation of Danish wind time series from a new global renewable energy atlas for energy system analysis. Energy 93(Part 1, 15):1074–1088
Arslan T, Bulut YM, Yavuz AA (2014) Comparative study of numerical methods for determining Weibull parameters for wind energy potential. Renew Sust Energ Rev 40:820–825
Aslani A, Wong K-FV (2014) Analysis of renewable energy development to power generation in the United States. Renew Energy 63:153–161
Atalay Y, Biermann F, Kalfagianni A (2016) Adoption of renewable energy technologies in oil-rich countries: explaining policy variation in the Gulf Cooperation Council states. Renew Energy 85:206–214
Azad AK, Rasul MG, Islam R, ShiShir IR (2015) Analysis of wind energy prospect for power generation by three Weibull distribution methods. Energy Procedia 75:722–727
Bezroukikh PP, Degtyarev VV, Elistratov VV, Pantshava ES, Petrov ES, Puzakov VN, Sidorenko GI, Tarnizhevsky BV, Shpak AA, Yampolsky AA (2007) Handbook of renewable energy sources in Russia and local fuels (territorial data). Moscow
Bhattacharya M, Paramati SR, Ozturk I, Bhattacharya S (2016) The effect of renewable energy consumption on economic growth: evidence from top 38 countries. Appl Energy 162:733–741
Bilir L, Imir M, Devrim Y, Albostan A (2015) Seasonal and yearly wind speed distribution and wind power density analysis based on Weibull distribution function. Int J Hydrog Energy 40:15301–15310
Boubaker K, Colantoni A, Marucci A, Longo L, Gambella F, Cividino S, Gallucci F, Monarca D, Cecchini M (2016) Perspective and potential of CO_2: a focus on potentials for renewable energy conversion in the Mediterranean basin. Renew Energy 90:248–256
Boute A, Willems P (2012) RUSTEC: Greening Europe's energy supply by developing Russia's renewable energy potential. Energy Policy 51:618–629
Ermolenko GV, Ermolenko BV, Gordeev IG (2013a) Prospective development of renewables as a tool for addressing environmental, economic, and social issues. Energy 2:2–8
Ermolenko GV, Ermolenko BV, Fetisova YA (2013b) Optimizing the plans for decentralized energy supply with the use of renewables and energy storage systems. Energy 12:11–19
Fathoni AM, Utama NA, Kristianto MA (2014) A technical and economic potential of solar energy application with feed-in tariff policy in Indonesia. Procedia Environ Sci 20:89–96
Ghafoor A, Rehman T, Munir A, Ahmad M, Iqbal M (2016) Current status and overview of renewable energy potential in Pakistan for continuous energy sustainability. Renew Sust Energ Rev 60:1332–1342
Government of the Russian Federation (2015) Decree of the Government of the Russian Federation. 28 July 2015, No. 1472-p
Gunderson I, Goyette S, Gago-Silva A, Quiquerez L, Lehmann A (2015) Climate and land-use change impacts on potential solar photovoltaic power generation in the Black Sea region. Environ Sci Policy 46:70–81
Hua Y, Oliphant M, Hu EJ (2016) Development of renewable energy in Australia and China: a comparison of policies and status. Renew Energy 85:1044–1051
IFC (2016) Russia Renewable Energy Program. http://www.ifc.org/wps/wcm/connect/RegProjects_Ext_Content/ifc_external_corporate_site/home-rrep. Accessed 27 Dec 2016
IRENA (2016) Renewable energy benefits: measuring the economics. http://www.irena.org/DocumentDownloads/Publications/IRENA_Measuring-the-Economics_2016.pdf
Islam MR, Saidur R, Rahim NA (2011) Assessment of wind energy potentiality at Kudat and Labuan, Malaysia using Weibull distribution function. Energy 36:985–992
Khare V, Nema S, Baredar P (2013) Status of solar wind renewable energy in India. Renew Sust Energ Rev 27:1–10
Kim K-T, Lee DJ, Park S-J, Zhang Y, Sultanov A (2015) Measuring the efficiency of the investment for renewable energy in Korea using data envelopment analysis. Renew Sust Energ Rev 47:694–702
Kiseleva SV, Ermolenko GV, Popel OS (2015) Atlas of renewable energy resources in Russia. Moscow

Kumar R, Agarwala A (2016) Renewable energy technology diffusion model for techno-economics feasibility. Renew Sust Energ Rev 54:1515–1524

Kumar S, Madlener R (2016) CO_2 emission reduction potential assessment using renewable energy in India. Energy 97(15):273–282

Lun IYF, Lam JC (2000) A study of Weibull parameters using long-term wind observations. Renew Energy 20:145–153

MacDonald A, Clack CTM, Alexander A, Dunbar A, Wilczak J, Xie Y (2016) Future cost-competitive electricity systems and their impact on US CO_2 emissions. Nat Clim Chang 6:526–531. https://doi.org/10.1038/nclimate2921

Mondal Md AH, Hawila D, Kennedy S, Mezher T (2016) The GCC countries RE-readiness: strengths and gaps for development of renewable energy technologies. Renew Sust Energ Rev 54:1114–1128

NASA (2015) NASA surface meteorology and solar energy, atmospheric science data center. https://eosweb.larc.nasa.gov/cgi-bin/sse/sse.cgi?skip@larc.nasa.gov. Accessed 10 Feb 2015

Nikolaev VG (2011) Resource and technical – economic basis for a wide development and deployment of wind energy in Russia. ATMOGRAF, Moscow

OECD/IEA (2013) Energy statistics of non-OECD countries. Paris

OECD/IEA (2015a) World energy outlook 2015

OECD/IEA (2015b) Renewable energy. Medium-term market report 2015. Market analysis and forecasts to 2020

Osmani A, Zhang J, Gonela V, Awudu I (2013) Electricity generation from renewables in the United States: resource potential, current usage, technical status, challenges, strategies, policies, and future directions. Renew Sust Energ Rev 24:454–472

Pristupa AO, Mol APJ (2015) Renewable energy in Russia: The take off in solid bioenergy? Renew Sust Energ Rev 50:315–324

Proskurina S, Heinimö J, Mikkilä M, Vakkilainen E (2015) The wood pellet business in Russia with the role of North-West Russian regions: present trends and future challenges. Renew Sust Energ Rev 51:730–740

Proskuryakova L, Filippov S (2015) Energy technology Foresight 2030 in Russia: an outlook for safer and more efficient energy future. Energy Procedia 75:2798–2806

Ramli MAM, Twaha S, Al-Hamouz A (2017) Analyzing the potential and progress of distributed generation applications in Saudi Arabia: the case of solar and wind resources. Renew Sust Energ Rev 70:287–297

REN21 (2015) Renewables 2015. Global Status Report. http://www.ren21.net/wp-content/uploads/2015/07/REN12-GSR2015_Onlinebook_low1.pdf. Accessed 17 Mar 2016

REN21 (2016) Renewables 2016. Global Status Report, Paris

REN21 (2017) UNECE Renewable energy status report 2017. http://www.unece.org/fileadmin/DAM/energy/se/pp/renew/Renewable_energy_report_2017_web.pdf. Accessed 17 Oct 2017

RusHydro (2016) Rushydro Power Generating Company. http://www.rushydro.ru/industry/biblioteka/14289.html#1. Accessed 17 Mar 2016

Russian Federal State Statistics Service (2015) www.gks.ru. Accessed 13 May 2016

Russian Federal State Statistics Service (2016) www.gks.ru. Accessed 17 Mar 2016

Russian Ministry of Energy (2015) Government policy to support renewables-based generation in the Russian Federation. Presentation at the international renewable energy congress – XXI: energy & economic efficiency. http://reencon-xxi.ru/wp-contenweuploads/2015/11/Гринкевич-Презентация-по-В⍰Э.pdf. Accessed 28 Mar 2016

Shepovalova OV (2015) Energy saving, implementation of solar energy and other renewable energy sources for energy supply in rural areas of Russia. Energy Procedia 74:1551–1560

UNDP (2010) Overview of renewables in Russia, thematic network "Energy efficiency and energy saving". http://solex-un.ru/energo/review/opyt-ispolzovaniya-vie/obzor-1-napravleniya-vie-ekonomika-effekty-sostoyanie-v-rf-0. Accessed 17 Mar 2016

UNESCO (2014) International Sustainable Energy Development Centre. Energy Bulletin No. 17

Boris Ermolenko is an Assistant Professor at D. Mendeleev University of Chemical Technology of Russia, Russian Federation. Dr. Ermolenko specializes in renewable energy and energy efficiency studies: resources, economics, policy, technologies and innovation. He authored over 50 research publications and took part in multiple national and international research projects and conferences on these topics.

Georgy Ermolenko is Head of Center for Renewable Energy at HSE Institute of Energy. Dr. Ermolenko specializes in renewable energy studies: resources, policy, technologies and innovation. He authored over 30 research publications and took part in over 15 national and international research projects on these topics. Dr. Ermolenko is Member of the Expert Council of the State Duma of the Russian Federation Committee on Energy; Deputy Chairman of the UNECE Expert Group on Renewable Energy.

Liliana Proskuryakova is Leading Researcher at HSE ISSEK S&T Laboratory. Dr. Proskuryakova specializes in energy studies, S&T policy and international cooperation in research and innovation. She authored over 40 research publications and took part in many national and international research projects on these topics. In 2003–2011 Dr. Proskuryakova served as member of the Board of Directors and adviser of the Partnership for Transparency Fund. She is member of the Advisory Board member of the "Lake Baikal" Foundation for the support of environmental research and development. Dr. Proskuryakova holds an MA degree in International Relations and a PhD in Political Science.

Global Market Creation for Fuel Cell Electric Vehicles 6

Alexander Sokolov, Ozcan Saritas, and Dirk Meissner

6.1 Introduction

Fuel cell electric vehicles (FCEVs) have been considered as a future vision for the automotive industry to replace the vehicles with petrol engines. Manufacturers in the industry have been working on concepts and prototypes for more than a decade now while having advances in addressing the main technical challenges like battery life. In parallel with the technological development, recent discussions about global warming and climate change caused by carbon dioxide emissions brought public support for emission-free vehicles, which are perceived as an asset. Despite of advancements and public support, the introduction of FCEVs is still not at the desirable levels. Car manufacturers frequently announce the near-time launch of FCEVs, which are postponed with the same frequency.

Numerous business models and market projections have been made for the launch of FCEVs resulting in promising roadmaps. But it is obvious that these roadmaps reflected the technology side quite realistically while lacking a fully-fledged consideration of the market side. The reason for this may be the fact that some of the actual and potential stakeholders are overlooked. Commonly market roadmaps make assumptions of customers' behaviour, take into account competing products, etc. but neglect the more or less "hidden" stakeholders, which become

The chapter builds on a previously published article "Saritas, O., Meissner, D. & Sokolov, A. A Transition Management Roadmap for Fuel Cell Electric Vehicles (FCEVs). J Knowl Econ (2018). https://doi.org/10.1007/s13132-018-0523-3."

A. Sokolov · O. Saritas · D. Meissner (✉)
National Research University, Higher School of Economics, Moscow, Russia
e-mail: sokolov@hse.ru; osaritas@hse.ru; dmeissner@hse.ru

D. Meissner et al. (eds.), *Emerging Technologies for Economic Development*, Science, Technology and Innovation Studies,
https://doi.org/10.1007/978-3-030-04370-4_6

obvious if one analyses the systemic impact of FCEVs as innovations within industry value chains in relation to the transportation sector in a broader sense.

The present chapter aims to make an attempt of a systemic analysis for a broader and more holistic analysis, which may portray the bigger picture and help to understand the industrial dynamics better. The key argument is that the stakeholder base in the transportation industry and thus for FCEVs is broader than usually thought, and there are a number of other issues to be addressed for a successful and widespread launch of FCEVs. The roadmap presented in the study, entitled "FCEV Global Market Creation", undertakes a broader analysis of society, technology, economy, environment and politics (STEEP) at two stages. First the STEEP systems are discussed, and the stakeholders for each item are analysed. From this analysis, some potential barriers for the diffusion of FCEVs have become clearer. It is suggested that a more in-depth analysis in a future study will yield more insights for the implementation of FCEVs.

Thus, the second section of the chapter provides a background for the study and clarifies the objectives of a broader stakeholder and issue analysis for the FCEVs. Approach and methodology used for analysis will be described in the third section. Following a systematic and holistic analysis of external drivers and stakeholders will be presented in the fourth section to uncover broader set of issues affecting the FCEVs. The final section will present a roadmap template to bring together demand and supply dynamics and strategies to be adopted for successful FCEV implementation.

6.2 Background and Objectives

Climate change, increasing pressures on natural resources and environmental hazards are among the indications that humanity is reaching to the end of the reliance on non-renewable sources of energy like oil, coal and natural gas. The scramble is now on to find renewable energy sources that will keep cars, homes and businesses running without destroying the environment. Among the alternatives for future energy sources, hydrogen comes as one of the first options. It is an infinitely renewable and relatively pollution-free fuel that scientists, policy-makers and society alike see as a viable alternative to fossil fuels.

In the automotive industry, hydrogen is a promising alternative for combustion engines. A hydrogen fuel cell in a car produces zero emissions with only water vapour and heat released through the tailpipe. Hydrogen is three times more efficient and provided that renewable sources are used for generation such as water and energy, and it does not release any pollution. Hydrogen can be used to power vehicles as well as to be used in fuel cells to generate and store energy. Both of these technologies can be used separately or on the same platform, where a car can be powered by a hydrogen combustion engine and the fuel cell as a source of energy to supply electric power in place of a conventional alternator.

On the technical side, the main issues are related to the engine technology and then related to recovery, storage and transfer of hydrogen into the car itself. However, there is a broader range of social, technological, economic, environmental and political (STEEP) issues, which have positive and negative implications for the widespread implementation of hydrogen in the automotive industry. For instance, whether hydrogen and fuel cell technologies can be clean and efficient is very much dependent to how hydrogen is produced. Generation of hydrogen from oil and natural gas is much cheaper but still puts pressures on natural resources. There are also struggles with storage and transportation under high pressure, which makes it bulky and impractical. As hydrogen has no smell, sensors must be used to detect leaks. In addition, a number of refuelling stations are required for hydrogen-powered cars. These are among the main barriers to the commercial development of hydrogen fuel cells.

The next few years appear to be critical for the commercialisation of FCEVs. There are some promising trends. For instance, total worldwide fuel cell shipments grew 20% between 2014 and 2015 and 200% between 2011 and 2015 (US Department of Energy 2015a, b). Adequate refuelling infrastructure will be required for FCEVs to be marketed as a credible and attractive alternative to conventional vehicles. Once hydrogen refuelling stations (HRSs) are available, the initial uptake of the FCEVs will be limited by the cost of buying and using the vehicles. The UK H_2 Mobility report (2013) predicts that sufficient early adopters should generate sales of approximately 10,000 vehicles per annum by 2020. As the vehicle costs become more competitive and refuelling network develops, FCEV uptake increases rapidly. The same report distinguishes three phases for implementing a viable business case for FCEVs:

1. Market seeding (2015–2020) with high capital costs and low number of stations—small at the beginning but expandable in the future. Revenues are also low because of low number of FCEVs on roads. Stations are planned to the nearest places, where a critical mass of consumers exists—such as close to the feet bases.
2. Investing in growth (2020–2025). The number of FCEVs will grow. New stations will be built, and the existing ones will be enlarged and upgraded. Revenues will increase gradually, and HRSs will become more attractive.
3. Developed network and market (2025–2030). Demand is high enough to grow revenues, and operations become profitable. A full network will be built by this time. Further HRSs will be built according to market demand.

No quick solutions are expected in the immediate future. At present, the total number of FCEVs has been estimated to increase by more than a factor of 10 from 2015 to 2020 and by a factor of 100 by 2030 (US Department of Energy 2015a, b). This is mainly due to a limited number of models on the market, limited infrastructure and higher costs compared to battery electric vehicles (BEVs) and plug-in hybrid electric vehicles (PHEVs) (Global EV Outlook 2013). First commercial introduction of FCEVs was in 2015, and more widespread HRSs are expected by 2020s. HRSs come to break even in the late 2020s. By the 2030s, 20% of vehicles are expected to be FCEVs and by the 2050s 50%. However, successful demonstrations in Germany, Scandinavia, Japan and California indicate that there are great potentials for the deployment and use of FCEVs. Further countries are expected to make significant efforts to exploit the potentials and avoid risks associated to the STEEP dimensions of the FCEVs. The countries which will prosper will be the ones, which do not only develop technologies but also implement them, develop necessary infrastructure, regulations and standards, as well as market and promote them to a range of consumers for a widespread use. As all these require skills and capabilities beyond the technical ones, the present study adopts an *eagle eye* view with a thorough analysis of overall context and framework conditions as well as future technologies, which may shape the future of mobility and transport and thus overhaul the automotive sector. The methodology used for the study is described in the next section.

6.3 Approach and Methodology

The process begins with a wider scanning and surveying phase (Miles et al. 2016), where the implications of key trends and drivers of change are examined with implications for the future FCEV markets. A framework based on the analysis of social, technological, economic, environmental and political (STEEP) systems was used to ensure that the topic is explored from multiple different but also interconnected lenses (Saritas 2013). The analysis was conducted at two steps.

A two-step STEEP analysis was used in the study process:

1. To identify "external drivers" through the analysis of social, technological, economic, environmental and political determinants of FCEVs
2. Stakeholder analysis and mapping to identify key players in each STEEP category

The external drivers (STEEP) are described in brief with emerging key issues under each of them. Following a basic outline of these sub-items, Wild Cards are suggested for each category to describe those events and developments with a small likelihood of occurrence but with big potential impacts on FCEVs. Three or four exemplary Wild Cards are presented under each category to provoke positive or negative disruptive thinking in the FCEV domain.

The main stakeholders are then described with an analysis of their potential influence on the diffusion and absorption of the FCEVs and a discussion on potential argumentation lines against FCEVs. At the current stage, this analysis is not exhaustive, and the stakeholders are listed in general terms with no direct relevance to a specific country or market.

The chapter concludes with a template of a roadmap to be developed for FCEVs in the second phase of the project. The roadmap template can be used for the purpose of developing multilevel and multidimensional strategic roadmaps, as exemplified by Saritas and Aylen (2010), for different countries or businesses operating in the FCEV domain.

6.4 Characteristics of External Drivers

6.4.1 Social Factors

Social factors involve ways of life (e.g., attitudes to work time, use of leisure time, family living patterns, work-life balance), demographic structures, social inclusion and cohesion issues (fragmentation of lifestyles, levels of (in)equality, educational trends). Public attitude towards hydrogen is considered to be one of the key factors towards the transition to FCEVs. The social perception of FCEVs can be assessed in terms of:

- Availability of a critical mass of customers
- Accessibility to infrastructure
- Aesthetics and convenience for making the vehicles attractive for the customers without compromising from quality

Availability

Building a critical mass of FCEV customers is a must for making the technology economically viable. For customers, initial purchase cost is one of the determinants of the selection of a vehicle. Without intervention or dedicated mechanisms, in the early years after market introduction, FCEVs will be significantly more expensive than the conventional vehicles, and refuelling will be limited with a low number of locations. These seem to be main barriers by the consumers. However, there are various ways of overcoming these initial barriers, which will be discussed in the Technology and Economy sections below (Sects. 6.4.2 and 6.4.3).

Accessibility

The accessibility to infrastructure is obviously raised as one of the key points for the widespread use of FCEVs. Gas stations need to invest in the ability to refuel hydrogen tanks before FCEVs become practical. Consumers require both local availability and national coverage for long-distance travels. Detailed spatial modelling studies can be done to identify those locations to deliver the greatest consumer benefit. The low number of customers will make it less likely for such investments at the early phase; however, after providing sufficient initial coverage, then the network could be developed in line with the demand by the vehicle owners/users. Accessibility should also be considered in terms of availability of the maintenance infrastructure and associated repair frequencies and costs.

Aesthetics and Convenience

Customers increasingly value new technologies, environmental friendliness and costs of a new car. Performance and driving behaviour of vehicles are also considered to be important when purchasing decisions are made. The dependency between total costs and the availability of hydrogen refuelling stations (HRSs) are the key determinants of buying a FCEV. A recent survey indicated that some 10% of new vehicle buyers showed themselves to be potential early adopters of FCEVs, being receptive to new technology and environmentally motivated (UK H_2 Mobility 2013). The expectation is that the cars may cost higher to purchase at the beginning. However, when lifetime costs are considered, it can be said that the operating costs will be considerably lower compared to existing internal combustion engines.

Various advantages of using FCEVs can be emphasised in the process of market creation, such as:

- Longer driving range as the electricity is generated in the car.
- Hydrogen charging time is usually quick.
- Like other electrical engines, there is no engine noise in the car—though, special effects can be created for motor fans.
- No carbon dioxide is pumped into the atmosphere, no noise is made, and the engine delivers better torque than most petrol cars, which translates to greater acceleration.

Considering all these factors, consumers appear to be receptive to FCEVs, especially in terms of vehicle performance and refuelling time. In the first 5–8 years of deployment, plug-in hybrid vehicles might be preferred given the fact that it is possible to use the existing petrol, diesel and electricity networks. However, as the HRSs expand, the advantage of longer driving range and quick refuelling time will be more attractive for car owners/users.

The UK H_2 Mobility (2013) report identified seven distinct consumer groups with defining characteristics under three categories as illustrated in the Fig. 6.1.

Being the 10% of the overall consumers, potential early adopters can be considered as the first target group. This group is willing to pay a premium prize for an

	Potential early adopters		Probable late adopters		Technology followers		
Segment	1 Well-off enthusiasts	2 innovative greens	3 High mileage luxury	4 Cautious pragmatists	5 Single car value	6 Younger sceptic	7 Uninterested rejecters
Attitudes to infrastructure	• High willingness to drive to find H2 • Medium concern over 1 HRS per city	• Strongest willingness to sacrifice convenience	• Below average willingness to drive to HRS • Strong concern over 1 HRS	• Medium willingness to drive to HRS	• Limited willingness to sacrifice convience	• Limited willingness to sacrifice convenience	• Will only drive short distance to find H2
Technology attitudes	• Strong tech enthusiast • Willing to pay for new tech	• Wants and might pay for new tech	• Low innovativeness • Well informed on car tech	• Not interested in new tech	• Very low interest in tech • Low innovativenes	• Low interest in new tech	• Low interest in tech
Green attitudes	• WTP for green • Tailpipe more important that WTW	• Strongest green attitude • Green hydrogen important	• Weak green motivation • Not willing to pay	• Slightly green, but no sacrifices	• Some concern about pollution, but not WTP for green cars	• Some concern about pollution but not WTP	• Lowest environmental concern
Cost sensitivity	• Payback not important • WTP for green and new tech	• Payback of low Importance • WTP for green and new tech	• Payback more important for new/green tech	• Low running costs important • No WTP for green/new tech	• Low running costs important • Value buyer	• No WTP for green/new tech	• No WTP for green/new tech • Capital cost primary fact
Proportion in sample	~10%		~40%		~50%		

Fig. 6.1 FCEV consumer segments. Source: UK H_2 Mobility Report (2013)

FCEV. This makes the introduction of FCEVs possible before cost parity is reached with existing vehicle technologies. The other groups would still require a discount to buy an FCEV.

Wild Card Thinking
Some Wild Card thinking can be applied for the acceleration of the speed of adoption of FCEVs:

- Strategies for attracting emerging young "technology-savvy society" would be a good group to target to expand the potential early adopter group.
- FCEVs can be presented with a new image by combining sleek design and technology, which may be used to create a new fashion. People are ready to pay 10 times higher for an Apple iPhone than others; similarly they may be ready to pay 1.2 times higher for a stylish car.
- Recently communications, electronics and photography sectors have converged to a large extent. It is expected that the automotive industry too will converge with those sectors. Integration and interoperability with information, communication, multimedia systems and social networking technologies with large touch screens will also attract a number of users.
- Mass media can be used to promote FCEVs. Media attitudes are equally important, especially mass media which has strong impact on public opinions. It has been frequently observed that media tends to report on accidents and failures of technologies most preferably instead of success stories. Here the impact of different media channels on consumer attitudes and behaviour could be given special attention.
- Society should be convinced about the safety, security and reliability of FCEVs, without any negative impacts on public and individual health.

6.4.2 Technological Factors

Technological factors focus on rates of technological progress, pace of diffusion of innovations, problems and risks associated with technology such as security and health problems. Some of the technological factors associated to FCEVs are:

- Reliability of supply
- Equipment (hardware)
- Vehicle technologies

Reliability of Supply
The supply of hydrogen is an important issue for the creation of FCEV markets. Necessary infrastructure should be established to ensure sound and stable supply of energy. Any lack of supply can be easily used as an argument against FCEVs.

When hydrogen is supplied, environmental implications should also be considered. Pure hydrogen can be industrially derived, but it takes energy. If that energy does not come from renewable sources, then fuel cell cars are not considered to be as clean as they seem. Ways should be found to generate hydrogen using renewable energy sources or integrated with carbon capture and storage (CCS) technologies.

Fuel storage and transport challenges seem to be solved at least technologically in principle. However there still remain concerns with respect to logistical issues, e.g. centralised or decentralised fuel production, infrastructure for fuel distribution and the appropriateness of the existing infrastructure for upgrading to the respective fuel distribution.

Equipment (Hardware)

The cost of buying and installing equipment for compressing, storing and dispensing hydrogen on site and the cost of financing the expenditure should be considered at this point. A hydrogen refuelling station (HRS) is considered to have a 20-year lifetime. The cost of land should also be factored in the cost calculations. Operation of the equipment and facilities will incur certain costs too such as the cost of maintaining the HRS, general operating cost and the cost of administration and sales and rental charge for the land used by the HRS—if it is not a mobile one.

Network development can be achieved with a number of small but widespread HRSs, which should be expandable with upgrades to medium- and large-scale HRSs. This option for expansion should be considered right from the beginning.

Vehicles

The vehicle technology is probably the most advanced and least problematic part of the overall FCEV business plans. A number of vehicle producers are ready for mass commercial production. Although FCEVs are more expensive than the cars with conventional engines at the moment, these costs are expected to go down following the "market seeding" phase between 2015 and 2020, when investments are made for growth and networks and markets are developed.

An important issue here may be the maintenance and repair infrastructure for FCEVs. As FCEVs are equipped with more sophisticated technologies, special attention will need to be given to repair facilities and equipment, which may look quite different than current garages for conventional vehicles. The transport and storage of hydrogen and spare electronic equipment will need special attention in terms of security and safety. Operating staff training for this new technology should also be taken into account.

Wild Card Thinking

Possible technology-related Wild Cards:

- New ways of extracting hydrogen. A team of Virginia Tech researchers has discovered a way to extract large quantities of hydrogen from any plant, a breakthrough that has the potential to bring a low-cost, environmentally friendly fuel source to the world.
- Mobile hydrogen refuelling stations can be used to provide further access to hydrogen in remote or congested areas or when a likely power cut starts effecting supply. Powertech company has pioneered the use of lightweight carbon fibre composite tanks for the high pressure bulk transportation of hydrogen, for mobile hydrogen fuelling station applications and for portable self-contained hydrogen

fuelling units. Transportable compressed hydrogen units can be custom-designed to meet customer needs, including transport trailers, mobile fuellers and portable filling stations.

- If the technology for BEVs does not develop significantly and if the process of producing hydrogen becomes easier and cheaper, then the case will be much stronger for FCEVs.

6.4.3 Economic Factors

Levels and distribution of economic growth, industrial structures, competition and competitiveness and market and financial issues are the sorts of factors to be considered under this category. As one of the key drivers for hydrogen cars, economic factors in this domain involve a number of points to be considered, including:

- Cost of FCEVs
- Demand
- Investments

Cost

Fuel cells (FCs) are still very expensive, even when compared to BEVs. FCs still remain to be ten times more expensive than internal combustion engines; however progress is made towards cost reduction. Another important aspect is hydrogen itself. Currently, there is no market price for hydrogen intended for alternative energy use comparable to that for gasoline. Hydrogen as an alternate energy carrier is in an early phase of development. Estimates of the cost of hydrogen per gallon of gas equivalent range from $2.10 to $10. Hydrogen produced from natural gas, the cheapest available method, is 3–4 times as expensive as gasoline, in terms of equivalent amounts of energy (US Department of Energy 2015a).

The Department of Energy in the USA aims to reduce the cost of hydrogen to $2 per gasoline gallon equivalents by 2020 (US Department of Energy 2015a, b). Its previous goal of $1.50, set before gasoline prices went up, was based on the use of natural gas as a source for hydrogen. The new goal is independent of the method of production, in response to questions about the environmental effects of using natural gas for hydrogen production.

Demand

Overall the market demand has three determinants:

1. Vehicle attributes including the price of the vehicle, performance and range and hydrogen consumption
2. Consumer attitudes in terms of different user segments presented above (Fig. 6.1) and the size of each group

3. Hydrogen refuelling stations with a widespread urban and national coverage and hydrogen process and emissions generated through the production process

Economic feasibility means a target of 200 people per pump at the annual depreciation target of 250 euros/year per customer to achieve the cost target of 500K euros over 10 years period (Hasegava 2013). This may be challenging in the first period between 2015 and 2020 but then will gradually become more realistic.

Considering the cost and demand factors, first and immediate clients for FCEV seem to be fleets, such as bus fleets in Europe. Large number of vehicles may create the economies of scale that justify the cost of building stations. Among the expectations of the fleet operators from the FCEVs are low emission, long range and fast fuelling vehicles. Van type of vehicles is more preferred by this group. Special rates or tax exemptions are expected for low- or zero-emission vehicles. Fleets are also desirable to begin with as they have more predictable driving patterns than other users. Filling stations might be located at the fleet base or close to it.

Investments

Venture capital and private equity investments in fuel cells and hydrogen increased worldwide by 9.2% between 2014 and 2015. Venture capital and private equity investments in US fuel cell companies grew by 96.2% during the same period (US Department of Energy 2015b). Coupling with the increasing investments, the FCEV industry may create further employment across the value chain from vehicle manufacture, development of new components, fuel production, distribution and supply, thereby bringing significant economic benefit.

The HRSs may not be profitable at the beginning. A break-even point may be expected in the late 2020s. Business cases should be created for the initial network of stations to ensure how first mover commercial advantage could be secured.

Wild Card Thinking

Possible Wild Cards in this category may include:

- The nearest-term application for FCs seems to be lift trucks (forklifts). Several industrial truck companies have announced commercial FC products that can replace battery-powered forklifts. These have been extensively tested and are available for commercial purchase today to be used in production plants, logistics and airports.
- A possible "electron economy" may replace the "hydrogen economy". In an electron economy, most energy would be distributed with highest efficiency by electricity, and the shortest route in an existing infrastructure could be taken. The efficiency of an electron economy is not affected by any wasteful conversions from physical to chemical and from chemical to physical energy. With an electron economy, attentions could be quickly turned to the energy storage technologies and upgrade to smart grids.
- On the more positive side, UK-based AFC Energy is confident it has identified a low-cost and sustainable source of hydrogen in the form of the waste gases

produced by the chlorine industry, and following successful trials at a chlor-alkali plant at Bitterfeld in Germany, the company is now working on a 50 kW commercial-scale version of its fuel cell technology. It should be considered that technology and economy should go hand in hand to achieve hydrogen breakthrough.

6.4.4 Environmental Factors

Pressures connected with sustainability and climate change and more localised environmental issues (including pollution, resource depletion, associated biodiversity and welfare concerns) are among the environmental factors to be considered. Environmental impact is one of the key arguments for supporting or contesting the transition process towards hydrogen-powered FC cars. There are two key factors to be considered:

- Emission reductions, which is highly desirable
- Hydrogen production, which is frequently used as a counter-argument to the hydrogen economy

Emission Reductions

The environmental benefits of hydrogen are a very positive attribute. When used in a fuel cell to power an electric vehicle, the emissions include only water and heat. But hydrogen is produced using energy from natural gas, coal, solar, wind or nuclear power, each of which has its own environmental effects. The UK H_2 roadmap indicates that hydrogen production mix in the roadmap for 2030 is 51% water electrolysis (WE), 47% steam methane reforming (SMR) and 2% existing capacities. WE, using renewable electricity, includes both on-site production at the HRS and centralised production with distribution to the HRS. In 2030, the roadmap shows that the UK national demand for hydrogen for FCEVs will be 254,000 tons p.a. FCEVs will help to meet long-term emission reduction targets by offering a practical mass market solution to help meet this objective.

Hydrogen Production

Transitional approaches relying on natural gas could facilitate the use of hydrogen technologies until production methods using other, more environmentally friendly resources become available. Therefore, one likely early path for the development of hydrogen could be using the wide availability of natural gas and its distribution pipelines to create hydrogen for on-site fuelling.

Hydrogen can also be produced from coal reserves. Most analyses show that the higher efficiency of hydrogen applications can result in lower greenhouse gas emissions, even when the hydrogen is produced from coal.

Other environmental benefits of FCEVs include improved air quality and reduction in noise pollution from traffic compared with conventional vehicles powered by conventional engines. These can be used as additional arguments to promote FCEVs.

Wild Card Thinking

Some of the associated Wild Cards from the environmental point of view may include:

- Hydrogen is widely accepted as a solution to global warming by the UN with a decree.
- A breakthrough in electric power storage occurs within a decade involving battery technology, fuel cells, new chemicals and materials with some nanotechnology applications. This breakthrough allows the integration of power systems in mobile equipment (cars and trucks) with stationary energy needs (back-up power and load management option for homes, offices and factories). These storage applications begin in the high-cost and high-value parts of the energy sector and as the technology matures, costs decline and zero-emission targets become closer to achieve.
- High corn prices driven by a bad harvest could hurt corn ethanol producers, which are suffering from a saturated market for ethanol. This may allow hydrogen to take off faster than expected as an alternative energy source.

6.4.5 Political Factors

Political factors involve dominant political viewpoints or parties, political (in)stability, regulatory roles and actions of governments, political action and lobbying by non-state actors (e.g. pressure groups, paramilitaries). Regarding the hydrogen field, the following political factors can be considered to be influential:

- Reduction in energy dependency
- Political incentives
- Joint action

Energy Dependency

The commercialisation of FCEVs and hydrogen has potential benefits in terms of reduction of carbon emissions, air quality improvements and energy security enhancements, in addition to wider economic benefits. For instance, switching from imported fossil fuels to hydrogen may bring £1.3 billion annual benefit to the UK economy by 2030. Energy cost reduction potential for the transport sector in the Japanese economy was estimated 23 billion euros for 2010 (Hasegava 2013).

Diversification of energy supply through hydrogen could help to reduce the reliance of imported fossil fuels for transport and thereby increase energy security in energy importing countries. The local production of hydrogen can also provide more of the process inputs to be produced locally by reducing the dependency on external energy markets.

Political Incentives

There are currently no production incentives for hydrogen. The political support often is limited to discussion about emission-free vehicles, but practical initiatives are missing. Limited public funds are available for research and demonstration projects.

As far as the cost of hydrogen is considered, a national pricing system will need to be introduced within countries. This will depend on the size of the country, distribution of hydrogen production facilities across the country and proximity to HRSs.

Tax exemptions are planned for hydrogen vehicles. However, the practical implementation of such tax exemptions might carry the danger of giving advantages to national or local manufacturers as was discussed after the crisis in 2008 when most European countries implemented such measures. Hence national (country-specific) measures have to be analysed in detail if these are applicable for fuel cell-powered cars.

Joint Action

In order to overcome the commercialisation challenge, a close cooperation is needed between vehicle manufacturers; equipment manufacturers in production of fuel cells, hydrogen refuelling stations and hydrogen technology components and subsystems; fuel retailers; hydrogen producers; energy utilities; and the government departments such as science, technology and innovation, transport and energy.

Wild Card Thinking

Some of the Wild Cards in the political sphere might include:

- Introduction of large-scale public procurement programmes for FCEVs
- Governments back zero-interest mortgage plans for hydrogen cars
- Massive movement of public transport vehicles and large fleets to FCEVs

6.5 Characteristics of Stakeholders

A number of roadmaps have been developed in course of the FCEV history. Although most roadmaps were professionally developed, they still lack an "eagle eye view" on the "hidden" stakeholders' attitudes, influence and argumentation lines. Developing effective roadmaps hence require knowledge and information about actual but also potential stakeholders, which follow their own strategies and might have the potential to block technologies and innovation diffusion or at least create obstacle and barriers to delay diffusion. It is important to underline that stakeholders vary in their influence on the diffusion of innovation, their power and their argumentation strategies.

6.5.1 Stakeholders: Societal

In the societal sphere, mainly customers are considered as stakeholders. However, they show special characteristics, in particular with reference to FCEVs. In the transportation, business car owners' associations also play an important role (Table 6.1).

Overall the individual stakeholders have only moderate influence on innovation diffusion, but associations representing car owners have considerable power at different levels.

6.5.2 Stakeholders: Technological

From the technological point of view, various stakeholders exist (Table 6.2). These refer to direct competitors who aim at similar application but using competing technological solutions and surrounding actors like infrastructure suppliers and fuel producers.

Table 6.1 Societal stakeholders

Stakeholders	Influence	Power	Argumentation strategies
• Traditional car owners	↗	↗	• Misses typical car features • Reluctant towards noiseless drive
• Young generation	↑	→	• Wish to differentiate from traditional drivers • Limited experience with infrastructure
• Car owners associations	↑	↑	• Adverse attitudes, mainly dominated by traditional drivers • Point on noise, danger of fuel supply, need to train traditional driver to adjust

Source: HSE
→ low, ↗ medium, ↑ high

Table 6.2 Technological stakeholders

Stakeholders	Influence	Power	Argumentation strategies
• Alternative FC (SOFC, PAFC, MCFC) producers	→	↗	• Similar application fields for FCs or at least potentially similar fields • Might point to dangers and environmental issues of membranes
• Infrastructure suppliers	↑	↑	• Decentralised infrastructures need to be build—investment cost • Existing infrastructure reshaped for fuel transport—opportunity cost
• Fuel producers (gasoline)	↗	↑	• Consequences of lacking demand for gasoline—refinery closures, job losses, impact on petrochemical industry

Source: HSE
→ low, ↗ medium, ↑ high

The analysis shows that competing technological solution providers have potentially stronger influence on the diffusion of innovation. The reason lies in their technology follower position in the FC development, polymer electrolyte membrane fuel cells (PEMFCs) have been the main FC type for transport applications for a while, but still other FC types might be used for mobile applications. Producers of these FCs might follow a strategy to point strongly on the inherent dangers of membrane technology, which are still manifold. Infrastructure suppliers have strong influence and power on the diffusion per se. As long as the infrastructure for fuel supply remains insufficiently developed, these actors need careful consideration and treatment. Other often neglected stakeholders are traditional fuel producers, who belong to downstream oil and gas business, eventually forming part of the petrochemical industry. Here the challenge is that the fuel is commonly produced in refineries from crude oil using different cracking technologies—fuel is only one product of cracking crudes. So far, the essentials for the chemical industry are produced together with traditional fuel, there is no either or. It follows that with a significant reduction of fuel demand, refineries need to lower output with respective impact on the subsequent industries. This gives reasonable arguments for the industry to block FCEVs.

6.5.3 Stakeholder: Economic

The most progressed FCEV producers are Asia-based companies. National producers in the Western markets are still lagging behind in technological terms; thus they will aim at influencing the national, regional and local communities. In a similar way, the petrochemical industry will act (see also technological stakeholders). Last but not least, the repair and maintenance infrastructure lobby is important. Currently any car can be fixed in emergency cases within the existing infrastructure, but there remains a challenge to (Table 6.3):

1. Upgrade the existing infrastructure on a broader scale for regular maintenance
2. To upgrade the emergency relief infrastructure

Table 6.3 Stakeholder diversity

Stakeholders	Influence	Power	Argumentation strategies
• Technology follower	↗	↗	• Technological leadership concentrated in Asia (Japan, South Korea), Europeans lagging, oppose with lobby work
• Petrochemical industry	↑	↑	• Job losses due to either refinery closure or high investment for new equipment
• Repair and maintenance industry	↗	↗	• Significant investment in equipment • No competences in new technologies, reluctant to accept dual system

Source: HSE
→ low, ↗ medium, ↑ high

Presumably the strongest opposition might come from the petrochemical industry, which is well aware of the investments needed in their refining facilities to compensate for the decreasing demand in gasoline and related products. The repair and maintenance industry will need physical investment in line with extensive training to assure adequate services to customers.

6.5.4 Stakeholders: Environmental

In technological terms FCEVs are not new at the small-scale production and operation. However environmental groups and also health- and safety-related interest groups might raise concerns about the reliability of manufacturing and operation of FCs in the broadest sense at large scale (Table 6.4).

It can be assumed that the concerns regarding Environmental Health and Safety (EHS) will not be announced and communicated by the interest groups to the end user directly; presumably the end user is not aware and interested in the manufacturing and handling of large-scale FC production units. Thus these interest groups are likely to influence the policy level to issue-related regulations and probably laws which might have reasonable impact on the business models. Also there is a possibility that EHS-related arguments and interest groups are used by other parties with the aim of delaying FC diffusion.

6.5.5 Stakeholders: Political

The mass introduction but also the pilot introduction of FCEVs will require physical infrastructural adjustments of the public transport system. Also there is a need at the policy level to implement complementary systems, for instance, by enforcing standards (Table 6.5).

The main policy actors will be at the level of municipalities and regions. National level policy-makers will have reasonable influence to design measures, which are

Table 6.4 Environmental stakeholders

Stakeholders	Influence	Power	Argumentation strategies
• Laws, legal regulations	↑	↑	• Especially important for EHS
• Environmental groups	↗	↗	• Long-term H2 impact not known
• Health, safety groups	↑	↑	• Unknown reliability of new standards and technologies, potential negative impact on safety and health of workers in all domains

Source: HSE
→ low, ↗ medium, ↑ high

Table 6.5 Political stakeholders

Stakeholder	Influence	Power	Argumentation strategies
• Municipalities	↑	↑	• Responsible for infrastructural decisions
• Regional	↑	↑	• Financial incentives for municipalities, regional standards, complementarities of standards between regions
• Federal	↗	↗	• Initiator and promoter role but less implementation power

Source: HSE
→ low, ↗ medium, ↑ high

supportive to FCEVs diffusion, still the implementation of any federal measures is at the regional and the municipalities' level. Especially municipalities have the responsibility for assigning and licensing respective space for related infrastructure, whereas regional authorities will be responsible for monitoring and quality/safety testing and certification.

6.6 Development of an Eagle Eye FCEV Roadmap

A roadmap for creation of a market for FCEVs is a tool for long-term complex planning, which allows setting strategic goals and estimating potential contribution of new technologies, products and services to build a competitive and sustainable FCEV market. The roadmap considers alternative ways to achieve the goals and choose the most efficient products and relevant technology applications. It provides decision-makers with estimates of future markets and prospects of innovative products and designs an innovative technological value chain from R&D to market entry. The FCEV roadmap presents estimated indicators of economic efficiency of the potentially prospective technologies and products, with the high demand potential and attractive consumer properties of FCEVs within the timescale also taking into account the elaborated stakeholder views. The proposed roadmap will be developed on the bases of both qualitative and quantitative methods, the expert community survey data and evidence-based analyses. In this stage the stakeholder analysis will be incorporated in the market dimension of the roadmap.

The roadmap template presented here is designed on the basis of a market-driven and technology-driven approach, which starts from the analysis of a market demand. The elaboration of the FCEV roadmap covers the analysis of key needs of the marketplace and customers, possible markets development within several scenarios, estimation of future demand for particular types of FCEVs and respective requirements including potential attitudes, measures and activities and argumentation lines of stakeholders. It will also require comprehensive analysis of technological innovation and product development based on identification of future dynamics. Thus the FCEV roadmap allows considering both technological and market sides providing a combination of market pull and technology push approaches. The process will generate a roadmap demonstrating new products and technologies, which are important in achieving the set goals and a business map containing economic appraisal and comparison of alternative paths of future development. In addition, the roadmap should provide a detailed analysis of market pull, including:

- Areas of product's application determining the demand for technological solutions
- Specificity of different segments of FCEV markets
- Balance between technological facilities and consumers' needs
- Economic estimation of technology trajectories
- Analysis of stakeholders
- Recommendations aimed at support of market-oriented technologies and products

It will also pay special attention to description of technology push factors:

- Technologies that provide competitive advantages for FCEVs
- Technological limitations
- Priority technological tasks
- Revealing of technological "forks"

The principal structure of the FCEV roadmap is presented on the Fig. 6.2.
The roadmap template includes four major layers:

1. *Technologies.* This layer contains the description of the prospective technologies within the identified time horizon. It provides a SWOT analysis of these technologies that summarises benefits and limitations of each technology. It also provides a forecast of target properties required to satisfy market needs and a set of the main technological tasks necessary to be done to reach these features. Finally, it gives an opportunity to estimate prospects for each technology in terms of readiness for implementation and potential effect.
2. *Products.* This layer provides a brief description of prospective products in terms of readiness for commercialisation and potential effects for researched area. It also estimates potential time of commercialisation and the most prospective market niches for each product.
3. *Markets.* There will be elaborated scenarios of potential FCEV market development based on the eagle eye view approach. The roadmap will provide a brief description of main market's features and possible strategies for each scenario and each market. Thus, all markets should be ranked from the most prospective down to the less ones.
4. *Alternatives.* The roadmap also reveals possible development of alternative products and solutions. It takes into account the dynamics of the main product properties, opportunities of export of these products and their cost.

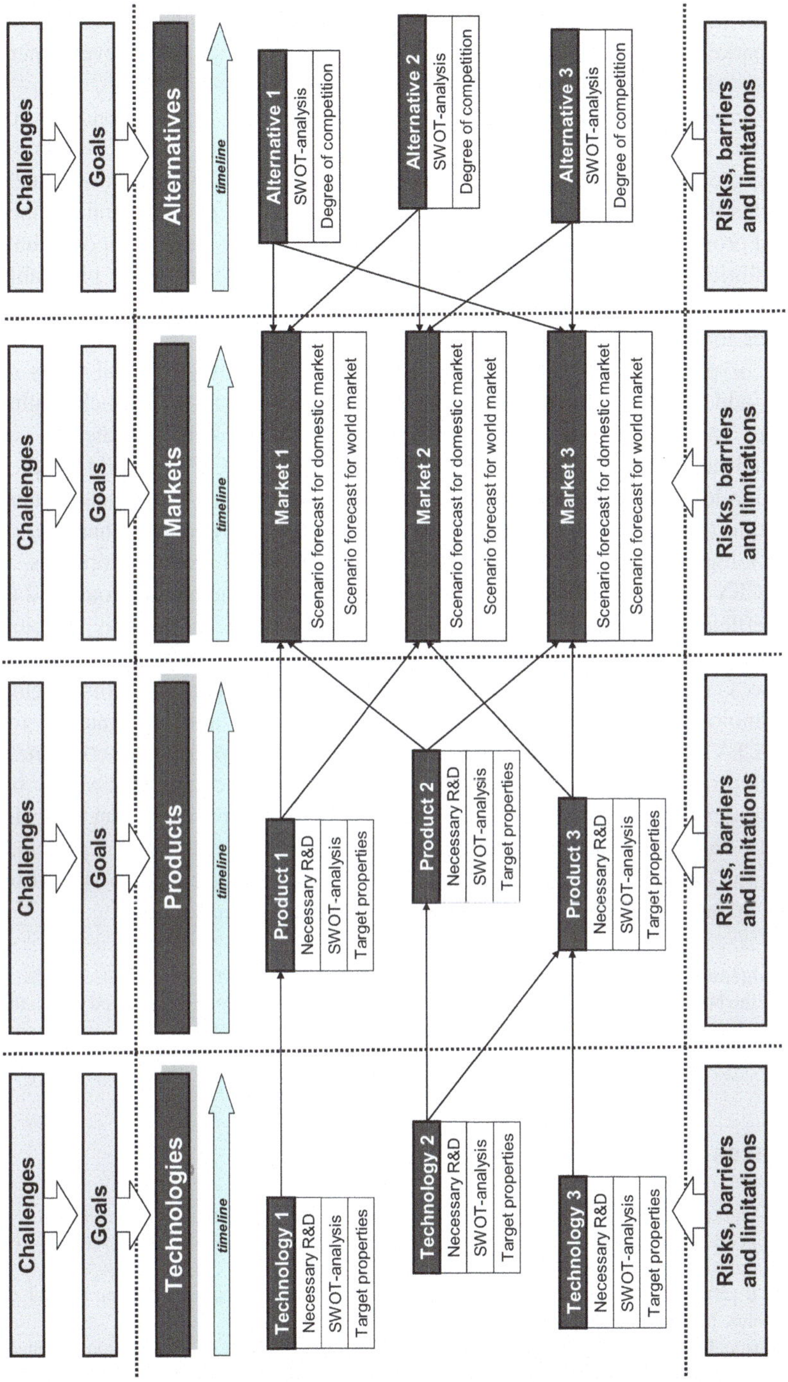

Fig. 6.2 Structure of roadmap. Source: HSE

For each layer, it is necessary to consider challenges and a set of relevant goals taking into account potential risks, the most significant particular challenges for FCEVs markets, to reveal obstacles that could hamper FCEVs market development and to assess key risks and threats.

The complete roadmap illustrates the links between the key technologies for FCEVs, the consumer properties of existing and advanced FCEVs, the most promising products and their respective market shares, volumes and growth rates. It highlights the structure of potential demand for innovative products and outlines their most prospective markets. The roadmap also provides an assessment of technical capabilities required for manufacturing of products with the most preferable consumer properties, which would allow generating the significant competitive advantages for FCEVs.

Being of practical value, the FCEV roadmap demonstrates optional paths of building added value chain "technologies—products—markets". Such paths/trajectories are aimed at detailed description of possible strategies of commercialisation on particular markets—what kinds of FCEVs should be produced; what level of their consumer properties will allow them to compete against other similar (conventional and new) goods at different time periods; what kind of new technologies should be introduced to obtain the required product properties.

The FCEV roadmap reveals alternative ways to achieve the market goals and to choose efficient allocation of resources. The roadmap takes into account manufacturing and market developments and prospects of technologies, products and services contributing to the design of complex innovation value chains ranging from technology to market entrance of FCEVs and allows building strategies for linking FCEVs development with other related industries (suppliers and consumers of related products/technologies). It integrates the expert community views on innovative development ways in FCEV and related areas, provide a set of well-grounded trajectories of innovation development and indicated principal "bifurcations" as points of the key decisions to be made. The roadmap should be regularly updated to enhance its practical value for decision-making.

Acknowledgements The book chapter was prepared within the framework of the Basic Research Program at the National Research University Higher School of Economics and supported within the framework of the subsidy by the Russian Academic Excellence Project '5-100'.

References

Global EV Outlook (2013) Understanding the electric vehicle landscape by 2020. https://www.iea.org/publications/freepublications/publication/GlobalEVOutlook_2013.pdf. Accessed 17 Feb 2017

Hasegava T (2013) Presentation on FCEV market development at Nissan. Higher School of Economics, Moscow

Miles I, Saritas O, Sokolov A (2016) Foresight for science, technology and innovation. Springer International Publishing, Switzerland

Saritas O (2013) Systemic foresight methodology. In: Meissner D, Gokhberg L, Sokolov A (eds) Science, technology and innovation policy for the future: potentials and limits of foresight studies. Springer, Berlin, pp 83–117

Saritas O, Aylen J (2010) Using scenarios for roadmapping: the case of clean production. Technol Forecast Soc Change 77(7):1061–1075

UK H2 Mobility Report (2013.) https://www.gov.uk/government/uploads/system/uploads/attachment_data/file/192440/13-799-uk-h2-mobility-phase-1-results.pdf. Accessed 17 Feb 2017

US Department of Energy (2015a) Hydrogen production, 2015 production section. https://energy.gov/sites/prod/files/2015/06/f23/fcto_myrdd_production.pdf. Accessed 17 Feb 2017

US Department of Energy (2015b) Fuel cell technologies market report 2015. Washington. https://energy.gov/sites/prod/files/2016/10/f33/fcto_2015_market_report.pdf. Accessed 17 Feb 2017

Alexander Sokolov is Deputy Director of HSE ISSEK and Director of HSE International Foresight Center. His main professional interests are related to Foresight, STI priorities, indicators and policies. Prof. Sokolov is tenure professor at HSE, he teaches Foresight for undergraduate and postgraduate students. He authored over 120 publications in Russia and internationally and managed many Foresight projects, including: Russian S&T Foresight: 2030; Foresight for Russian ICT sector (2012); Innovation Priorities for the Sector of Natural Resources (2008–2010); Russian S&T Delphi Study: 2025 (2007–2008); Russian Critical Technologies (2009) et al. Prof. Sokolov is member of several high-level working groups at OECD and other international organizations, serves for advisory boards at several international conferences and journals.

Ozcan Saritas is Professor of Innovation and Strategy at the HSE and editor-in-chief of "Foresight"—the journal of future studies, strategic thinking and policy. He worked as Senior Research Fellow at the Manchester Institute of Innovation Research, The University of Manchester. His research focuses on innovation and policy research with particular emphasis on socio-economic and technological Foresight. With a PhD from the "Foresight and Prospective Studies Program," he introduced the "Systemic Foresight Methodology", and has produced a number of publications on the topic. Dr. Saritas has extensive work experience with the international organizations, including the United Nations, OECD, and the European Commission. He has been involved in large scale national, multinational and corporate research and consultancy projects on sectors including energy, climate change, agriculture, food, water, transportation, and ICT among others; published a number of articles in respected journals; and have delivered keynote speeches in more than 50 countries across the world. Besides his research and publication activities, he designs and delivers academic and executive education courses on Foresight and Strategic Planning. He has recently co-authored a book, entitled "Foresight for Science, Technology and Innovation" published by Springer, which has become one of the key readings in the field.

Dirk Meissner is Deputy Head of the Laboratory for Economics of Innovation at HSE ISSEK and Academic Director of the Master Program "Governance for STI". Dr. Meissner has 20 years experience in research and teaching technology and innovation management and policy. He has strong background in policy making and industrial management for STI with special focus on Foresight and roadmapping, funding of research and priority setting. Prior to joining HSE Dr. Meissner was responsible for technology and innovation policy at the presidential office of the Swiss Science and Technology Council. Previously he was management consultant for technology and innovation management with Arthur D. Little. He is member of OECD Working Party on Technology and Innovation Policy. He is Associate Editor of Technological Forecasting and Social Change, IEEE Transactions on Engineering Management, Journal of the Knowledge Economy, member of Editorial Review Board at Small Business Economics and Journal of Knowledge Management. He guest edited Special Issues in Industry and Innovation journal, Journal of Engineering and Technology Management, Technological Analysis and Strategic Management among others.

Technology Roadmaps: Emerging Technologies in the Aircraft and Shipbuilding Industries

7

Marina Klubova, Lubov Matich, Vladimir Salun, and Natalia Veselitskaya

7.1 Introduction

Recently the emergence of technologies and development of industries in the world have been inextricably linked to the response to grand challenges (EC 2010). In this regard, a clearly established scientific and technology policy should promote the opportunities and ways to overcome threats associated with said grand challenges (Chaminade and Edquist 2006; Bergek et al. 2008; Georghiou 2011; Keenan et al. 2012; Edquist 2011; Gokhberg 2013).

In this context, the development of the aircraft and shipbuilding industries has multiplier effects on the economy and society, on a domestic and international level. The speed and volume of traffic flows are clearly correlated with the development of the global economy. Furthermore, emerging technologies in the field of aircraft and shipbuilding will effectively respond to various types of global challenges: economic, environmental, scientific and technological, and social. Such technologies form the basis for the improvement of existing and the emergence of new types of aircraft and ships. Such improvements and innovations eventually lead to:

- The transformation of global value chains
- An increase in energy efficiency through the use of alternative energy sources
- A reduction in negative environmental impacts (carbon dioxide and other harmful emissions, noise pollution, etc.)
- The prevention of terrorist acts
- An acceleration of knowledge and technology transfer

M. Klubova (✉) · L. Matich · V. Salun · N. Veselitskaya
National Research University, Higher School of Economics, Moscow, Russia
e-mail: mklubova@hse.ru; lmatich@hse.ru; vsalun@hse.ru; nveselitskaya@hse.ru

D. Meissner et al. (eds.), *Emerging Technologies for Economic Development*, Science, Technology and Innovation Studies,
https://doi.org/10.1007/978-3-030-04370-4_7

An integrated analysis of the emerging technologies in aircraft and shipbuilding is justified due to similar features of both industries in their products' life cycles. In particular, these industries are characterized by the high duration of technological and production processes, high capital expenditures, etc.

Despite the wide opportunities that emerging technologies offer, the high capital intensity of the aircraft and shipbuilding industries makes it necessary to set scientific and technological priorities. Depending on which global challenges are recognized as the most significant (energy efficiency, environmental friendliness, safety, etc.), alternative variants (forks) of the development of emerging technologies can be selected and implemented by the country. The forks form the basis for building scenarios for emerging technologies in the aircraft and shipbuilding industries.

7.2 Background

Global foresight experience demonstrates that for the successful identification and further implementation of industry development priorities, it is important to determine "the future shape" of those developments. During this process, long-term scientific, technological, and socioeconomic developments are considered based on the factor influence system, including joker events, global challenges, trends, threats, drivers, barriers, and restrictions (Saritas and Proskuryakova 2017).

Over the course of the last few decades, state and regional foresight exercises for the transport system in general, as well as for individual vehicle types, have been actively carried out. They are initiated by associations of organizations and unions, large companies producing transport equipment, companies operating on transportation markets, etc. (Vishnevskiy and Yaroslavtsev 2017).

The information base of this research includes the following studies in the field of aircraft construction and shipbuilding (Fig. 7.1):

- Leading global forecasts and foresight studies
- Technology foresight studies at the national level
- Technology roadmaps
- Strategic documents of leading companies in the aircraft and shipbuilding industries
- Scientific publications

During this study, more than 50 documents describing the scientific, technological, and socioeconomic development of the transport sector were analyzed (Fig. 7.1), and a bibliometric analysis of more than 2000 scientific articles (the Horizon 2005–2016) on "Emerging technologies" topics was conducted.

Global forecasts and foresight studies call one's attention to the development of technologies in the design, production, testing and repair of vehicles (EREA 2012; EC 2012), global trends and challenges (ACEA 2011; NISTEP 2014; OECD 2012),

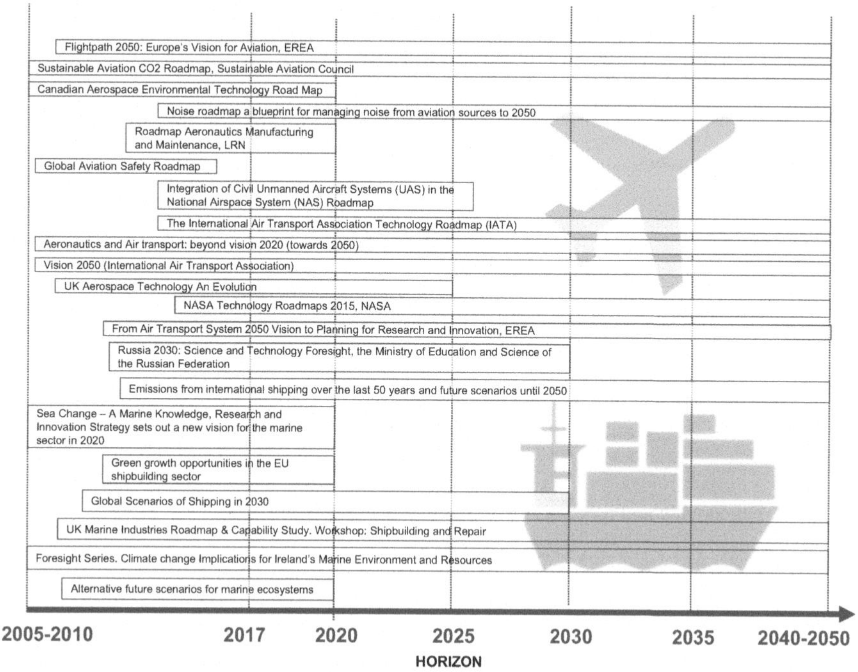

Fig. 7.1 Foresight and roadmaps for the shipbuilding and aircraft industries (examples). Source: HSE

important problems, and prospects for the development of air and water transport (IATA 2011; IEA 2012; OECD 2015).

National technology foresight studies focus on the problems of the energy efficiency of transportation, the reduction of negative environment impacts (EC 2011, 2015a; Marine Institute 2005; CE Delft 2012; Pinnegar et al. 2006), transport safety (EC 2015b), and strategy development for improving product competitiveness (UK Marine Industries Alliance 2011; GenerationBalt 2011). Studies also refer to traffic management technologies, the interaction of vehicles with one another and with the existing infrastructure (EC 2014; Federal Ministry of Transport and Digital Infrastructure of Germany 2015).

International and national roadmaps analyze the trajectories of technology implementation aimed at reducing emissions (IATA 2013; NRC 2011; Sustainable Aviation Council 2012), noise pollution (Sustainable Aviation Council 2013), the cost of production and operation of transport equipment (Marine Industries Roadmap & Capability Study 2011), and security problems (ICAO 2006). Furthermore, there are roadmaps for individual types of vehicles, which are formulated on the basis of an analysis and the prioritization of individual technologies and technological solutions (NASA 2013).

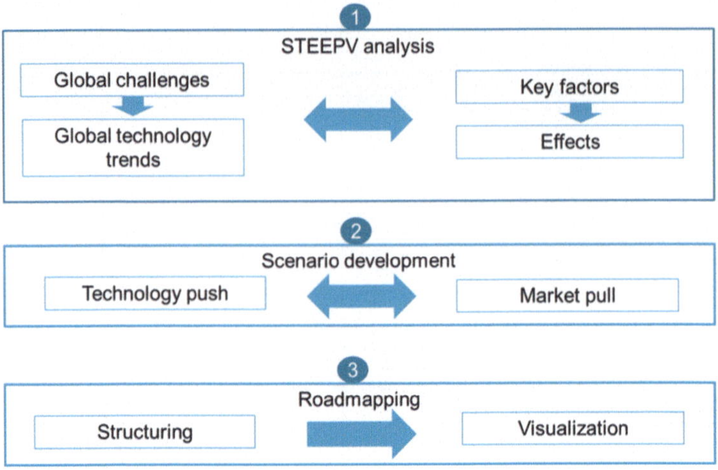

Fig. 7.2 Methodology and process visualization. Source: HSE

Forecasts were also prepared by Boeing (2016), Airbus (Airbus Group 2016), Bombardier, Embraer, Hindustan Aeronautics (India), Safran (2016), and others. Companies formulate foresight-based strategies for the development of their own businesses in order to identify technological priorities, opportunities, and potential threats that might limit the companies' future growth.

The list of key shipbuilding companies producing scientific and technology forecasts includes Mitsubishi Heavy Industries (2017), Hyundai Heavy Industries (2016), Daewoo Shipbuilding & Marine Engineering (NVIDIA Corporation 2014), Samsung Heavy Industries (2012), and China State Shipbuilding Corporation (2017).

With a help of the Web of Science database, an analysis of scientific publications on emerging technologies was conducted: a significant number of studies were devoted to the development of transportation infrastructure (including the aircraft and shipbuilding industries). The most interesting research on emerging technologies in the aerospace industry concerns self-renewing materials for aircraft construction (Yang et al. 2013). Shipbuilding and aircraft engineering papers are devoted to the management of unmanned vehicles (Wahlstrom et al. 2015; Vachtsevanos et al. 2016), the design of electric motors for ships (Pestana 2014), etc.

This chapter presents a three-stage approach (Fig. 7.2) involving the parallel and consistent use of various foresight methods, including a STEEPV analysis of grand challenges, factors and trends, scenario building, and roadmapping.

The *STEEPV analysis* involves a comprehensive study of the research object in six key categories: social, technological, economic, environmental, political, and value (Table 7.1). This method has been used for over 10 years in Europe (FOREN 2001) and other countries, having proved an effective tool for analyzing the external environment.

Table 7.1 STEEPV

Category	Description
Social	Lifestyle (e.g., the use of free time, the trends of family relations), demographic structure, social integration, and cohesion (disparities in lifestyle, equality/inequality, education)
Technological	The level of technological progress, the speed of innovation diffusion, problems and risks associated with technology (including safety and health risks)
Economic	The level and rates of economic growth, the structure of industries, competition and competitiveness, market and financial aspects
Environmental	Ensuring sustainable development and combating climate change, more localized issues related to the environment (including pollution, depletion of resources, loss of biodiversity, etc.)
Political	Dominant political structures, political stability/instability, regulating the roles and actions of governments, lobbying for the interests of nongovernmental groups (e.g., advocacy groups, military groups)
Value	Attitude toward work [e.g., entrepreneurship, career development, dependence on authorities, demand for mobility (work, workplaces)], preferences for recreation, culture, social relations, etc.

Source: HSE based on Loveridge (2002)

The construction of scenarios provides the option of "crosscutting" promising areas that have the potential to implement a complete innovation cycle: from basic research to the commercialization of the final product.

The development of roadmaps makes it possible to move from the identification of promising areas to the organization of specific innovative projects and a list of activities necessary for their implementation. Roadmaps are often used by large companies (often on a regular basis).

The roadmap is a generalizing document that reflects a multilevel system of strategic development within a subject area on a single time scale. It contains indicators of the economic effectiveness of advanced technologies and products with high demand potential and attractive consumer properties (Matich 2017).

The concept of integrated roadmaps includes two interrelated components: a technological roadmap reflecting the prospective directions of scientific and technological development and a business roadmap that provides an economic evaluation of these areas (Vishnevskiy et al. 2016).

Foresight research involves the integration of two approaches: technology-driven (technology push) and market-oriented (market pull). Based on the application of "technology push" methods, the elements of the innovative technological chain are formed (such as research with the potential for practical application, innovative technologies, high-tech products, and services). The "market pull" approach implies an analysis of demand for innovative products and individual technologies used in their production. The integration of these approaches is particularly effective for high-tech industries that establish a link between the application of promising products and the technological capabilities for their production, which are identified on the basis of research and development results (Dekhtyaruk et al. 2014).

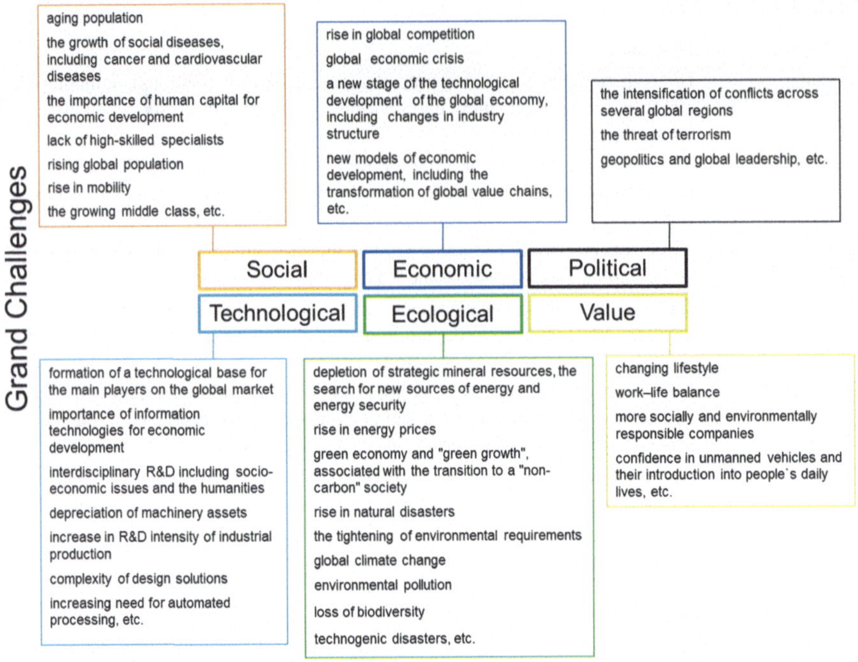

Fig. 7.3 Grand challenges. Source: HSE

Despite the advantages of roadmaps and scenarios, their isolated application often yields a fragmented picture of the future.

Therefore, over the last few years, the process of roadmap development has often been combined with a scenario approach (Saritas and Aylen 2010). Scenarios can be used to verify the reliability of technological roadmaps. Scenarios assume the emergence of disruptive innovations and wild cards. An integrated approach including both roadmaps and scenarios allows for overcoming the limitations of the individual methodologies.

7.3 Global Trends and Emerging Technologies

7.3.1 Key Determinants

The identification of global trends that determine the future is one of the key objectives for scientific, technological, and innovation policies. During this study, more than 150 grand challenges were revealed. The most significant ones are shown in Fig. 7.3. They are classified by STEEPV.

The technological development of the shipbuilding and aircraft industries is influenced by the following macroeconomic trends:

- Sustainable traffic growth
- Energy efficiency as a key factor determining transportation preferences
- Globalization and the liberalization of transport
- The uneven development of air transportation according to regions of the world, the increasing importance of emerging markets
- Customized, personalized transport services, combined with or supplemented by information and communication devices
- The consolidation of the world market: mergers, acquisitions, and alliances
- The concentration of industries and formation of global corporate leaders
- The emergence of new large corporate players and regional production centers

Grand challenges include trends in the shipbuilding and aircraft industries and in related fields. For example, the emerging trends on the markets for passenger and cargo transportation determine the prospects for the technological development of aircraft construction because they create demand for certain technologies and aircraft components. In general, grand challenges and macroeconomic trends in the development of the transportation market create the following effects:

- An extension of the route network
- An improvement in flight safety
- An improvement of air transport's environmental friendliness (reducing the environmental impact of air transport)
- The complication of the operating conditions of aviation equipment
- The diversification of the ground infrastructure
- A rise in operational efficiency and the diversity of aviation business models
- The increased intermodality of traffic
- An expansion of operating conditions for ships
- The development of technologies in maritime areas (e.g., the biotechnology of the seas)
- The intensive development of the shelf, particularly the Arctic and the Antarctic
- The proliferation of new economic models on the goods delivery market

Based on these trends, we can identify the groups of key factors that have the greatest impact on emerging technologies (Fig. 7.4).

Technological capabilities (primarily horizontal technologies) represent the factors contributing to the development of emerging technologies on the part of technology push (for more detail, see Sect. 7.3.2 Global Technology Trends). The majority of other factors characterize the socioeconomic and environmental needs, thus forming the requirements for the various parameters of vehicles. Therefore, we can consider these factors as those from the market pull side. Groups of factors "efficiency," "mobility," "individualization," "ecology," and "security" are used as a basis for scenario analysis and are discussed in more detail in Sect. 7.3.3, Scenarios.

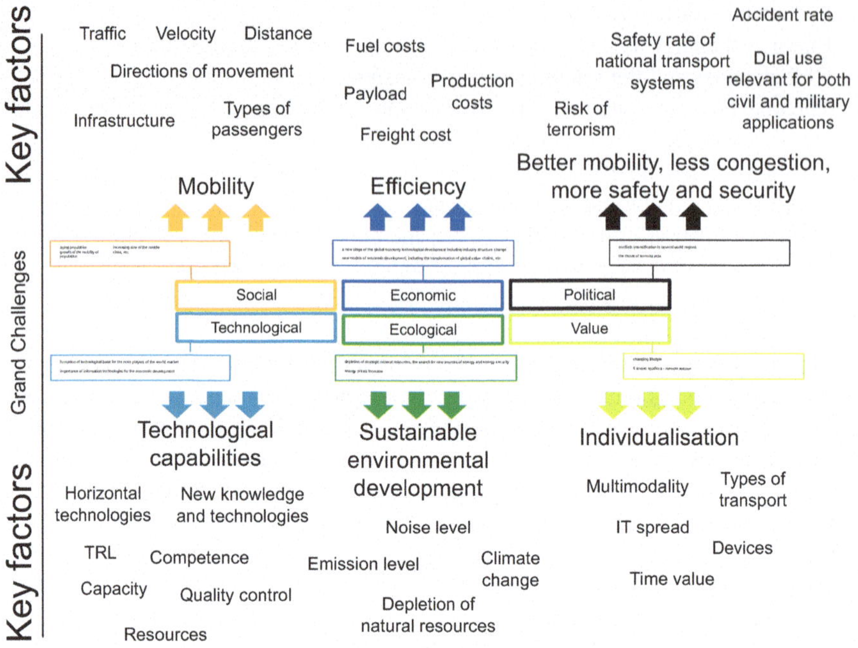

Fig. 7.4 Key factors impacting emerging technologies individualization. Source: HSE

7.3.2 Global Technological Trends

The study of the socioeconomic development from the perspective of global challenges has led to the development of the concept of emerging technologies, which are designed to respond to said challenges. A bibliometric analysis of the above topics confirms this thesis. So, to the query "emerging technologies," the following results were obtained; in the last decade, growth periods (2005–2008 and 2011–2015) and a certain decline of interest in the topic have been noted. The total number of publications during the period under review exceeded 2000 articles (Fig. 7.5).

A further semantic analysis of the publications allowed us to identify the thematic categories.

Figure 7.6 presents the key areas of publication activity on "emerging technologies" (engineering, electronic):

- Engine arrangement
- Control devices
- ICT technologies
- Infrastructure

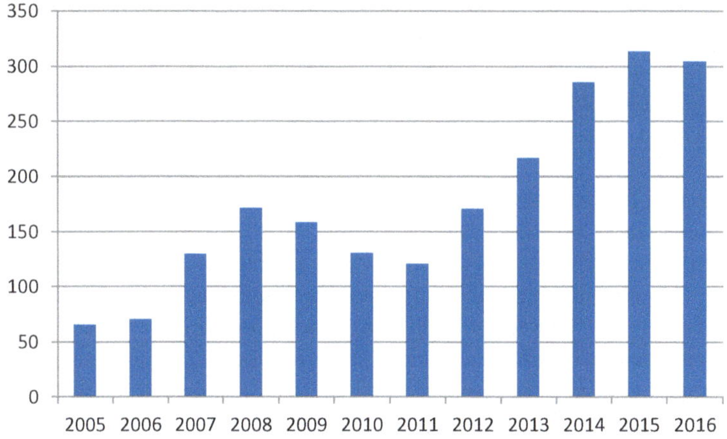

Fig. 7.5 Publications on "emerging technologies" (engineering, electronic). Source: HSE based on Web of Science

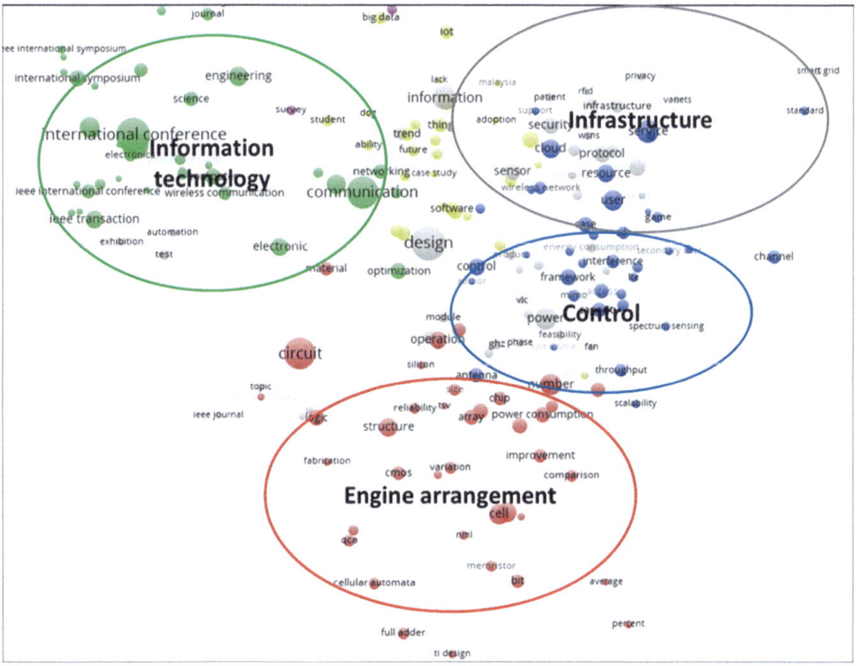

Fig. 7.6 Semantic analysis of the publications on "emerging technologies" (engineering, electronic). Legend: for each item, the size of the item's label and the size of the item's circle depend on the prominence (weight) of the item. The distance between two terms provides an indication of the number of co-occurrences of the terms. The colors divide items into clusters (VOSviewer Manual 2016). Source: HSE based on Web of Science

The bibliometric analysis on the field of emerging technologies helps one form an idea of the key research issues and the areas of their influence.

Global technology trends have a crosscutting impact on all areas, including the aviation and shipbuilding industries, by creating opportunities for innovative development. The main global technology trends identified in the study include:

1. Research in the field of alternative energy sources
2. The transition to advanced manufacturing
3. The digitalization of production (industrial Internet of things, "factories of the future," big data, augmented reality, cloud technologies, etc.)
4. The move to the "Internet of Everything" (Bandyopadhyay and Sen 2011)
5. The development of information analysis technologies
6. The transition to new interfaces (neuro-, bio) and augmented reality
7. The development of materials with new properties
8. The use of additive technologies

For example, we can examine the influence of additive technologies on the aircraft, shipbuilding, and other industries in more detail. Additive production implies direct manufacturing of three-dimensional parts without the use of intermediate operations for processing blanks (raw material savings of more than 70%). The demand for additive production is attributable to industry needs for the simplification and acceleration of the process to create individual parts. Layer-by-layer processing allows for creating three-dimensional products according to a given individual pattern. This technology refers to a new industrial stage with a modified production chain.

Today, additive production processes (AP processes) are actively used in rapid prototyping. AP technologies are also used for the rapid production of tools for printing and molds.

With the improvement of technology quality, the scope of additive production is substantially expanded. Currently such technologies are used in the production of goods and products, including elements of mechanisms, parts, and assemblies for the aerospace, defense, shipbuilding, automotive, and electronics industries.

The simultaneous appearance and complementarity of these trends enhance the positive effects and contribute to the formation of such scientific and technological areas as:

- Robotics
- Sensor systems
- Systems of artificial intelligence
- 3D printing
- New materials
- ICT
- Microelectronics
- Storage and power converters
- Holographic technology

Table 7.2 Important facts concerning technological areas for the aircraft and shipbuilding industries

More than 1000 parts of the A350 are now made by 3D printing—more than on any other commercial aircraft
Parts created by 3D printing are 30–55% lighter and use 90% fewer raw materials than those made by traditional methods
3D printing will affect the technologies used by the Navy for the modeling and design of ships, submarines, aircraft carriers, and of everything else that is on board
3D printers produced by the German company BigRep are already used in the production of screws
The launch in 2018 of the Zephyr T drones by Airbus, powered only by solar energy

Source: Benthien et al. (2015), Hiufu Wong (2016), Airbus (2016)

As presented in Table 7.2, robotics and 3D printing technologies have already found a wide application in the aircraft and shipbuilding industries. Further, it is likely that horizontal technologies will be used increasingly often in these industries.

The most popular horizontal technological areas used in different parts of aircraft and ships and during their production processes include:

- Sensor systems and 3D printing used during equipment, design, and production processes
- Nanotechnology and new materials used in engine production, materials, and design

7.3.3 Scenarios

The key factors have a different impact on the development of the transport sector. An analysis of the possible areas of influence of separate factors (Fig. 7.8) allows one to group the emerging technologies in three scenarios.

The impact of key factors (Fig. 7.4) varies across different scenarios of the transport sector's development. The construction of the scenarios is based on the main groups of transport sector development factors: efficiency, ecology, security, and individualization (Table 7.3). An analysis of the separate factors' possible influence (positive or negative) (Fig. 7.8) allows one to group the emerging technologies into three scenarios:

- Scenario A "National industries"
- Scenario B "Green energy"
- Scenario C "Transport individualization"

For example, the growth of such a factor as "traffic and distance" contributes to the development of scenario A and scenario B, while, other things being equal, this development prevents the emergence of scenario C or postpones the time of its implementation.

Table 7.3 Application of horizontal technologies in aircraft and shipbuilding industries

Horizontal technology/ applications	Engine	Equipment	Materials	Design (construction, etc.)	Production technologies
Robotics		+			+
Sensors		+		+	+
Artificial intelligence		+			+
3D printing		+		+	+
Nanotechnology and new materials	+		+	+	
IT		+			+
Microelectronics	+				+
Accumulators and energy converters	+				+

Source: HSE

Information presented in Table 7.4 is complemented by Fig. 7.7 and gives the reasoning behind the selection of key technologies for the development of each scenario. The key factors encompass the requirements for the parameters of future vehicles.

Expected technological breakthroughs will improve the design and construction of vehicles, testing, etc. However, the pace of social, environmental, and value changes poses an ever-increasing number of challenges for the transport industry. To answer such challenges, advanced technologies are needed. The improvement of transport efficiency by means of optimizing fuel consumption and reducing the cost of maintenance, etc., may contradict targets for improving environmental impacts, for example, the reduction of aviation noise levels. In this regard, there are alternative technological forks, which are distributed between different scenarios.

Scenario A: National Industries

The *main tasks* under this scenario are balanced economic growth, the increase of high-tech product exports, and safe operation of the transport system. The *key concept* of transport sector development in this scenario is technology push. State policy incentivizes the development of shipbuilding and aircraft building as system-forming industries of the economy. The main technological objectives financed by the government are to increase the safety of passenger transportation, cargo transportation, the functioning of transport infrastructure, increasing production efficiency, and competitiveness of products. *Main players are* state scientific centers, large aircraft and shipbuilding corporations, and public authorities.

Table 7.4 Matrix of main groups of factors/scenarios

	Scenario A National industries	Scenario B Green energy	Scenario C Transport individualization
Efficiency	■■■	□□□	■■□
Ecology	■■□	■■■	□□□
Security	■■■	■■□	■■□
Individualization	■□□	□□□	■■■

Source: HSE

Fig. 7.7 Matrix key factors/scenarios. Source: HSE

Examples of technologies and technological solutions:

- Open rotor
- Advanced turbofan
- Virtual engineering
- Robotics for smart welding
- Intermetallic alloys for combustion chimneys
- New polymers, thermoplastic

Scenario B: Green Energy

Main tasks under this scenario are improving the ecological situation, preserving biodiversity, and reducing the likelihood hazardous natural phenomena and man-made environmental catastrophes. The *key concept* for transport sector development is the search for solutions to environmental problems. The main technological areas are the reduction of the level of emissions of CO_2, NOx, etc., a decrease in noise pollution caused by vehicles, and the reduction of the negative impact transport infrastructure has on the ecosystem and solving problems of vehicle utilization. *Main players* are public organizations, state scientific centers, and large aircraft building and shipbuilding corporations.

Examples of technologies and technological solutions:

- Hybrid vehicles
- Electric engines
- Research on alternative fuels
- Biopolymer membranes
- Floating power plants
- Wind turbine-powered sea vessels

Scenario C: Transport Individualization

The *main tasks* under this scenario are raising attention to individual needs when designing vehicles, providing convenience and accessibility of transport for the public and businesses. The *key concept* of transport sector development in this scenario is market pull. The rising need for multimodal transport systems on the one hand and the emergence of opportunities to overcome the same distance in about the same time by different vehicles on the other contribute to the emergence of new types of transport and personalization of services. The management of individual mobility will be carried out using new information and communication technologies. Accordingly, the *main players* are small and medium businesses, vehicle users, and software developers.

Examples of technologies and technological solutions:

- Flying cars (Toyota 2017)
- Craft-to-craft communication

- Jetpacks
- Vertical takeoff aircraft
- Bionic structures
- Self-healing, multifunctionality, a new generation of rolled metal fibers, and smart materials
- Multi-criteria multiparameter optimization and virtual testing technologies

7.3.4 Technology Roadmap "Emerging Technologies in the Aircraft and Shipbuilding Industries"

Our roadmap was developed on the basis of the approaches of the world's leading centers in the field of strategic forecasting and planning, taking into account the results of foreign strategic documents in the field of aircraft and shipbuilding development, and also based on the results of expert analysis.

To specify the format of the visual representation of the roadmap, we created a hierarchical structure of the roadmap, which defines the relationships between global trends and emerging technologies (Fig. 7.8).

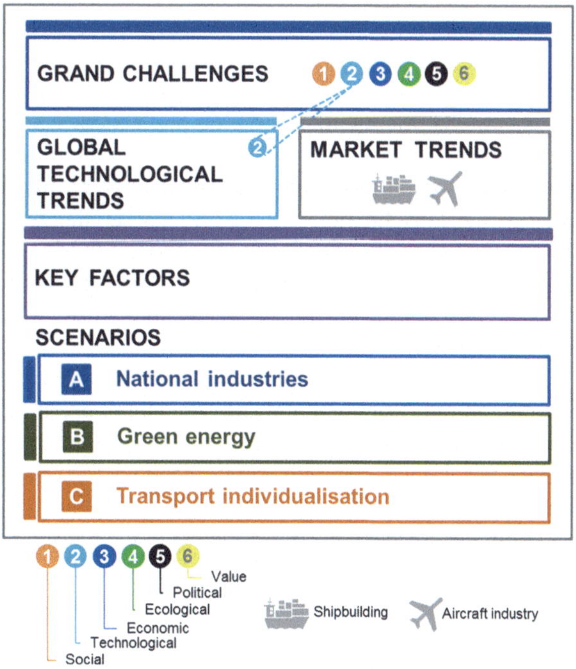

Fig. 7.8 Structure of the technology roadmap "emerging technologies in the aircraft and shipbuilding industries." Source: HSE

7.4 Conclusions

The aircraft and shipbuilding industries have a multiplier effect on the socioeconomic development of countries. Grand challenges, including technological, social, economic, and others, affect the vision of the future for these industries. This influence is expressed in the emergence of new technologies and products, modifications of business models, etc. As a consequence, one of the priority tasks of states is to determine the directions of development as well as the emerging technologies in these sectors.

The study also included a comprehensive analysis of the grand challenges affecting the development of the aircraft and shipbuilding industries in the medium and long term. The construction of scenarios was based on the main groups of factors affecting the development of the transport sector (efficiency, ecology, security, individualization), which in their turn are influenced by grand challenges. According to the results, the roadmap "Emerging Technologies in the Aircraft and Shipbuilding Industries" was developed. The strategic documents in the field of aircraft and shipbuilding were used as the background for this paper. The methodology used is based upon a comprehensive analysis of STEEPV factors, scenarios, and roadmap development.

As a result of this study, several interesting conclusions may be drawn. First, the STEEPV analysis of the key factors helped us identify the prospects for the development of the aircraft and shipbuilding industries in light of the social, technological, economic, environmental, political, and value aspects. During the research, more than 150 challenges were identified and classified by STEEPV. Some of them created market trends, such as sustainable traffic growth; energy efficiency as a key factor determining transportation preferences; transport globalization and liberalization; uneven development of air transportation; customized, personalized transport services combined with or supplemented by information and communication devices; the consolidation of the world market: mergers, acquisitions, and alliances; the concentration of industries and the formation of global corporate leaders; and the emergence of new large-scale corporate players and regional production centers. These trends significantly affect the technological development of the shipbuilding and aircraft industries.

Second, the study revealed that grand challenges create trends in the shipbuilding and aircraft industries and in related fields. For example, the trends emerging on the passenger and freight transport markets determine the prospects for aircraft technological development as they create demand for certain aircraft technologies and components.

Third, the effects of the development of the transport market that were spurred on by grand challenges and macroeconomic trends were defined in this paper. These include the extension of the route network; the improvement in flight safety and the environmental friendliness of aircraft; the increased complexity of aviation equipment's operating conditions; and the diversification of ground infrastructure.

The bibliometric analysis identified the main technology trends and nine key technology areas determining the technological development of the aircraft and shipbuilding industries which are robotics, sensor systems, artificial intelligence systems, 3D printing, new materials, ICT, microelectronics, storage and power converters, and holographic technology.

On the basis of the aforementioned results, three scenarios were developed: Scenario A "National industries," Scenario B "Green energy," and Scenario C "Transport individualization." Each scenario assumes the need for the priority development of various groups of technologies addressing grand challenges. The scenario development is especially important in view of the high capital intensity of the aircraft and shipbuilding industries.

As a result of the research, we developed the "Emerging Technologies in the Aircraft and Shipbuilding Industries" roadmap. It demonstrates the relationships between the impact of grand challenges and the possible directions for the technological development of the aircraft and shipbuilding industries, which are grouped according to the three scenarios. In addition, the roadmap shows how each factor affects the feasibility of implementing each scenario (whether it facilitates or hinders the implementation of each scenario). The roadmap provides an opportunity to transform priorities in the aircraft and shipbuilding industries into specific technology development projects.

The results of this study, including the developed roadmap, can be used as an instrument for priority setting, the selection of innovation projects, and their implementation by policy-makers and other users of the results of foresight studies. It stipulates scenario forks within the set of emerging technologies, which contribute to bringing the industries to a desired future state.

Acknowledgments The book chapter was prepared within the framework of the Basic Research Program at the National Research University Higher School of Economics and supported within the framework of the subsidy by the Russian Academic Excellence Project '5-100'.

References

ACEA (2011) Global trends in transport routes and goods transport: influence on future international loading units. http://www.acea.be/uploads/publications/SAG_16_Global_Trends_in_Transport_Routes__Goods_Transport.pdf

Airbus Group (2016) Global market forecast airbus 2016–2035. http://www.airbus.com/company/market/forecast/

Airbus (2016) Zephyr, the high altitude pseudo-satellite. https://airbusdefenceandspace.com/our-portfolio/military-aircraft/uav/zephyr/

Bandyopadhyay D, Sen J (2011) Wireless Pers Commun. 5 internet of things: applications and challenges in technology and standardization, 58:49. https://link.springer.com/article/10.1007/s11277-011-0288-5

Benthien H et al (2015) Manufacturing of compounds of a vehicle using additive layer manufacturing. US Patent&Trademark Office United States Patent Application. http://appft.uspto.gov/netacgi/nph-Parser?Sect1=PTO1&Sect2=HITOFF&d=PG01&p=1&u=%2Fnetahtml%2FPTO%2Fsrchnum.html&r=1&f=G&l=50&s1=%2220160136891%22.PGNR.&OS=DN/20160136891&RS=DN/20160136891

Bergek A, Jacobsson S, Carlsson B, Lindmark S, Rickne A (2008) Analyzing the functional dynamics of technological innovation systems: a scheme of analysis. Res Policy 37(3):407–429

Boeing (2016) Current market outlook 2016–2035. http://www.boeing.com/resources/boeingdotcom/commercial/about-our-market/assets/downloads/cmo_print_2016_final.pdf

CE Delft, Ecorys, IDEA Consult (2012) Green growth opportunities in the EU shipbuilding sector 2012. http://www.cedelft.eu/publicatie/green_growth_opportunities_in_the_eu_shipbuilding_sector/1291

Chaminade C, Edquist C (2006) From theory to practice. The use of the systems of innovation approach in innovation policy. In: Hage J, Meeus M (eds) Innovation, science and institutional change. A research handbook. Oxford University Press, Oxford

China State Shipbuilding Corporation (2017) Strategy, SWOT and corporate finance report. https://www.businessresearchstore.com/product/190938-china+state+shipbuilding+corporation+-+strategy++swot.html

Dekhtyaruk Y, Karyshev I, Korableva M, Velikanova N, Edelkina A, Karasev O, Klubova M, Bogomolova A, Dyshkant N (2014) Foresight in civil shipbuilding—2030. Foresight-Russia 8 (2):30–45

EC (2010) Facing the future: time for the EU to meet global challenges, IPTS.

EC (2011) Sustainable Aviation Council, UK, Flightpath 2050: Europe's vision for aviation. https://ec.europa.eu/transport/sites/transport/files/modes/air/doc/flightpath2050.pdf

EC (2012) Aeronautics and Air transport: beyond vision 2020 (towards 2050), 2010. http://www.gppq.fct.pt/h2020/_docs/brochuras/transportes/A%20Background%20Document%20from%20ACARE.pdf

EC (2014) Long-term perspectives. Communicating transport research and innovation. http://www.transport-research.info/sites/default/files/thematic-analysis/20150430_175035_36562_TRS24_fin.pdf

EC (2015a) Horizon 2020 Smart, green and integrated transport work programme 2014–2015. http://ec.europa.eu/research/participants/data/ref/h2020/wp/2014_2015/main/h2020-wp1415-transport_en.pdf

EC (2015b) LeaderShip 2015. Defining the future of the European shipbuilding and repair industry. http://eur-lex.europa.eu/legal-content/EN/TXT/?uri=CELEX%3A52003DC0717

Edquist C (2011) Design of innovation policy through diagnostic analysis: identification of systemic problems (or failures). Indus Corp Change 20(6):1–29. https://doi.org/10.1093/icc/dtr060

EREA (2012) From Air Transport System 2050 vision to planning for research and innovation. http://www.onera.fr/sites/default/files/Departements-scientifiques/DPRA/ats_2050-phase2_en.pdf

Federal Ministry of Transport and Digital Infrastructure of Germany (2015) Forecast of transport interconnectivity 2030. http://www.bmvi.de/SharedDocs/EN/publications/flyer-2030-transport-interconnectivity-forecast.html?nn=212250

FOREN (2001) A practical guide to regional foresight. http://foresight.jrc.ec.europa.eu/documents/eur20128en.pdf

GenerationBalt (2011) Foresight of the South Baltic maritime labour market 2017. http://www.generationbalt.eu/data/user/documents/Foresight_Maritime_Labour_Market_SouthBalt.pdf

Georghiou L (2011) Final evaluation of the lead market initiative, Publications Office of the European Union, Luxembourg. doi: https://doi.org/10.2769/29882

Gokhberg L (ed) (2013) Long-term priorities of applied science in Russia. Higher School of Economics, Moscow

Hiufu Wong M (2016) Five cool technology leaps about to change aviation for good. http://edition.cnn.com/2016/02/25/aviation/aviation-emerging-tech-singapore-airshow/

Hyundai Heavy Industries (2016) Pioneering innovative technology. http://pdf.nauticexpo.com/pdf/hyundai-heavy-industries/electro-electric-systems/31139-89635.html

IATA (2011) Vision 2050. Report 2011. http://www.iata.org/pressroom/facts_figures/Documents/vision-2050.pdf

IATA (2013) IATA Technology Roadmap, 2013. http://www.iata.org/publications/Pages/technology-roadmap.aspx

ICAO (2006) Global Aviation Safety Roadmap. http://cfapp.icao.int/fsix/_Library%5CGlobal_Aviation_Safety_Roadmap_Part_2.pdf
IEA (2012) Global transport outlook to 2050. https://www.iea.org/media/workshops/2013/egrdmobility/DULAC_23052013.pdf
Keenan M, Cutler P, Marks J, Meylan R, Smith C, Koivisto E (2012) Orienting international science cooperation to meet global 'grand challenges'. Sci Publ Policy 39:166–177
Loveridge D (2002) The STEEPV acronym and process-a clarification, Ideas in Progress. Paper, 29
Marine Institute (2005) Foresight series. Climate change implications for Ireland's marine environment and resources 2005. http://oar.marine.ie/bitstream/10793/560/3/Foresight%20Series%202%20Climate%20Change%20Implications.pdf
Marine Industries Roadmap & Capability Study (2011) Workshop D: shipbuilding and repair. http://www.ifm.eng.cam.ac.uk/uploads/Research/CTM/Roadmapping/marine_wksp_4.pdf
Matich L (2017) Roadmaps as a tool for modeling complex systems. Working Paper Series Science, Technology and Innovation, Higher School of Economics, No. 73/STI/2017
Mitsubishi (2017) Lifting, powering and accelerating the world Mitsubishi heavy industries forklift, engine and turbocharger holdings. Mitsubishi Heavy Industries Technical Review 54(1). http://www.mhi-global.com/company/technology/review/pdf/e541/e541002.pdf
NASA (2013) Integration of Civil Unmanned Aircraft Systems (UAS) in the National Airspace System (NASA) Roadmap. http://www.faa.gov/uas/media/uas_roadmap_2013.pdf
NISTEP (2014) Science and technology trends—quarterly review. Science and Technology Foresight Center, NISTEP, Tokyo. http://www.nistep.go.jp/achiev/ftx/eng/stfc/stfc_all-e.html
National Research Council Canada (2011) Canadian Aerospace Environmental Technology Road Map v2.0 (CAETRM). http://www.gardn.org/
NVIDIA Corporation (2014) Empowering world-class marine designers. https://www.nvidia.fr/content/grid/resources/case-study-daewooshipbuilding.pdf
OECD (2012) Strategic transport infrastructure needs to 2030. Paris. http://www.oecd-ilibrary.org/economics/strategic-transport-infrastructure-needs-to-2030_9789264114425-en
OECD (2015) ITF Transport Outlook 2015. http://www.oecd.org/environment/itf-transport-outlook-2015-9789282107782-en.htm
Pestana H (2014) Future trends of electrical propulsion and implications to ship design. In: 2nd International Conference on Maritime Technology and Engineering, Lisbon, Portugal. Maritime Technology and Engineering 1 & 2, pp 797–803
Pinnegar JK, Viner D, Hadley D, Dye S, Harris M, Berkout F, Simpson M (2006) Alternative future scenarios for marine ecosystems. http://www.meece.eu/meetings/bologna/documents/AFMEC%20technical%20report.pdf
Safran SA (2016) Electrical systems and power electronics: towards more electric aircraft. https://www.safran-group.com/innovation/our-technologies/electrical-systems-and-power-electronics/electrical-systems-and-power-electronics-towards-more-electric-aircraft
Samsung Heavy Industries (2012) Global leader in ocean & power. http://pdf.nauticexpo.com/pdf/samsung-heavy-industries/global-leader-ocean-power/31569-54099.html
Saritas O, Aylen J (2010) Using scenarios for roadmapping: the case of clean production. Technol Forecast Soc Change 77(7):1061–1075
Saritas O, Proskuryakova L (2017) Water resources – an analysis of trends, weak signals and wild cards with implications for Russia. Foresight 19(2):152–173
Sustainable Aviation Council (2012) Sustainable aviation CO2 roadmap. http://www.sustainableaviation.co.uk/wp-content/uploads/2015/09/SA-Carbon-Roadmap-full-report.pdf
Sustainable Aviation Council (2013) Noise road-map a blueprint for managing noise from aviation sources to 2050. http://airlinesuk.org/wp-content/uploads/2013/04/SA-Noise-Roadmap-Summary-April-2013.pdf
Toyota (2017) Patenting the design of a flying car. http://www.freepatentsonline.com/20160176256.pdf
UK Marine Industries Alliance (2011) A strategy for growth for the UK Marine Industries. https://www.gov.uk/government/uploads/system/uploads/attachment_data/file/31813/11-1310-strategy-for-growth-uk-marine-industries.pdf

Vachtsevanos G, Georgoulas G, Nikolakopoulos G (2016) Fault diagnosis, failure prognosis and fault tolerant control of aerospace/unmanned aerial systems. In: 24th Mediterranean Conference on Control and Automation, Athens, Greece, 21–24 June 2016
Vishnevskiy K, Karasev O, Meissner D (2016) Integrated roadmaps for strategic management and planning. Technol Forecast Soc Change 110:153–166
Vishnevskiy K, Yaroslavtsev A (2017) Russian S&T Foresight 2030: case of nanotechnologies and new materials, foresight. pp. 198–217
VOSviewer Manual. Nees Jan van Eck and Ludo Waltman, 28 Sept 2016
Wahlstrom M, Hakulinen J, Karvonen H (2015) Human factors challenges in unmanned ship operations – insights from other domains. In: 6th International Conference on Applied Human Factors and Ergonomics
Yang J, Zhang H, Huang M (2013) Mingxing emerging technology in aerospace engineering polymer-based self-healing materials. In: Aerospace materials. Handbook advances in materials science and engineering. Boca Raton, pp 531–606

Marina Klubova is Research Fellow at the International Foresight Center, HSE ISSEK. She obtained her Master's degree in Economics from the Lomonosov Moscow State University. Ms. Klubova was involved in projects related to roadmapping and market analysis, including "Technology Foresight of Civil Shipbuilding to 2030", "Roadmaps of S&T Development of Civil Aviation" among others. Her research interests cover S&T and innovation policy, strategic planning, roadmapping and scenario analysis.

Lyubov Matich is the Head of Project Management Unit at the HSE ISSEK. She obtained her Master's degree in Economics from the Lomonosov Moscow State University. Ms. Matich prepared a wide number of corporate and industry-level projects on innovative development programs for Russian companies; technology roadmaps for the space, petro, and aircraft industries among others. Her professional interests refer to company strategies, management of innovations, and technology roadmapping.

Vladimir Salun is Director of the Center for Industrial and Corporate Projects at the HSE ISSEK. The main activities of the Center are focused on innovative development of the key sectors of the Russian economy, introduction of advanced methods in strategy planning and project management at state-owned and private companies, improvement of their competitiveness through better selection and implementation of STI priority objectives. Dr. Salun has an excellent record of managing numerous consulting assignments for leading Russian corporate companies, in particular related to developing their innovation strategies and technology roadmaps. He has PhD degree in Physical and Mathematical Sciences.

Natalia Veselitskaya is Senior Researcher at the HSE ISSEK Foresight Center. Natalia holds a PhD in Innovation Management. She has worked on strategic foresight studies for the support of research and innovation and for corporate innovation management. Ms. Veselitskaya focuses on the implementation of foresight methods for the resolution of different technological and socio-economic challenges. In particular, she is engaged in developing technology roadmaps for companies, regions and industries.

Part III

Living Systems and Environment

Water Treatment and Purification: Technological Responses to Grand Challenges

8

Ozcan Saritas and Konstantin Vishnevskiy

8.1 Introduction

Nanotechnologies have been increasingly used in wide-ranging application areas. The present chapter sheds light the potential uses of nanotechnologies for supplying clean water. Water is one of the Grand Challenges facing humanity. The chapter begins with a discussion on the increasing demand for fresh water across the globe and technological challenges and opportunities associated with water supply. It is shown that new technologies are needed for effective and resource-efficient water treatment and nanotechnologies may offer affordable solutions.

Overall, the chapter advocates that supplying drinking water to billions of people, while at the same time protecting water resources, is the best strategic guideline for the water supply industry and for applying new techniques and materials including nanotechnology. Developing new technologies for deep purification of water with nanotechnology will increase healthy water supply by removing both visible and invisible impurities hazardous to human and animal health. Thus, the chapter discusses nanotechnology solutions for water treatment and purification. How nanotechnologies can significantly increase the efficiency of certain traditional water purification processes such as coagulation, sorption and flotation is discussed. Next technological, market and institutional aspects are considered regarding nanotechnology solutions. The chapter is concluded with future scenarios and strategic steps to be taken for the implementation of nano-based water treatment and purification.

While investigating in nanotechnology and its uses for water treatment and purification, the chapter focuses on the case of Russia. More recently, Russia is

O. Saritas (✉) · K. Vishnevskiy
National Research University, Higher School of Economics, Moscow, Russia
e-mail: osaritas@hse.ru; kvishnevsky@hse.ru

D. Meissner et al. (eds.), *Emerging Technologies for Economic Development*, Science, Technology and Innovation Studies,
https://doi.org/10.1007/978-3-030-04370-4_8

engaged in improving its energy and water infrastructures, as well as more efficient use of its natural resources (Saritas 2015). However, the country still experiences a set of challenges related to the protection and use of water resources, water purification and networks, consumption patterns, discharge, treatment and reuse (Saritas and Proskuryakova 2017).

8.2 Background

Grand Challenges are defined as global problems in the long-term that are potentially capable of changing the world. They are challenges faced by everyone (Saritas and Miles 2012). Coming up with adequate solutions would increase well-being and quality of life and improve the environment—and this requires raising awareness and fostering cooperation among key stakeholders (Miles et al. 2016). Grand Challenges can emerge in economic, social and research spheres, but for any such challenge, it requires responses at regional, national and global levels. Furthermore, the potential consequences of Grand Challenges should be made clear, and scientists should be able and willing to propose adequate responses which have to meet stringent criteria, including the practical applicability of relevant research, and a potential for increasing efficiency as well as the availability of adequate economic and social investments. In order to develop effective technology and innovation-driven responses, the principles and efficiency of interindustrial priorities need to be analysed which might not be directly connected with a specific science and technology (S&T) area and application of relevant results. Also there is consensus that the objectives have to be achieved at international and national levels (i.e. those set to meet global challenges) which will eventually lead to increased well-being and quality of life, improved environment and stronger internal and external competitiveness of countries and regions. Consequently, these challenges are seen as key global subject areas thus regularly included in strategies and forecasts developed by relevant government agencies and international organisations (e.g. environment protection, information technologies, etc.). It is broadly understood that these challenges may turn into opportunities associated with applying innovative products and serve as an initiator for overcoming established 'bottlenecks' of innovation systems.

Water is one of the most tangible and fastest-growing challenges the humanity faces today. The water security is a grand challenge which is strongly tied to survival, social well-being and economic growth (Inayatullah and Elouafi 2014). It is also a crucial element of the water-energy-climate-food nexus. As an example Maggio et al. (2016) discuss the role of water for global food security. Sustainable energy policies require taking water into consideration (Horner et al. 2016). Moriarty and Honnery (2014) highlight ongoing climate change and impacts on water. The world is expected to face a 40% global shortfall between estimated demand and available supply by 2030 (UNwater 2015).

Water is not just a 'life source' but also a source of many diseases. The 'shortage of drinking water' is associated with the growth of sickness rates as well as environmental pollution. These are among the other global challenges to which scientists all over the world are seeking adequate responses. Polluted water causes up to 80% of all human diseases in the world and accelerates ageing by 30%. More than a billion people experience a shortage of safe drinking water, and 2.4 billion do not have access to water which meets established sanitary norms. This means that almost a third of the planet's population lives in areas experiencing a shortage of pure drinking water. According to certain forecasts, by 2025 this share will increase to two thirds (Prüss-Üstün et al. 2008).

The low quality of drinking water supplied is frequently caused by several natural, anthropogenic and technology-related reasons. In some regions, the quality of water is affected by natural concentrations of pathogenic microorganisms in the atmosphere, soil, water, flora and fauna. They can engender epidemics of infectious and parasitic diseases such as cholera, typhoid fever, salmonellosis, dysentery, amoebiasis, lambliasis, viral hepatitis or poliomyelitis. There are areas with naturally increased background radiation level, and in certain regions water has a significantly increased concentration of microelements.

Among the reasons for low water quality are inadequate drinking water purification technologies and secondary pollution in the course of treatment and transportation through pipeline networks. A significant trigger of deteriorating water quality in densely populated areas, and the subsequent growth of oncological and other serious diseases, is untreated or inadequately treated waste water dumped into natural reservoirs after industrial, communal or agricultural use. This sewage contains vast amounts of mineral and organic pollutants (contaminants) ranging from heavy metals to oil products, which negatively affect the ecosystem. Such pollutants are a result of faulty industrial waste treatment and sewage systems' application. They may also be caused by the dissemination of algae from increased concentration of phosphorus and nitrogen in waste products, detergent powders and fertilisers. Dangerous substances are also generated during soil erosion, when pesticides, chemical reagents and heavy metals enter the water supply. Another cause is pollution of subterranean waters during intense and inefficient irrigation.

A list of industrial waste water contaminants would number millions. Many of these substances have carcinogenic and mutagenic properties. For example, benzopyrene causes oncological diseases and is classified as a Class 1 dangerous substance. It is a product of burning organic fuels and is commonly present in areas with heavy automobile traffic; particularly high concentrations are also observed near fuel-burning power plants.

Substances which can cause serious diseases are widely used in agriculture. These include mineral fertilisers which on the one hand increase crop yield while on the other increase concentration of nitrates, phosphates and potassium in soil and subterranean waters—and thus in vegetables, fruit and drinking water. Up to 65% of nitrates introduced into the human digestive tract subsequently find their way into blood and tissue, hindering their ability to take in oxygen. Pesticides are also dangerous. There are approximately a thousand of such substances which do not degrade for many years. Rather, they accumulate and migrate in the environment and find their way into water and then into the human body.

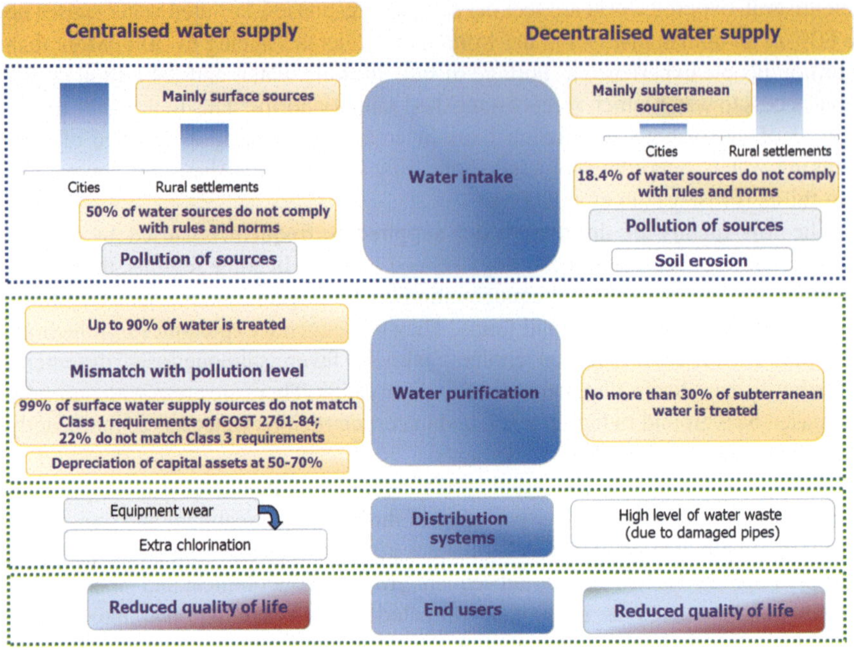

Fig. 8.1 Industry-specific water supply challenges in Russia. Source: HSE

Thus the current state of water purification and treatment industries indicates that a whole system of challenges exists on their agenda. In summary, these challenges can be grouped into three main types:

1. *Global*—due to the growth of industrial and anthropogenic pollution of water sources, shortage of drinking water, spreading of serious diseases, ageing of population
2. *Regional*—due to more stringent requirements for the quality of consumed water as 50% of surface water sources do not match official standards and only about 30% of subterranean water is treated
3. *Industrial*—due to depreciation of capital assets of the water supply and sewage industry

Obsolete and worn-out equipment of the water supply systems increase the number of accidents. These, in turn, cause loss of water and soil erosion and damaged roads and building foundations. Figure 8.1 shows the chain of events in the water supply in the case of the Russia.

It is increasingly clear that meeting these challenges requires radically innovative S&T solutions (Saritas and Proskuryakova 2017). Past few decades have seen the emergence of nanotechnologies as one of the key pillars of industrial development. A number of forward-looking studies have been undertaken at various levels of governance, including national (Saritas et al. 2007; Vishnevskiy and Yaroslavtsev 2017), regional (Battistella and Pillon 2016) and sectoral levels (Aydogdu et al. 2017) to investigate the ways of exploiting nanotechnologies in various S&T areas. Other studies have been undertaken to explore the social, economic and environmental aspects of nanotechnologies within the framework of foresight studies (Loveridge and Saritas 2009, 2012). The present chapter takes a closer look at the nanotechnology solutions for water treatment and purification. With a forward-looking methodology, the study makes use of frequently used foresight methods including, reviews, SWOT analysis, forecasting, scenarios and strategy planning (Saritas and Burmaoglu 2015).

8.3 Nanosolutions and Alternatives for Water Treatment

Water treatment, as well as sensible management and use of water resources, is a major area of S&T activities. The scientific community clearly realises the importance of developing new technologies for deep purification of water to remove visible and invisible impurities hazardous to human health. Supplying drinking water to billions of people, while at the same time protecting water sources, are the best strategic guidelines for the water supply industry and for applying new techniques and materials including nanotechnology (Bottero et al. 2006; Theron et al. 2008).

Applying nanotechnologies can significantly increase the efficiency of certain traditional water purification processes such as coagulation, sorption and flotation (Draginsky et al. 2005). Membrane-based technologies that are used to make molecular nanostructured materials, including baro-membrane and electromembrane processes, membrane bioreactors and membrane degassing already exist (Desiatov et al. 2008; Nikolayev 2008; Poliakov and Vidiakin 2009). Technological solutions based on nanoproducts are available, including carbon nanotubes and fullerenes, dendrimers, zeolites and catalysts. At present, the following technologies are in the process of being launched: extremely efficient carbon nanotube-based sorbents, organic-mineral composites containing highly selective nanoparticles in an organic matrix and innovative molecular mesh sorbents, among the others. However, the priorities and prospects for applying these materials by interested parties remain rather ambiguous.

There is a broad consensus that it is currently quite challenging to determine which water treatment and purification techniques should be classified as nanotechnology. Sorption technology and the use of catalysts are most frequently named as solutions based on nanotechnology. All remaining traditional water treatment techniques are not considered nanotechnology-related. Table 8.1 presents the advantages and disadvantages of traditional technologies, with potentials of applying nanotechnologies, and alternative technologies, where nanotechnologies are largely not applicable.

Table 8.1 Advantages and disadvantages of traditional and alternative technologies

Name	Description	Advantages	Disadvantages
Traditional technologies where nanotechnologies can be applied			
Sorption	Solid or liquid substances, with sorption happening on the surface of its particles. Silica gel, activated carbon, white carbon, certain oxides, resins, etc. are used as adsorbents	An adsorbent can absorb only one substance, i.e. they act selectively (something membranes are not capable of). This technique can be efficiently applied if water is purified to a very low concentration of pollutant: adsorbent consumption would not be high, which makes perfect economic sense	After a while the adsorbent becomes saturated and needs to be replaced In certain cases, the costs of applying this technology can be quite high
Catalysis	Substances that accelerate water purification process	The main advantage of using catalysts is that they only accelerate the process and do not actually take part in the process	The catalyst remains in the water and may harm people and the environment
Alternative (non-nano) technologies			
Chlorination	Treating water with chlorine and chlorine compounds. The most common technique for drinking water disinfection; based on free chlorine and the ability of its compound to suppress microbes' enzyme systems by catalysing redox processes	Efficiently kills pathogenic germs Chlorination is necessary; without it, unchecked breeding of microorganisms would make water unsuitable for drinking	If a certain concentration is exceeded, water acquires an unpleasant odour Chlorinated water has an adverse effect on human skin Interaction between chlorine and certain organic compounds produces their chlorine derivatives which are hazardous to human health. Some such compounds have carcinogenic properties
Ozonisation	Based on application of ozone gas. After interaction with chemical and microbiological pollutants, ozone transforms into ordinary oxygen	Ozonisation kills germs and does not leave an unpleasant odour. No hazardous by-products remain in the water	A large amount of sediments may accumulate during chemical treatment Ozone production is energy-intensive; accordingly, it is quite an expensive product
Coagulation, flocculation	Introduction of aluminium or iron salts		Application of this technique requires

(continued)

Table 8.1 (continued)

Name	Description	Advantages	Disadvantages
	or polyelectrolytes into treated water. This enlarges suspension and makes colloid particles stick together, become larger, and sediment	This technique is relatively inexpensive and quite efficient	subsequent additional treatment as traces of reagents remain in the water
Flotation	This method is based on removing impurities with the help of air bubbles. Coming to the surface, they capture impurity particles including oils and oil products, creating a film or a foam layer on the water surface, which is subsequently removed with the help of special foam-gathering mechanisms	Can treat large volumes of water and is quite inexpensive	Flotation is not particularly effective compared with membranes in terms of removing impurities. A certain amount of impurities remain in the water. If we try to achieve better results through flotation, it becomes a very lengthy process
Distillation	Distillation—evaporation of liquid with subsequent cooling and condensation of the vapour. When water is evaporated through phase transition, it is purified of almost all impurities	Distillation is quite a simple and very efficient technique	Distilled water in effect is unsuitable for drinking Distillation is quite an expensive technique

Source: HSE

Alternative technologies are believed to have the following major disadvantages:

- Use of chemical reagents
- Short reaction time
- Saturation of water with traces of reagents
- High costs
- Insufficient purification of water
- Need to use bulky equipment

However, nanotechnologies discussed across this chapter should not be seen as competing alternatives; rather, they should be merged together to *develop integrated technologies*. For example, the application of micron-grade carbon powder requires special devices for its subsequent removal. Therefore, a combination is used (i.e. carbon powder plus membrane), because submicron-sized particles get through regular filters but not through such fine filtering elements as membranes.

Nanofiltering and reverse osmosis membranes offer significant advantages but also have some disadvantages (Karelin 1988). The main difficulty is that such purification in effect produces a distillate unsuitable for drinking. Therefore, to produce regular water, membrane-based techniques must be applied in combination with others. In addition, membrane nanotechnologies generate about 30% waste. Ion-exchange technology was proposed to utilise this waste. They work together well, which reduces the amount of water dumped into sewage. Application of ion-exchange technology to treat waste water generated through membrane nanotechnologies-based water purification significantly reduces the cost of ion-exchange installations (at least threefold). In future, an integrated approach may be used which takes into account various technologies' specific features and input water properties. It is also believed that in future *selective membranes* will be developed, filtering out only specific components. This would enable water to be purified from all kinds of pollutants while retaining all necessary elements. Such membranes can be self-structuring, working in a similar way to biological analogues. They would enable natural water resources to be restored, which would ultimately solve the problem of providing adequate drinking water supply for all.

In addition, to the above technologies, promising water treatment technologies mentioned by experts include electric pulse water treatment. Under certain conditions, alternative products such as various filters and sterilisers can also be used. Their advantages and disadvantages deserve a closer look.

Sediment Filters

Sediment filters are used primarily to remove mechanical and organic suspensions, sand, rust particles and colloid substances. The main drawback of this class of filters is its relatively low take-up capacity of solids or heavily polluted water as this means they require frequent cleaning when actively used. In such situations, automated particulate filtering systems seem more appropriate. The filter medium is dry aluminium silicate, which allows particles above 20 microns in size to be filtered (Lysov et al. 2002). To remove larger particles (20–50 microns), mesh or disk coarse filters are used. To remove small particles (5–20 microns), special ceramic (Macrolite) particulate filters are applied.

De-ironing Filters

The main purpose of this filter class is to remove dissolved iron and manganese from water and some other compounds. Certain filters of this type also quite efficiently remove dissolved hydrogen sulphide. The filter medium is manganese dioxide-containing substances such as Birm, Filox, Greensand, etc., which act as catalysts for the oxidation reaction. As a result of the reaction, unwanted impurities dissolved in water (iron or manganese) transform into insolubles and sediment into the filtrate layer; they are then washed out during backwashing. A specific feature of this technology is that it does not require special reagents for backwashing. Some of the filter media do require regeneration (with potassium permanganate).

Softening Filters

This class of filters is designed to soften water, i.e. remove hardness salts like calcium and magnesium (Alekseyev and Gladkov 1994). Using special particulate agents also removes a certain amount of organic compounds, heavy metal salts, manganese, iron, nitrates and sulphates. The filtering medium is cation-exchange resins which have high hardness salt take-up capacity. Such filters require periodic regeneration with saline solution, which implies a need for special (saline) vat to make and use such solutions.

Charcoal Filters

Charcoal filters are designed to remove dissolved gases, residual chlorine and certain organic compounds from water, get rid of unpleasant odours and generally improve the water's organoleptic properties. The filtering medium here is fibrous activated charcoal which, due to its high absorptive capacity (1 g of such material can absorb up to 200 mg of pollutants), means it can efficiently absorb dissolved gases and compounds. To increase service life and prevent 'volley' dumping of pollutants into an output channel, activated charcoal made of coconut shells is used in charcoal filters, which has a high absorptive capacity (four times higher than traditional charcoal). This reduces the need to frequently change the filtering medium (necessary if traditional materials are used because organic compounds accumulating in the filtrate are quite hard to extract via backwashing). To prevent biologic overgrowing, special charcoals with bacteriostatic additives are used in charcoal filters.

Ultraviolet Sterilisers

Ultraviolet sterilisers are used to disinfect water and to neutralise germ pollution—microbes, bacteria and viruses in water. Ultraviolet irradiation is the most common way of neutralising germ pollution. This method is different from other methods (ozonisation or chlorination) because it lacks hazardous impurities. Chambers containing ultraviolet lamps are used as sterilisers, installed in hard cases. Water flows through such sterilising chambers where it is constantly treated with ultraviolet radiation.

The experience of international water purification shows that traditional technologies that incorporate innovative components combined with membrane-based processes can significantly increase the efficiency of water treatment and purification.

8.4 The Future of Nanotechnology-Based Water Supply

The future of water treatment and purification is currently in the process of being shaped towards incorporating nanotechnologies into the water supply industry to significantly increase the quality of drinking water and reduce the incidence of diseases caused by polluted water sources and pipeline networks. The present section discusses various nano-based technologies, products and processes and explores their feasibility of application.

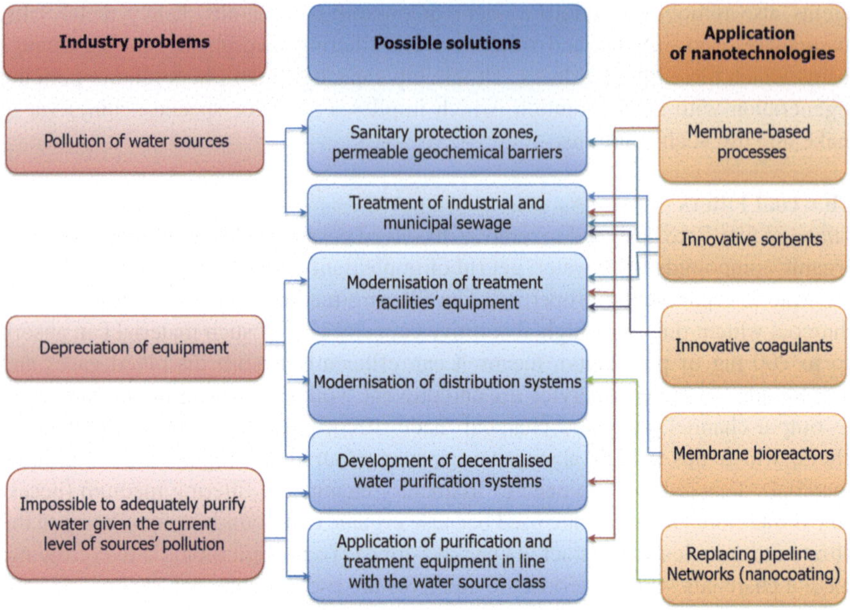

Fig. 8.2 Possible applications of nanotechnologies in industrial water treatment and purification. Source: HSE

8.4.1 Application Areas for Nanosolutions

At this stage, the largest area of application for nanotechnologies is water treatment for municipal and industrial use (which constitutes 67% of water use), with a major share of that (up to 95%) belonging to industry. In the near future (before 2020), given overall market growth, it is likely that we will see a shift in the share of different segments in favour of the following three areas: municipal waste water treatment will grow to 10%, applications in major technological processes to 12% and treatment of industrial waste water to 25%. These segments will show accelerated growth, with the remaining areas growing at more modest rates. Figure 8.2 illustrates the possible applications for nanotechnologies in industrial water treatment and purification.

Membrane-based nanotechnologies have good prospects in industrial, as well as municipal, water treatment and purification areas—including nuclear and fuel-burning power generation (Sterman and Pokrovsky 1991), radio- and microelectronics and food and chemical industries.

Many industries require water of a particular quality and composition, which creates stable demand for membrane technologies by various industrial companies.

Thus water treatment and purification combine three segments: water for the general population, water for industrial purposes and wastewater treatment. Each of these areas can be further divided into smaller segments, with specific problems and demand for new technologies, products and components.

8.4.2 Market Prospects for Nanoproducts and Nanotechnologies in Russia

As already noted, water treatment and purification market is growing in Russia quite quickly (Proskuryakova et al. 2018). There is a high demand for household water filters, despite the low quality of some of those products. The growing demand is driven by greater attention to a healthier lifestyle, including the consumption of only natural and pure food and high-quality water. Meanwhile the market development is not uniform in all regions. In some regions, demand is increasing rapidly for new water treatment technologies: for example, Moscow and other large cities are seeing greater popularity of membrane-based techniques. At the same time, rural areas which frequently lack any mains water supply have practically no market for water treatment and purification, e.g. in rural areas which are occupied by about 30% of the total population. Moreover, industrial characteristics also influence the water treatment market development such as:

- Significant depreciation of capital assets and need for upgrading and replacement of obsolete equipment in enterprises
- Modernisation of municipal installations
- Numerous newly built production facilities and commercial real estate
- Active construction of housing and social infrastructure

Water purification technologies are actively applied in the mass production of industrial and household filtration systems and are well represented even in the retail sector. However, there is practically no R&D in Russia that aims to advance drinking water purification processes; only various components of the water supply system are manufactured which lack mass demand. Neither are there any companies capable of implementing all stages of the technological chain in the international market. Russia does not occupy a leading position. Accordingly, 'Russian-made' water purification and treatment systems are in effect just assembled from individual (imported) components.

Remarkably, the prospects for this market segment depend not just on the level of public funding but also on private investments. The commercial segment shows the highest growth rate today, so the long-term private-public partnership (PPP) mechanisms (including long-term contractual agreements) may play a major role. Another important condition of the growth in the water purification market is a close interdependency between all stages of the technological life cycle. It is important to note that this interdependency appears in various production segments individually and if they are brought together under a single management system. However, production of water purification solutions must be divided between various participants of this process: research should be done at specialised research centres, development of technologies and the manufacture of water purification system components by specialised public-private corporations and end products by engineering companies. Furthermore, developed countries demonstrate examples of efficient production of equipment for water treatment systems by single companies

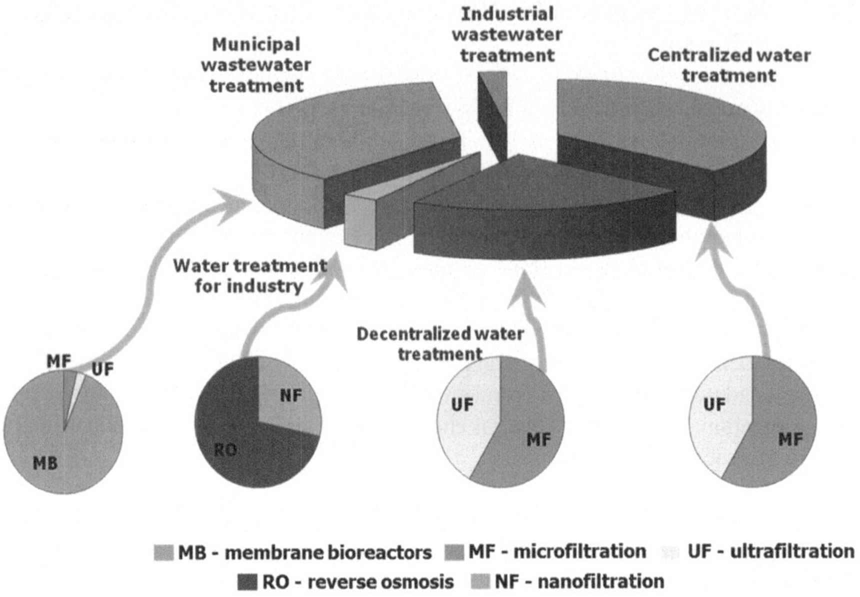

Fig. 8.3 The possibilities of applying nanotechnologies in various market segments. Source: HSE

such as Dow Chemical, whose activities include research, technology development and production of membranes, filters, water purification systems, etc.

Thus, the rapid growth of this market segment makes any assessments of 'anticipatory' technological, socio-economic and organisational steps even more uncertain. Therefore, the distribution of players in this market, the potential to produce innovative products and the long-term demand for such products will be particularly important. Figure 8.3 gives some estimates of demand for membrane-based technologies in various market segments and the supply structure of products used in baro-membrane processes for each segment.

Membranes and installations for microfiltration and ultrafiltration and for use in membrane bioreactors in combination with non-membrane technological solutions show the highest potential demand (Judd 2006). Next, each of these technologies are described briefly followed by a SWOT analysis for the feasibility of their application.

Microfiltration membranes used in tubular modules installations have proved to be an efficient rough filtration technique applied at the initial stages of the drinking water production process. Microfiltration membranes filter out small suspensions, fine and colloid impurities, algae and unicellular microorganisms larger than 0.1 microns. Microfiltration is widely used in medicine, food industry (alcoholic and soft drinks, wine, beer, vegetable oil and other products) and water treatment installations. A SWOT analysis of the aforementioned membrane is given in Table 8.2.

Ultrafiltration membrane process which is mostly used in installations with hollow-fibre, flat, roll and tubular modules is a process of separating solutions and

Table 8.2 SWOT analysis of microfiltration membranes

Strengths (compared with alternative products)	*Weaknesses (compared with alternative products)*
• Compact equipment • Easily increased output • Water treatment process can be automated	• Short service life • Only some pollutants (within a specific range) can be removed • Require periodic washing and cleaning
Opportunities (external factors contributing to product promotion)	*Threats (external factors hindering product promotion)*
• Need to modernise existing equipment • Increased requirements for waste water treatment • Extremely fast growth of water consumption • Development of special-purpose water treatment segments	• Conservatism of major consumers—water supply companies • Budgetary limitations
Alternative products and processes	
Particulate filters, aeration, chemical treatment, disinfection	

Source: HSE

Table 8.3 SWOT analysis of ultrafiltration membranes

Strengths	*Weaknesses*
• Efficiently remove large organic molecules, colloid particles, bacteria and viruses, without filtering out dissolved salts • No need for preliminary chlorination • More efficient coagulation and sedimentation with reduced consumption of coagulants and incomplete coagulation	• Need to be used in combination with other membrane methods to efficiently remove all pollutants • Require washing to clean the membrane from pollutants
Opportunities	*Threats*
• Additional purification of tap water • Household application	• Conservatism of major consumers—water supply companies • Budgetary limitations
Alternative products and processes	
Sand granular filters	

Source: HSE

colloid systems using semipermeable membranes. Ultrafiltration is applied in the following areas:

- Clean water surfaces at water intake stations.
- Perform additional purification of water for city mains.
- De-iron and improve the quality of subterranean water.
- Treat water for industrial use.8.3.

A SWOT analysis of the aforementioned membrane is given in Table

Table 8.4 SWOT analysis of nanofiltration membranes

Strengths	*Weaknesses*
• High level of purification • Produce physiologically perfect drinking water with good organoleptic properties	• Require thorough preliminary cleaning from chlorine and other substances • Roll modules with nanofilter membranes are quite unreliable
Opportunities	*Threats*
• Can be applied in local additional purification in buildings • High export potential	• Conservatism of major consumers—water supply companies • Budgetary limitations
Alternative products and processes	
Deioniser water-softening installations	

Source: HSE

Nanofiltration membrane process for application in roll modules installations is a fractional membrane process which removes only ions that are multiply charged (such as calcium, magnesium, iron, etc.) from the water, not all salts. Nanofiltration membranes also efficiently remove low-molecular compounds. Due to lower energy consumption than reverse osmosis, nanofiltration is successfully applied abroad as the most commonly used process to produce drinking water from surface sources. This process provides particularly pure high-quality water for use in medical, pharmaceutical, electronic, glass-making, food and other industries. Water passed through a system of membrane filters is purified of hazardous bacteria, viruses, microorganisms, colloid particles (including pesticides), molecules of heavy metal salts, nitrates, nitrites and other harmful impurities. A SWOT analysis of the aforementioned membrane is given in Table 8.4.

Reverse osmosis membranes are applied in flat and roll modules installations. The reverse osmosis method is based on filtering solutions through semipermeable membranes which let the solvent's molecules through but fully or partially filter out molecules and ions of dissolved substances. This type of membrane is used in various industries which need very pure water (galvanic production, printed circuits manufacturing, instrument making, electronics, applying noble metal coatings, bottled water and drinks production, food industry, pharmaceuticals, etc.). A SWOT analysis of the aforementioned membrane is given in Table 8.5.

Ion-exchange membranes used in electrodialysis installations and membrane bipolar electrolysers are based on the application of a particular class of substances—natural or synthetic ionites. Synthetic ionites (deioniser resins) are 'mesh' polymers with various linkage degrees of the gel micro- or macroporous structure, covalently bonded with ionogenic groups. There are two ways to apply the ion-exchange method for water treatment and purification: using deioniser resins (ionites) and via ion-exchange membranes—in electrodialysis and electrodeionisation processes. Unlike baro-membrane techniques (where pressure is applied to separate mixtures through membranes), electromembrane processes separate mixtures by applying an electric field. Electrodialysis has more potential applications than reverse osmosis. It can be used for desalination of sea water and at the finishing stage of deeply desalted water production. Its main advantage over reverse osmosis is its ability to treat water with a

Table 8.5 SWOT analysis of reverse osmosis membranes

Strengths	*Weaknesses*
• Up to 99.9% selectivity • Unique quality of treated water • Remove low-molecular humic compounds • All-purpose: efficiently remove pollutant mixtures such as heavy metals, calcium and magnesium ions, phosphates, sulphates and chlorides • No secondary pollution of water • Convenient for transportation and installation • Long service life of the system, with periodic backwashing of membranes • Simple to run, reliable • Installation can work in automatic mode • Very environmentally safe	• Expensive • High energy consumption • Too much purification for drinking water supply—some of the useful elements are removed from the water
Opportunities	*Threats*
• High export potential: desalination is a relevant objective in many countries • More stringent requirements for environmental safety as a market growth factor	• Desalination is not an issue in most of Russia • Conservatism of major consumers—water supply companies
Alternative products and processes	
Distillers and evaporator systems	

Source: HSE

high concentration of salts, when reverse osmosis becomes unprofitable because of the low degree of conversion. The major spheres of application are in electronics and fuel-burning power plants. Electromembrane processes are also used in membrane bipolar electrolysers to produce 'chlorine water'. A SWOT analysis of the aforementioned membrane is given in Table 8.6.

Membrane bioreactors (MBR) combine membrane and biological purification. One of the most promising ways to treat waste water involves using activated sludge bioreactors in combination with ultra- or microfiltration membrane modules, capable of processing and utilising significant volumes of pollutants. Membrane bioreactors can be applied at various water treatment stages (e.g. pretreatment before nanofiltration and reverse osmosis and prefinishing treatment before the disinfection stage). Most commonly, they are used to treat sewage. A SWOT analysis of the aforementioned membrane is given in Table 8.7.

Innovative coagulants include partially hydrolysed salts, aluminium dihydroxy sulphate, aluminium dihydroxochloride, aluminium pentoxychloride and aluminium oxychloride sulphate, with the following functionalities (Tkachev et al. 1978, 1988; Gerasimov 2001; Getmantsev 2003):

- Highly efficient and effective coagulation process
- Low reagent consumption
- High environmental protection properties

Table 8.6 SWOT analysis of ion-exchange membranes

Strengths	*Weaknesses*
• High-quality water treatment with minimum consumption of reagents • Stable performance	• Certain kinds of membranes have low hydrophilic properties • Require regeneration to restore their ion-exchange potential
Opportunities	*Threats*
• Thorough purification by removing calcium and magnesium ions, as well as iron and manganese ions (in dissolved state) • No limits for increasing productivity of ion-exchange membrane-based installations	• Limited application area (desalination, de-ironing) • Limited water purification from certain reagents and a wide range of pollutants • Susceptible to pollution by organic substances • Require washing for regeneration because they can be a breeding environment for bacteria
Alternative products and processes	
Distillation, common oxidation, catalytic oxidation	

Source: HSE

Table 8.7 SWOT analysis of membrane bioreactors

Strengths	*Weaknesses*
• Long life cycle of the activated sediment • Create conditions for breeding specific bacteria • No overflow of sediment • High-quality treatment of output sewage • More efficient way to treat sewage than classic bioreactors	• Require pretreatment of sewage containing large suspension particles
Opportunities	*Threats*
• No need for additional filtration and disinfection	• Application area limited to pretreatment of water and treatment of municipal and industrial waste water
Alternative products and processes	
Mechanical filtration and sedimentation, coagulation and flocculation	

Source: HSE

- Soft impact overtreated objects
- Potential to produce sediments with specified properties

A SWOT analysis of the aforementioned membrane is given in Table 8.8.

Innovative sorbents (including carbons) offer improved efficiency compared with traditional ones. Innovative sorbents (most of which belong to the nanotechnology sphere) can be divided into carbon-based and others. New-generation sorbents are used to remove cations (of copper, iron, ammonium, vanadium, manganese, aluminium, lead, zinc, phosphates) and anions (including sulphides, fluorides and nitrates) from water. Innovative sorbents are also used to absorb oil products and ether-soluble

Table 8.8 SWOT analysis of innovative coagulants

Strengths	*Weaknesses*
• Easy to use • Can be used to treat a wide range of waste waters • Can be used in a wide range of pH and alkalinity values; do not change the pH value of treated water	• Coagulant elements (aluminium) remain in the water after treatment • Can be applied only at pretreatment stage; subsequent treatment required
Opportunities	*Threats*
• Scientific progress opens up the prospects of creating even more efficient coagulants	• Development of alternative treatment techniques • Need to treat purified water to get rid of new pollutants (e.g. radiation)
Alternative products and processes	
Pressure flotation, flocculation, filtration	

Source: HSE

Table 8.9 SWOT analysis of innovative sorbents

Strengths	*Weaknesses*
Compared with other carbon-based sorbents	
• Suitable for a wide range of industries • High abrasive resistance	• Activated carbons (AC) must be replaced/regenerated using chemical, thermal or biological techniques
Compared with alternative products	
• Convenient for loading/unloading, storage and transportation (do not generate dust) • Fire-resistant • Suppress germ growth	• Timber-based AC have lower mechanical strength than charcoals produced from other materials • Alternative products have higher ash content and offer higher cost-efficiency (such as coal coke, pitch, electrode pitch coke, oil coke) • Less strong • More expensive
Opportunities	*Threats*
• AC can be used to absorb organic substances of artificial origin • Wide range of application areas in the chemical, food, pharmaceutical, power generation and metallurgic industries, as well as in oil and gas production and refinery	• Risk of environmental pollution due to AC's ash content
Alternative products and processes	
Mechanical filtration, sedimentation, coagulation and flocculation, flotation, chlorination and ozonisation, plus baro-membrane-based purification techniques (depending on pollution rate and type)	

Source: HSE

substances, to deeply purify water of various microorganisms (bacteria and viruses), including for the production of drinking water from swamp water sources. In addition to the above-mentioned new-generation sorbents, there are innovative sorbents capable of removing arsenic, cadmium and zinc from water and sewage. A SWOT analysis of the aforementioned membrane is given in Table 8.9.

Table 8.10 Innovative water treatment and purification products

Key and promising products	Beginning of mass production in Russia	Market segment
Membranes		Reduced cost, increased productivity, reduced energy consumption, increased temperature and chemical reagents resistance
Microfiltration membranes	Short term	
Ultrafiltration membranes		
Nanofiltration membranes		
Reverse osmosis membranes		
Ion-exchange membranes		
Membrane bioreactors	Medium term	
Membranes with dendrimers	Long term	
Membranes with fullerenes	Long term	
Nanoreactive membranes	Medium term	
Nanocomposite membranes	Medium term	
Membranes based on molecular zeolite mesh	Long term	
Nanocomponents in traditional technologies		Reduced cost, increased productivity, reduced energy consumption
Innovative sorbents	Short term	
Innovative coagulants	Short term	
Activated nanocatalysts inbuilt into membrane systems	Long term	
Nanosize biopolymers with adjustable properties, for removing pollutants	Long term	

Source: HSE

Table 8.10 provides a summary with a benchmark about the market segments for promising technologies, as well as the expected start date of their mass production in Russia.

Next, Fig. 8.4 describes the development prospects for various water treatment and purification technologies incorporating nanocomponents.

8.4.3 Market Segments for Nanotechnology Water Treatment and Purification

Following the focus on alternative technologies, this section exemplifies various market segments for nanotechnology-based water treatment and purification.

8.4.3.1 Water Treatment for Industrial and Municipal Use

Many industries require water with special properties, though each of them has particular criteria. Furthermore, there is a constant stable demand for water softening

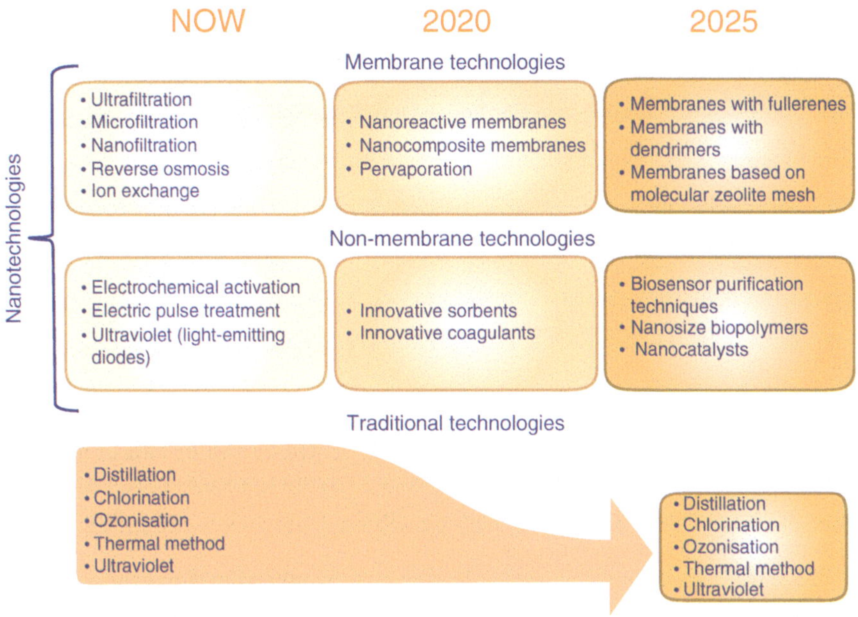

Fig. 8.4 Development prospects of nanotechnologies for water treatment and purification. Source: HSE

(removal of Ca and Mg salts) and desalination (removal of a significant proportion of all dissolved components). Water treatment is also required in nuclear and fuel-burning power generation, radioelectronics and microelectronics, food industry and biotechnology, chemical industry, housing and communal services.

8.4.3.2 Drinking Water Production and Quality Improvement

Promising areas in this segment include the purification of subterranean and surface waters, desalination of brackish and sea water and upgrading of water supply installations (Koroteyev and Desiatov 2003). Dozens of major drinking water production and municipal and industrial sewage treatment projects have been implemented across the world. These projects' installation capacities range from 1000 to 100,000 cubic metres per day. In France, a unique project for a one-step drinking water production from a river source using a nanofiltration membrane-based system was implemented. The equipment guarantees safety (no chemicals used to disinfect water) and a high quality of output water, including its softening; the installation supplies a region with 800,000 residents.

8.4.3.3 Household Membrane Water Purifiers

A membrane module is the basic element of a water purifier. The main consumers are manufacturers of household water purification systems. Water treatment and drinking water production segment dominates the market across the world.

Table 8.11 Structure and dynamics of the provision of nanotechnologies for water supply in Russia (%)

Segment	2015	2020
Membrane-based technologies—nanofiltration and reverse osmosis	85	80
Water treatment (for municipal and industrial use)	50	40
Production of drinking water	40	40
Household water purifiers	8	15
Industrial waste water treatment	1	1
Municipal sewage treatment	0	1
Industrial application (specialised)	2	3
Other (and innovative) water treatment technologies	15	20

Source: HSE

However, it is expected that by 2020 its growth in Russia will be slower than the global average (from 15 to 20% globally, from 39 to 40% in Russia). Nevertheless, these figures suggest that the segment has good prospects and looks attractive to investors (Table 8.11).

The sewage treatment segments still remain at the initial development stages. In Russia, the application of nanotechnologies for treatment of waste water is very rare. However, domestic and international developers already offer equipment with nanostructured corrosion-resistant filtering elements based on steel, titanium, zirconium, nickel and ceramics. Applying this equipment for waste water treatment means they can stay within the maximum permissible concentration limits. When it comes to Russian water treatment technologies, we should primarily mention electric pulse treatment (the equipment is manufactured in Tomsk). In addition, several other water treatment techniques may be relevant in terms of identifying various groups of consumers, manufacturers and products in the course of further research in this field. Table 8.12 summarises various water treatment and purification techniques with relevant market segments in Russia.

8.5 Future Scenarios and Strategies for Nanotechnology-Based Water Treatment and Purification

8.5.1 Forecasts for the Membrane Technologies Market

Membrane technologies are widely perceived a significant and promising area of water treatment and purification nanotechnologies. Figures 8.5 and 8.6 show the likely development prospects for water purification nanotechnologies markets in Russia and the world. The forecast includes three scenarios: optimistic, moderate and pessimistic.

The *optimistic scenario* of global market growth suggests an annual growth rate of about 10–11%. The main growth factors include the increasing shortage of drinking water, development of membrane technologies and a significant increase

Table 8.12 Market segmentation of water treatment technologies in Russia

	Drinking water purification			Industrial water purification	Waste water treatment	
	Centralised water supply	Collective use	Household filters		Industrial waste water treatment	Municipal sewage treatment
Membranes						
Baromembrane processes						
Ultrafiltration	+	+		+	+	+
Nanofiltration	–/+			+	–/+	
Reverse osmosis	–/+			+	–/+	
Microfiltration	+	+	+	+	+	+
Electric membrane processes						
Electrodialysis				+		
Electrodeionization				+		
Membrane bioreactors (MBR)					+	+
Membrane degassing				+		
Nanosorbents	+	+	+	+	+	+
Nanocoagulants	+			+	+	+
Traditional technologies						
Ion exchange	–/+			+	–/+	
Chlorination	+			+	+	+
Ozonisation	+			+	+/–	+/–
Coagulation, flocculation	+			+	+	+
flotation	+			+	+	+
Distillation				+		

Note: The green pluses (+) strong positive correlation; (+) positive correlation; (∓) negative correlation
Source: HSE

of membrane technologies' supply from China, with a large growth of membrane module production (ITA 2005). The optimistic forecast is supported by the growing popularity of complex natural and waste water purification technologies, coupled with a growing shortage of water. The emergence of competitive closed-cycle technologies, and increased public concern in Europe and the USA about the state of affairs in this field, could also be the drivers of the optimistic scenario in the near future. The *pessimistic forecast* envisions low market growth rates—no higher than 5–6% a year. The *moderate scenario* assumes that average market growth rate would amount to 7–8% a year in the long term.

Russia's global market share is small (according to various estimates, about 3% of physical volume and less than 1% in monetary value). This is due to low water costs and the systemic reasons why Russia is technologically lagging behind.

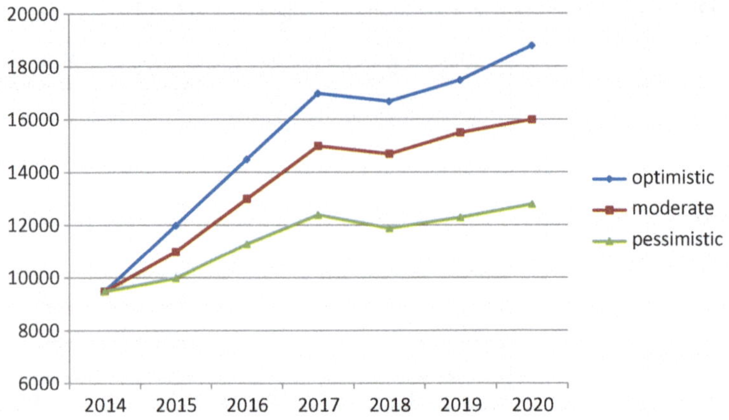

Fig. 8.5 Actual and expected global market growth (million USD). Source: HSE

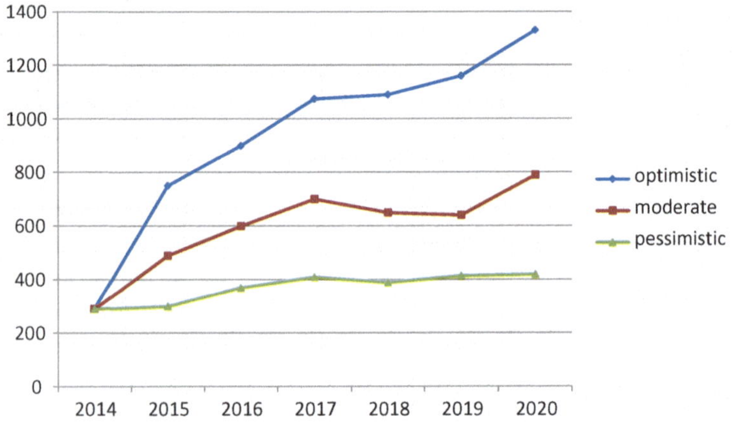

Fig. 8.6 Actual and expected Russian market growth (million USD). Source: HSE

The *optimistic scenario* for Russian market growth is based on a high growth rate of the overall water treatment market (for industrial and municipal purposes). The optimistic scenario assumes continued government support and the active popularisation and promotion of relevant technologies. According to the optimistic scenario, by 2020, Russia's market share will reach to **7%**, or **1.3** billion USD. According to the *pessimistic scenario* where the government does not provide any support, the annual growth rate of the membrane technologies market will not exceed 4%. The main barrier hindering the application of nanotechnologies is traditional water treatment techniques; however, their efficiency may be increased by introducing new designs unrelated to nanotechnologies. The *moderate scenario* assumes limited government support, making constrained growth of about 11% feasible.

8.5.2 Strategic Options

8.5.2.1 Federal Strategies

An aggressive strategy envisages supporting segments with the maximum market potential and might contribute to achieving the optimistic development scenario. The need for an aggressive strategy is conditioned by the current situation in the industry and by the understanding that high-quality water treatment has strategic importance to preserve the nation's health and environmental balance. Such a strategy is based on the assumptions that the prospects for the membrane technologies market are optimistic assuming that water purification nanotechnologies have numerous application areas, ranging from industrial waste water treatment and water purification to production of drinking water, and that the demand for membranes will continue to grow globally, with Russia becoming an exporter of nanomembranes and one of the industry's 'regulators'.

The core of such a strategy is to support the market segments centralised and decentralised water purification and water purification for industrial use. For implementing this strategy, a combination of various measures is required. First, government support to industry should be provided by means of direct funding at the initial stage of upgrading technology and equipment, adopting relevant legally binding documents (including national standards, terms of reference, end product specifications which may include nanomembranes) and supporting demand via preferential treatment of consumers who opt for these products. Furthermore, improving customs rules for the industry and supporting the participation of Russian membrane manufacturers in technology chains comprising leading companies and countries are essential measures. A significant market breakthrough can be achieved by building an advanced nanomembrane factory, which would require government participation to go ahead. Second, market players must adopt a more aggressive approach in areas such as the active promotion of nanomembrane-based products which has many potential consumers who're not planning to switch to such products because of a lack of awareness of their efficiency and performance, the promotion of water purification nanotechnologies among end-product users and extending cooperation between Russian manufacturers and developers and launching cooperation with foreign partners (in particular, a more active search for potential consumers).

Implementing an active strategy corresponds with the moderate development scenario and would be possible under the following conditions:

- Provision of government support for the industry, including through the targeted federal programme 'Pure Water'
- Active participation of several stakeholders in the programme (such as Rusnano, Rostechnologies, etc.) to increase the production and consumption of nanoproducts
- Retaining existing research and human potential and training highly skilled researchers for this field
- Promoting the application of nanomembranes

- Setting up cooperation with manufacturers and consumers in various countries, including joint ventures with leading producers

8.5.2.2 Regional Strategies

Strategies to promote innovative water purification technologies in various market segments are determined by the application potential of the new technologies in centralised and decentralised water supply systems. Making the most of this potential depends on the growth in water consumption segments, their market shares, the upgrading or replacement of worn-out pipeline networks and how well water treatment installation equipment meets official Russian standards for the quality of source water. To design technology diffusion promotion strategies, we grouped similar regions together based on similar regional features such as the nature of problems they face in water treatment and purification. Depending on the availability of a centralised water supply, we classified regions into two groups: Group A comprises regions with a higher availability of mains water supply; and Group B consists of regions with a lower availability of mains water. Figure 8.7 illustrates the two groups of regions, distinguished by the availability of centralised water supply, the quality of centralised and decentralised water sources and the quality of 'output' water.

The regions included in these groups are characterised by reasonably different features. Group A-1 and A-2 regions possess relatively high-quality central water supply resources, but A-1 regions and 7 of 17 regions show relatively low quality of decentralised water supply sources, whereas A–2 regions feature a low quality of output water and 6 out of 9 regions have a relatively low quality of decentralised water supply sources. A-1 regions accordingly need to modernise water treatment and purification equipment and pipeline networks. This might be done by the application of new technologies which is basically possible because most of these regions have a relatively high economic development level. However, the relatively low quality of decentralised water supply sources in some of the regions is a motivation to promote individual and collective water purification technologies. A-2 regions operate worn-out pipeline networks which makes it first priority to replace them. Introduction of new technologies, including membrane-based ones, will have no economic or social effect without upgraded pipeline networks. If the problem is with obsolete water treatment installation equipment, the application of new technologies may be justified. The relatively high level of economic development in regions with low-quality decentralised water supply sources opens up opportunities to promote technologies for individual and collective water purification. The quality of source water is rather low in A-3 regions, however, a relatively high quality of water is delivered to end users. This implies a favourable condition of pipeline networks and equipment at centralised water supply installations. Applying membrane bioreactors and other nanotechnologies for treating industrial and municipal waste water is possible, which subsequently would have a positive effect on the water source. The good prospects of the waste water treatment market offer a high

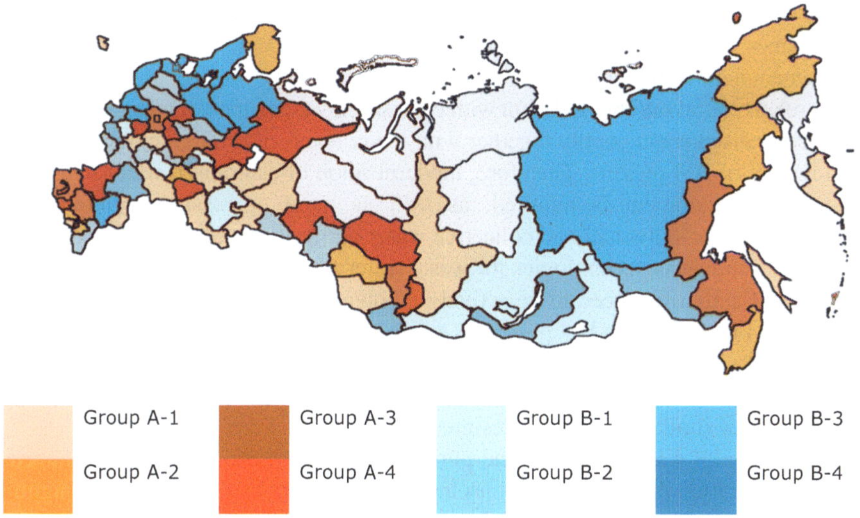

Fig. 8.7 Region-specific strategies to promote water treatment and purification nanotechnologies. Source: HSE

demand potential for relevant nanotechnologies. Like in A-3 regions, the source water and output water is of low quality in A-4 regions, e.g. five of ten regions have low levels of economic development. In the other five regions, a relatively low quality of decentralised water supply sources is combined with a relatively high economic development level. Therefore, the first priority is upgrading water treatment equipment and pipeline networks. The federal government needs to be actively involved in solving the problems of those regions with relatively low development levels. Subsequently, it might be possible to promote nanotechnologies for individual and collective water purification in the more economically developed regions with low quality of decentralised water supply sources.

Relatively problem-free regions are grouped in B-1, where a high quality of water in centralised water supply systems is combined with a relatively high quality of decentralised water sources, but most of the regions in this group have low levels of economic development. It's typical for this group that planned replacement of water treatment and purification equipment is required which potential for nanotechnologies application for water treatment and purification can be applied both for water supply and sewage treatment purposes. Depending on the perspective—medium or long term—these can be supplementary or alternative strategies. In B-2 regions a relatively high quality of decentralised water sources is combined with an unfavourable situation with centralised water supply which is also a typical feature less developed of economic regions (9 out of the 12 regions). Accordingly, two potential modernisation strategies emerge: (1) extending the centralised water supply system, while at the same time modernising the equipment and infrastructure, and (2) a more active application of innovative technologies for individual and collective water supply. The eventual choice of strategy requires feasibility studies in any case. The strategies can also be applied in combination and in various timeframes. Lack of adequate financial support can become a major obstacle for implementation of both strategies.

A relatively favourable situation with centralised water supply, and a poor quality of decentralised water supply together with lower levels of economic development, is a feature of B-3 regions. Therefore, modernisation of centralised water treatment installation equipment is required, as well as more aggressive promotion of technologies for individual and collective water purification, membrane bioreactors and other innovative technologies for waste water treatment. In these constellations the federal authorities need to be actively involved to successfully deal with these problems. Finally, Group B-4 possesses a low quality of decentralised water sources which is combined with centralised water supply problems. There, the modernisation of water treatment and purification equipment is the first priority, and strategies similar to the ones designed for Group B-2 can be applied.

The future of water treatment and purification is in the process of being shaped. The incorporation of nanotechnologies into the water supply solutions would significantly increase the quality of drinking water and reduce the incidence of diseases caused by polluted water sources and pipeline networks. Hence, new application fields for nanosolutions are emerging. To take advantage of these opportunities, governments follow different strategies. The present chapter discussed that there are no uniform strategies by governments both at the federal and regional levels. Instead, these strategic approaches vary and take regional and national characteristics into account.

Acknowledgements The book chapter was prepared within the framework of the Basic Research Program at the National Research University Higher School of Economics and supported within the framework of the subsidy by the Russian Academic Excellence Project '5-100'.

References

Alekseyev LS, Gladkov VA (1994) Improving quality of soft waters. Stroyizdat, Moscow

Aydogdu A, Burmaoglu S, Saritas O, Cakir S (2017) A nanotechnology roadmapping study for the Turkish defense industry. Foresight 19(4):354–375. https://doi.org/10.1108/FS-06-2017-0020

Battistella C, Pillon R (2016) Foresight for regional policy: technological and regional fit. Foresight 18(2):93–116

Bottero J-Y, Rose J, Wiesner MR (2006) Nanotechnologies: tools for sustainability in a new wave of water treatment processes. Integr Environ Assess Manag 2(4):391–395

Desiatov AV et al (2008) Application of membrane technologies for water purification and desalination. In: Koroteyev AS (ed) Chemistry, Moscow

Draginsky VL, Alekseyeva LP, Getmantsev SV (2005) Coagulation in natural water purification technology. Moscow

Gerasimov GN (2001) Application of coagulation – flocculation processes to treat surface waters. Water supply and sanitary equipment 3

Getmantsev SV (2003) Current state of aluminous coagulants' production and import in Russia. Water supply and sanitary equipment 2

Horner N, Oliviera AGP, Silberglitt R, Poppe MK, Rocha BB (2016) Energy foresight, scenarios and sustainable energy policy in Brazil. Foresight 18(5):535–550
Inayatullah S, Elouafi IA (2014) The alternative futures of the International Biosaline Agriculture: from salinity research to greening the desert. Foresight 16(5):389–409
ITA (2005) Water supply and wastewater treatment market in China. U.S. Department of Commerce International Trade Administration, Washington, DC. http://www.ita.doc.gov/media/publications/pdf/chinawater2005.pdf. Accessed 31 July 2017
Judd S (2006) Membrane bioreactor process fundamentals/Orange: Microfiltration 4. The National Water Research Institute
Karelin FN (1988) Application of reverse osmosis for water desalination. Stroyizdat, Moscow
Koroteyev AS, Desiatov AV (2003) Application of space technologies for desalination of sea and brackish water. In: Sea water desalination technologies and equipment for Caspian region. Proceedings of international theoretical and practical conference. Aktau, pp 14–26
Loveridge D, Saritas O (2009) Reducing the democratic deficit in institutional foresight programmes: a case for critical systems thinking in nanotechnology. Technol Forecast Soc Change 76(9):1208–1221
Loveridge D, Saritas O (2012) Ignorance and uncertainty: influences on future-oriented technology analysis. Technol Anal Strat Manag 24(8):753–767
Lysov VA, Vilson YV, Butko AV, Butko DA (2002) Application of silica-alumina flocculants for water treatment and purification. Water supply and sanitary equipment 11
Maggio A, van Criekinge T, Malingreau J-P (2016) Global food security: assessing trends in view of guiding future EU policies. Foresight 18(5):551–560
Miles I, Saritas O, Sokolov A (2016) Foresight for science, technology and innovation. Springer International Publishing, Switzerland
Moriarty P, Honnery D (2014) Future Earth: declining energy use and economic output. Foresight 16(6):512–526
Nikolayev V (2008) Membrane-based technologies in water management. Water Magazine 5:16–22
Poliakov AM, Vidiakin MN (2009) Membrane bioreactor equipment and technology market. Sanitary Engineering 4
Proskuryakova L, Saritas O, Sivaev S (2018) Global water trends and future scenarios for sustainable development: the case of Russia. J Clean Prod 170:867–879
Prüss-Üstün A, Bos R, Gore F, Bartram J (2008) Safer water, better health: costs, benefits and sustainability of interventions to protect and promote health. World Health Organization, Geneva
Saritas O (2015) SPIEF 2014: technology, energy and Russia's future development agenda. Foresight 17(3):233–239
Saritas O, Burmaoglu S (2015) The evolution of the use of Foresight methods: a scientometric analysis of global FTA research output. Scientometrics 105(1):497–508
Saritas O, Miles I (2012) Scan-4-Light: a Searchlight function horizon scanning and trend monitoring project. Foresight 14(6):489–510
Saritas O, Proskuryakova L (2017) Water resources – an analysis of trends, weak signals and wild cards with implications for Russia. Foresight 19(2):152–173
Saritas O, Taymaz E, Tumer T (2007) Vision2023: Turkey's national Technology Foresight Program: a contextualist analysis and discussion. Technol Forecast Soc Change 74 (8):1374–1393
Sterman LS, Pokrovsky VN (1991) Physical and chemical techniques for water treatment at fuel-burning power plants. Energoatomizdat, Moscow
Theron J et al (2008) Nanotechnology and water treatment: applications and emerging opportunities. Crit Rev Microbiol 34(1):43–69
Tkachev KV, Kisil Y, Zapolsky AK (1978) Coagulant technology. Chemistry, Leningrad
Tkachev KV, Trifonova LA, Bormontova SV (1988) Inorganic coagulants technology. Moscow

UNwater (2015) World Water Development Report. http://www.unwater.org/publication_categories/world-water-development-report/. Accessed 27 July 2017

Vishnevskiy K, Yaroslavtsev A (2017) Russian S&T foresight: case of nanotechnologies and new materials. Foresight 19(2):198–217

Ozcan Saritas is Professor of Innovation and Strategy at the HSE and editor-in-chief of "Foresight"—the journal of future studies, strategic thinking and policy. He worked as Senior Research Fellow at the Manchester Institute of Innovation Research, The University of Manchester. His research focuses on innovation and policy research with particular emphasis on socio-economic and technological Foresight. With a PhD from the "Foresight and Prospective Studies Program," he introduced the "Systemic Foresight Methodology", and has produced a number of publications on the topic. Dr. Saritas has extensive work experience with the international organizations, including the United Nations, OECD, and the European Commission. He has been involved in large scale national, multinational and corporate research and consultancy projects on sectors including energy, climate change, agriculture, food, water, transportation, and ICT among others; published a number of articles in respected journals; and have delivered keynote speeches in more than 50 countries across the world. Besides his research and publication activities, he designs and delivers academic and executive education courses on Foresight and Strategic Planning. He has recently co-authored a book, entitled "Foresight for Science, Technology and Innovation" published by Springer, which has become one of the key readings in the field.

Konstantin Vishnevskiy is Head of the HSE ISSEK, Department for Digital Economy Studies. He holds a PhD from Moscow State University, Faculty of Economics. Dr. Vishnevskiy has long-standing experience in the development of technology roadmaps, the elaboration of Foresight methodology and corporate innovation development programs, the integration Foresight into government policy as well as financial and econometric modeling. He participated in many high-level research projects on S&T Foresight on national and regional level both in Russia and abroad. Dr. Vishnevskiy authors about 70 scientific publications on long-term planning and Foresight, roadmapping, digital economy, macroeconomic regulation and government policy, innovation strategies for businesses. He presented about 100 reports on academic and professional conferences and workshops concerning Foresight, roadmapping and innovations.

Emerging Technologies Identification in Foresight and Strategic Planning: Case of Agriculture and Food Sector

9

Leonid Gokhberg, Ilya Kuzminov, Pavel Bakhtin, Anton Timofeev, and Elena Khabirova

9.1 Introduction

In conditions of global challenges for sustainable development and attempts to reduce global threats driven by complex issues (such as climate change, ageing population, natural resource scarcity, water security, human health and wellbeing threats) (Kallhauge et al. 2005; Keenan et al. 2012), the global and national governance systems are faced with extremely difficult missions. In particular, the global food problem is far from solution, as several hundred million people in less developed countries face undernourishment and even famine (FAO 2009), while global population growth (United Nations 2015), which is far from plateauing, puts additional demand pressure on the global food production-distribution systems. The global food problem is aggravated by clearly expressed negative environmental trends threatening to decrease gross A&F output in the future (World Bank 2007). It is admitted that A&F sector is one of the largest greenhouse gas emitters globally (O'Mara 2011; Tubiello et al. 2013), and the situation grows worse because of global shift towards consumption of animal products, more resource intensive and environmentally unsustainable. The negative trends also include degradation of bioproductivity of agricultural land, namely, soil erosion (Montgomery 2007), soil

L. Gokhberg · I. Kuzminov (✉) · A. Timofeev
National Research University, Higher School of Economics, Moscow, Russia
e-mail: lgokhberg@hse.ru; ikuzminov@hse.ru; Aatimofeev@hse.ru

P. Bakhtin
Information and Analytical Systems Unit, HSE ISSEK, Moscow, Russia
e-mail: pbakhtin@hse.ru

E. Khabirova
Unit for Digital Economy Studies, HSE ISSEK, Moscow, Russia
e-mail: etochilina@hse.ru

D. Meissner et al. (eds.), *Emerging Technologies for Economic Development*, Science, Technology and Innovation Studies,
https://doi.org/10.1007/978-3-030-04370-4_9

compaction (Hamza and Anderson 2005), negative net nutrients flaw and fertility fall (Pimentel 2006). Among the other environmental threats, there are World Ocean bioproductivity loss due to overfishing (Srinivasan et al. 2010), contamination with harmful anthropogenic substances (Aarkrog 2003), climate change and acidification (Hoegh-Guldberg and Bruno 2010).

The solutions to the existing global challenges, as many researches and international think tanks see it, lie in wide-scale adoption of new technologies (Omenn 2006). At the same time, many organizational innovations, such as new business models, citizen vigilance schemes and governance mechanisms, which could possibly alleviate some of the global issues, are also becoming feasible solely because of the development of some universal, or platform, or enabling technologies (Gokhberg et al. 2013).

Therefore, governance and management systems have to acquire technology-awareness capabilities (Spitsberg et al. 2013; Momeni and Rost 2016; Bildosola et al. 2017). This means that effective STI policy becomes a more crucial success factor for governance both on global and national levels, as well as for corporate strategic management. For the policies to become effective, they need to be both evidence-based (Smith and Haux 2017) and proactive (Mani 2004; Aghion and Griffith 2008). This, in turn, formulates the necessity of constant monitoring of *the technologies that are emerging*, as they are important drivers of efficiency of the human activity.

The concept of emerging technology, its scope and definition is a highly discussed topic in social sciences. While a number of impactful publications concentrate on conceptualizing the term "emerging technology" (among others see Rotolo et al. 2015; Halaweh 2013), it is sufficient for the scope of this paper to understand the emerging technology as a new technology that might have a significant impact on the economic activity in certain sectors of the economy.

In the era of explosive growth of diversity of S&T and of quantity of available information (Hidalgo 2015), technologies identification and mapping becomes less feasible without the use of modern data science techniques. This necessity is caused not only by the "information push" but also by significant drawbacks of human-performed analytics. While spotting critical technologies and existing and emerging trends, as well as forecasting future development and setting STI priorities, sectoral experts often demonstrate subjectivity, incompleteness of their knowledge and bias of views favouring familiar subject areas (see, e.g. Winkler and Moser 2016). Reconciliation of ideas among the large expert groups could lead to overextended periods of *foresight* studies—one of the main sources of evidence for modern STI policy. This means that some technologies might already transform from the emerging stage to the stage of commercially viable products before a dedicated foresight report on the emerging technology landscape is published. These factors along with budget limitations drive governments and private companies towards at least partial automation of foresight and strategic planning activities.

One example of early attempts at solving these problems is the tech mining (Porter and Cunningham 2004; Madani 2015; Huang et al. 2015; Bakhtin and Saritas 2016). It strongly relies on large expert validation and manual filtering and cleaning of data outputs. For the purposes of emerging technologies identification, the text mining in combination with semantic analysis tools seems to be most appropriate, as a task of identification of new man-made phenomena of known nature (technologies in this case) can be reduced to designation of new syntactic constructions describing them.

To demonstrate the potential of the text-mining techniques for technologies identification and mapping, we conducted the case study of emerging technologies identification in the agriculture and food sector by applying an intelligent Big Data Text Mining system iFORA, developed at HSE ISSEK featuring semantic analysis of heterogeneous textual data sources. Moreover, there are proposed new analytical tools for technologies' sequential analysis. For the identified prospective technologies with the synthesis of the materials of expert foresight exercises, desk research and Big Data Text Mining analytics, there were aggregated (1) parameters of technologies' potential to bring answers to sectoral and national challenges and (2) corresponding future markets forecasts. Based on this analysis, prospective S&T development areas and technological solutions for the Russian A&F sector were highlighted and STI policy strategies outlined in the context of two scenarios of the sector development.

9.2 Methodology

Current study is based on a set of quantitative, qualitative and expert-based S&T Foresight techniques. The main information source for the study is the ample material S&T Foresight for Agriculture and Food Sector in the Russian Federation until 2030 (HSE 2017; Gokhberg et al. 2017). With the participation of representatives of about 400 organizations (leading research organizations, agricultural companies, universities, industrial unions and associations and development institutes), there were undertaken 25 in-depth expert interviews, and 15 panels and workshops were organized. These structured discussions were dedicated to the main S&T, socio-economic and environment trends along with the most promising technologies and markets' main directions of future policy in the sector. Understanding the danger of experts' biased opinions (which is particularly true for lobbyist business associations), described data collection methods were counterbalanced with the desk research and specialized texts automatic analysis.

A&F sector-relevant texts were processed with Big Data Text Mining system iFORA. The composition of data sources of the text-mining system at the time of the exercise includes:

- Stratified random sample of summaries and metadata of 1,885,624 internationally top-cited research papers for 10-year period from 2005, representing various research areas and acquired from the open citation indexes and other open data sources
- Stratified random sample of summaries and metadata of 2,277,869 international patents for 10-year period, acquired through open-access sources of the World Intellectual Property Organization Patent Cooperation Treaty (WIPO PCT)

- 7,007,033 newsfeed items from Alexa and SimilarWeb[1] tops of global news portals with science and technology flavour, for the period since the inception of the WWW
- 171,247 analytical reports available from international and national organizations and agencies

At the time of the study, the system featured more than 12 million individual documents, with several hundred million individual sentences, of which up to 3 million documents were at least partially relevant to A&F sector and adjoining sectors, such as biotechnology and bioenergy.

Synthesis of the materials of expert foresight exercises, desk research and Big Data Text Mining analytics allowed preparing and validating of *lists of technologies, markets, trends and challenges* for Russian A&F sector. Among all identified technologies, the emerging ones were classified by text-mining analysis of dynamics of their presence intensity in the discourse during the last years (for details on applied computational procedures and measures, see in Bakhtin et al. 2017) with visualization in the form of trend maps. They consist of two-dimensional plots with one axis representing the popularity of a term (the so-called significance axis) and the other showing the year-by-year dynamics of the normalized popularity (relative frequency of use) (the so-called trending axis). Interpretation of the results of this analysis was done based on the assumption that currently unfolding technology trends (including the development and adoption of emerging technologies) are characterized by the growth of interest towards them at least in one of the corpora of documents processed (patents and news/blogs).

For balanced classification of identified technologies, semantic mapping is executed with the use of combination of machine learning methods, based on co-occurrence term graph clustering and regularization of mutual information and topic monopolism of terms. The principal results of semantic mapping are clusters of principal terms that give insights into optimal *groupings of technologies* by semantic similarity.

After technologies identification and mapping stages, it is proposed to implement the following analytical tools for STI policy evidence gathering by combination of expert Delphi procedures and text-mining techniques:

- Assessment of technology potential to bring answers to global and national challenges through an expert evaluation using special questionnaires and further validation of the results with computation of textual parameters of the semantic proximity (for details on measurement, see in Bakhtin et al. 2017) between two lists: technologies and challenges. These two ontologies were preliminarily composed during desk research, supplemented with the materials of in-depth expert interviews and workshops, and finally verified with text-mining instruments on the extensive corpus of A&F related documents.

[1]URL: https://www.alexa.com/ and URL: https://www.similarweb.com/

- Aggregation of connected technological markets forecasts on the basis of lists of prospective markets composed by expert participants of the study and with an extraction of forward-looking market volume estimates from up to a million market report press releases relevant to agriculture[2] from key global players in the field.

The analysis of these additional parameters for identified technologies can serve as an instrument of strategic priorities setting in the field of technologies that are currently emerging through the ever-intensifying global S&T development process.

9.3 Emerging Technologies Extraction and Mapping

As a result of proposed text-mining analysis, the *ontology of global agriculture and food technologies* is developed. It consists of several thousands of technologies, which were classified by dynamics of intensity of their presence in the discourse during the last years (see details in Sect. 9.2). The results of this analysis were visualized by trend maps on media resources and international patents (Figs. 9.1 and 9.2, respectively[3]). It is suggested that *emerging technologies* with strong potential of surviving and upscaling to the global production systems receive an increasing public awareness in wide audience information resources (media) and growing presence in intellectual property protection documents (patents). *The upper right quadrant* of the trend maps consists of the strongest topics shaping the future agenda of the sector; they are popular and gaining traction. In media they exemplified by CRISPR technologies, agroforestry and aquaponic technologies, precision agriculture and microalgae utilization, etc. In patents this group also consists of several genetic technologies but moreover includes fertigation and hatchery technologies. *The lower right quadrant* of maps contains the so-called weak signals: they are highly trending but underrepresented in discourse yet. They can contain the *emerging technologies*. This group presented in media by smart irrigation technologies, molecular breeding and zinc finger nuclease technologies, etc. At the same time, the patent map shows, for instance, that bionematicides, mutagenesis and synthetic seed technologies are increasingly gaining popularity. Among the popular topics losing their significance in media are fertilization, pruning, antifouling technologies and many more. In patents, topics with declining popularity are composting and horticultural technologies, among others.

For *dynamic classification* of identified technologies, which is an essential task of technology landscape mapping (Bakhtin and Saritas 2016), semantic map instrument was used (see details in Sect. 9.2). It is a clustered co-occurrence graph, which consists of thousands of connected vertices, each of which represents an important

[2]Identification of market reports lexically relevant to agriculture is done by the advanced methods of topic modelling and term-document biclustering.

[3]In the illustrative visualization, only few points can be labelled without overlap.

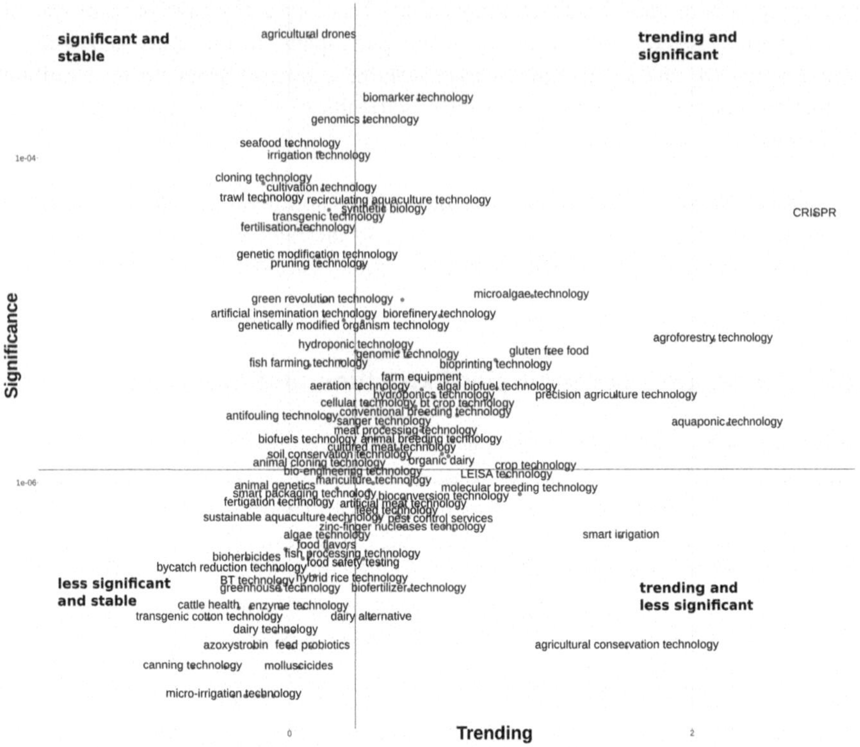

Fig. 9.1 Trend map of agricultural technologies on media resources. Source: HSE

term[4] (Fig. 9.3). Technologies were automatically grouped in the following large coloured clusters of the most promising S&T development areas:

- Urban agriculture technologies (robotic greenhouses, vertical farms, recirculating aquaculture systems, aquaponics, hydro-/aeroponics, artificial lighting, sensors and control systems, etc.)
- Advanced precision agriculture technologies (geographic information systems, autonomous vehicles, GNSS and remote sensing, etc.)
- LEISA (Low External Input Sustainable Agriculture) technologies and waste processing technologies (waste processing, soil management, organic agriculture, etc.)
- Biotechnologies (genetic engineering, vaccines, antibiotics, probiotic production technologies, embryo transfer, DNA sequencing)
- Advanced food technologies (biochemical, enzyme technologies, nanotechnologies for food industry, active packaging, nutrient additives)
- Smart bioenergy convergent technologies (biofuel production, solar energy, etc.)

[4]In the illustrative visualization, only few of the vertices can be labelled without overlap.

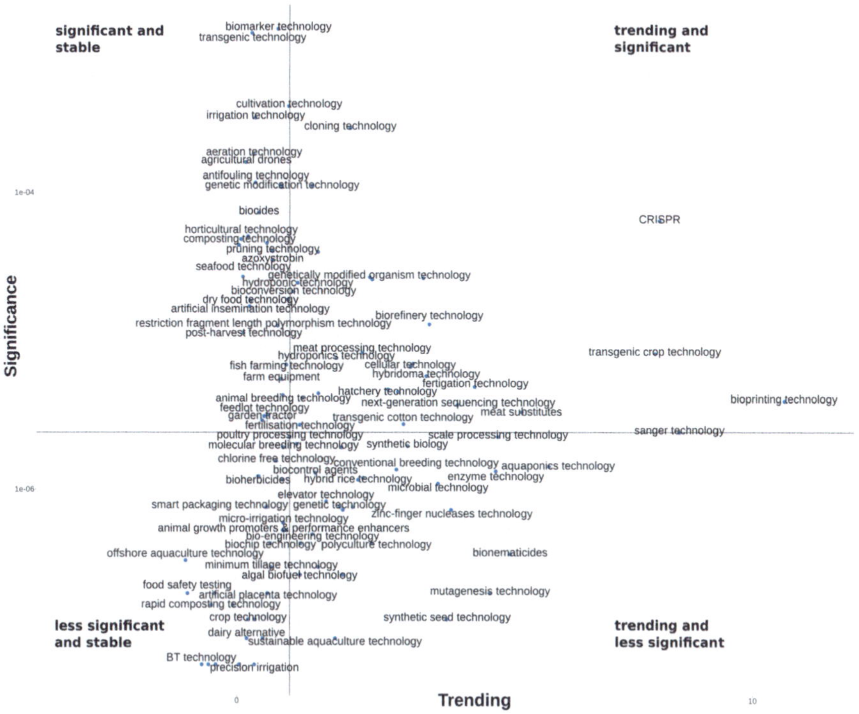

Fig. 9.2 Trend map of agricultural technologies on patent resources. Source: HSE

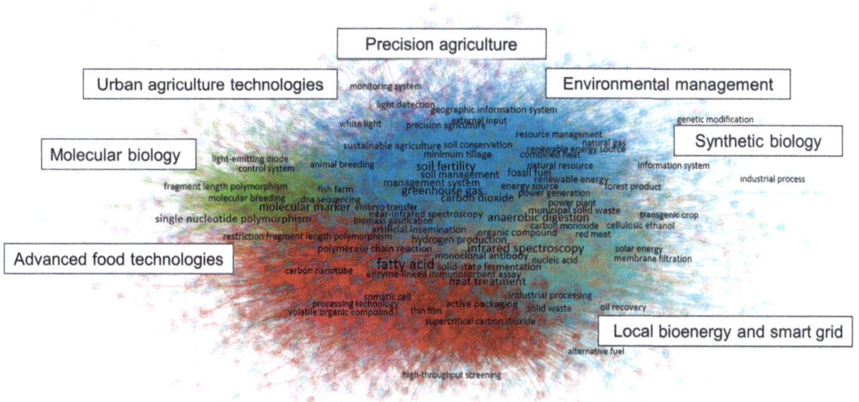

Fig. 9.3 Semantic map of identified A&F sector technologies and related topics. Source: HSE

- System integration technologies for managing logistics in agricultural sector (big data, machine learning, robotic storage facilities, etc.)
- Advanced technologies for the fisheries sector

Dynamic terms' classification based on clustering mechanisms provides mapping of the identified technologies, thematic linkage among them and with non-technological concepts and topics in the sphere. It provides an ad hoc ontology for fast analytics and for expert discussions in the absence of official taxonomies. This allows mapping even *emerging* fields, which have not yet been categorized for the purposes of official statistics or legal regulation. Technologies of identified thematic groups were further analysed on the next step to make a conclusion about their potential from strategic priorities setting viewpoint.

9.4 Implications to Strategic Priorities Setting in the Field of Emerging Technologies

During the preparation of Science and Technology Foresight for Agriculture and Food Sector in the Russian Federation until 2030 (HSE 2017) on the basis of the presented results, prospective S&T areas were highlighted which are supposed to play a critical role in stabilizing country's position on traditional markets and entering the emerging innovative ones.[5] Nine prospective S&T areas are envisaged, comprising 68 specific innovative technological groups[6] (which are presented in Annex), with 47 related global challenges and 37 markets for advanced production technologies, equipment and consumer agricultural products.

A list of STEEPV (social, technological, economic, environmental, political and value) *sectoral challenges* identified with the experts' participation and automatic knowledge extraction procedures was reviewed by Saritas and Kuzminov (2017). Through text-mining analysis of their mentions in the contexts of previously identified technologies, their semantic proximity was evaluated (for measurement details, see in Bakhtin et al. 2017). The potential of a technology to bring answers to global and national challenges can be seen as the evidence that it is a prospective one that worth a particular attention during strategic priorities setting.

[5]For profound survey of the status and prospects of S&T development of agriculture and food sector in Russia, see Gokhberg and Kuzminov (2017).

[6]Taking into account the broad definition of technology concept from *Oxford English Dictionary*, where it is explained as "the codified and applicable pieces of knowledge that can be used for production and distribution of goods and services, other purposeful economic and non-economic, but socially impactful activities", the main concept of operationalization problem may occur. However, deep theoretical questions about adjustable parameters of technology atomicity (what level of technical knowledge formalization should be seen as a technology) remain out of the scope of the paper and need additional comprehensive research.

The most promising S&T development areas for the A&F sector are often related with the emergence of new *technological markets*. With the means of syntactic parsing (Socher et al. 2013), we aggregated forward-looking global markets' volume estimates for 2030 for previously identified emerging technologies directly from market reports' press releases from key global players in the field (for full tables of A&F markets forecasts, see in HSE 2017). Based on the analysed forecasts, it should be mentioned that to position Russia on global markets of A&F sector, choose adequate strategies for market expansion; an understanding of existing and future image of both traditional commodity markets and perspective unconventional products is required. Attention should be paid not only to the export of traditional and modernized means of production and a wide range of products, including biochemical ones, but also to the export of *new technologies*, information systems and smart services for system integration based on domestic information and space technologies. Platform solutions, which will be to some extent applied in all branches of A&F sector, are worth special consideration. The largest prospective markets are *emerging markets* for agricultural lands with improved characteristics, unmanned systems in A&F sector, and smart management systems for technological processes and automated regulation of economic processes. Until 2030, Russia will be able to achieve strong presence on such prospective global markets as the new-generation melioration systems market, home hydro- and aeroponic systems, the market for a new nutrient medium for plants and various biodegradable plastics. Other prospective niches include the following:

- Smart agriculture, including high-tech plant growing and animal farming products, based, among other things, on new technological solutions
- Functional foods, with unique useful properties
- New varieties, hybrids, breeds and cross-breeds created through accelerated selection
- Balanced standardized forage for high-yield animal farming and aquaculture
- Highly efficient and safe active ingredients for vaccines, antibiotics, and antivirus preparations for animal farming and weed and pest killers
- Biotechnology- and synthetic biology-based food production systems, including new strains of useful microorganisms, bioreactors and enzyme complexes
- Climate-adaptive production systems, including next-generation irrigation complexes

Summing up the two outlined parameters from expert and text-mining evaluations for identified technologies—naming connection to global challenges and related markets forecasts, illustration for one example of prospective technological area—advanced precision agriculture technologies are presented in Table 9.1. It is assumed that these two main characteristics get reliable first approximation of social, political and economic value of the identified emerging technologies on national scale. Therefore, it is strongly proposed to consider them in strategic priorities setting.

Table 9.1 Additional parameters' summary for prospective technological area example

Prospective technologies in a group	Connection to global challenges and trends	Global markets forecast volume in 2030
Advanced precision agriculture technologies		
– Big data and Internet of Things technologies for agriculture – New electronic technologies, wireless networks and microsensors – Advanced robotic technologies based on artificial intelligence, swarm intelligence and machine learning – Application of unmanned aerial vehicles in agriculture – Application of nano-and pico-satellites in agriculture – Fully automated operation of agricultural machinery – Technologies for deep processing of agricultural materials at the place of their harvesting (mobile semiautomated mini-plants)	– Abandoning full-coverage irrigation in favour of precision underground irrigation techniques, to achieve significant water saving – Abandoning regular full-coverage fertilizer application in favour of dynamically adjusted techniques – Replacing conventional fertilizers with composite capsular ones – Development of distributed production of advanced hardware based on additive and telecommunication technologies – Moving on from manned agricultural machinery to automated equipment, based on micro-geopositioning and self-learning robots – Replacement of traditional agricultural labour markets with skills and competencies markets based on cloud technologies, controlled robots and remote employment	– Unmanned systems in A&F sector—$250 bln – Systems for precise agriculture, telematics and unmanned technologies in crop farming—$65 bln – Robotized systems for urbanized and industrial crop farming—$30 bln – Self-driving machinery based on micro-geopositioning—$4 bln – GPS/GLONASS sensors and RFID chips for the logistics in A&F sector—$4 bln

Source: HSE

Vital characteristic of sectoral S&T innovation process—rates of emerging technologies' adoption and diffusion—determines, from the one side, and relies upon the sector economic and institutional development, from the other side. Depending on the future evolution of global and national economy, oil price dynamics and investment spending and the most likely policy choices made at key threshold points—macroeconomic, institutional and political ones—*two possible development scenarios* for the period from 2020 to 2030 were developed in S&T Foresight for A&F sector in the Russian Federation until 2030 (HSE 2017).

In the "local growth" scenario, the gradually strengthening oil prices and the increasing volume of the export of hydrocarbons and other raw material resources will be the main drivers for the development of A&F sector technologies. Due to the expecting economy enhancement, GDP will annually grow by 1.3% between 2017 and 2019. By 2020 this indicator's growth rate may achieve 3–4% per year in a medium-term perspective because of the expansionary monetary and fiscal policy aimed primarily at the improvement of the investment environment. Structural and institutional changes in the economy focused on the sustainable growth will enable the Russian economy to develop more rapidly than the world economy.

The basic conditions for "Global Breakthrough" scenario realization include the accelerating growth of the Russian economy in a medium term by 1–2 percentage points due to the increasing public investment spending. This scenario assumes that the monetary policy will soften more quickly in the next 2 years and that additional investments will be made to develop science, support agricultural products' export and foster consumer demand for domestic products. Owing to the fiscal and monetary policy, investment growth rate in the scenario "Global Breakthrough" may exceed the figures of the "local growth" scenario 2–2.5 times after 2020. The average annual growth rate of consumption in this scenario will be by 0.7 percentage points higher. GDP will increase faster by 1.6 percentage points than in the "local growth" scenario.

Russia can become a global supplier of high-value-added products, technologies and services. This is the goal of the "Global Breakthrough" scenario. Another option is less ambitious and easier to achieve yet also desirable: this entails saturating the domestic market with competitive domestic products and technologies—the "local growth" scenario. Both trajectories are possible and may start at the same point in time from identical external conditions. The difference between the higher and the lower trajectories is determined by the quality of the institutional framework and economic mechanisms by the year 2020; the gap between these trajectories will grow over time, making the leap from the less favourable scenario to the more ambitious one more and more resource intensive (Fig. 9.4).

If Russia will heavily invest in the outlined emerging and prospective technologies, it can become a global supplier of high-value-added products, technologies and services. Advancing outlined promising S&T areas and putting into action the innovative "Global Breakthrough" scenario together will help to increase the Russian agricultural sector's global competitiveness and contribute to the following goals:

- Russian producers' entering prospective food markets and Russia's integration into the global agricultural production system
- Significant reduction of ready-made food product imports and increased added value generated by agricultural industries
- Ensuring food security, including efficiently meeting demand for high-quality, affordable foods in line with medically recommended consumption norms
- Creating new highly productive jobs in the A&F sector, increased employment and improved quality of life in rural areas
- Increasing investment appeal of the A&F sector
- Saving the country's hard currency reserves and promoting overall economic growth

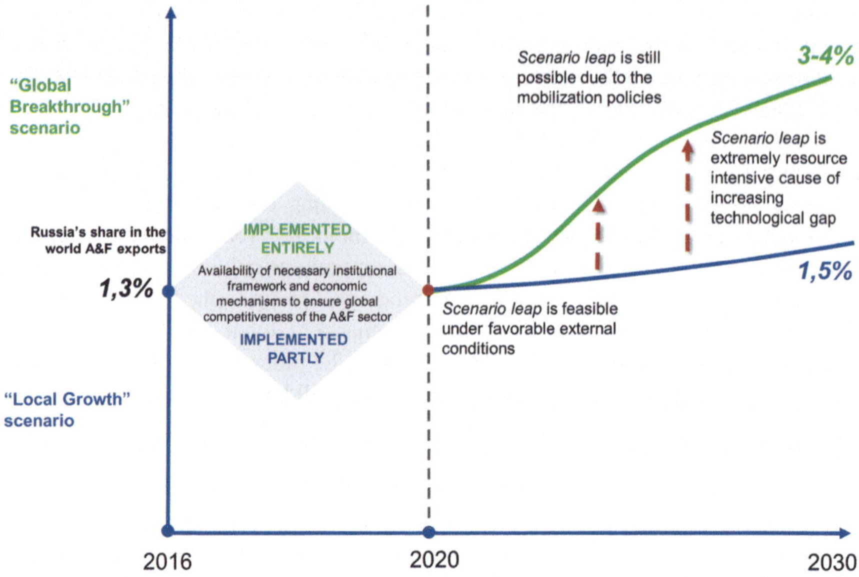

Fig. 9.4 STI development scenarios for the Russian A&F sector. Source: HSE

The development of the national technological potential depends on better institutional solutions for innovation activities and proper support mechanisms for technology transfer (Thurner and Zaichenko 2015). Research on various technology sectors that have an agricultural application would require intensification and support measures for knowledge dissemination through technology platforms and regional clusters. Especially emerging cross-sectoral technological solutions require a supportive environment for innovative companies, including access to venture capital funding. The Russian government targets the establishment of spin-offs and start-ups at leading technical and agricultural universities and research centres. Naturally, financial resources' allocation during this process should rely upon identified structure of prospective technological areas with stronger monitoring tools at the sectoral ministry.

9.5 Discussion and Conclusions

Developed big-data-augmented approach to technologies identification and mapping allows proposing the ontology of currently emerging technologies in global agriculture and food sector. The raw results of the emerging technologies identification show that the algorithms used (syntactic construction extraction) gain an appropriate phenomenon indeed: an expert validation of extracted lists of terms shows their relevance to the sphere of interest. The high results' accuracy is achieved through the comprehensive information source coverage and automatic analysis with objective criteria. Thereafter two additional parameters were evaluated for the list of identified technologies: (1) their potential to bring answer to global and national challenges and (2) future technological market forecasts. Based on such a characterization of technologies, prospective S&T development areas for the Russian A&F sector were highlighted.

For deepening the analysis of the emerging technologies' potential value for national economies, the current study can be continued with big-data-augmented approaches to an assessment of Russian comparative scientific and technological level in identified S&T areas. It can be done by an integration of expert evaluation procedures, bibliometry and estimation of complex parameters on the corpus of sectoral scientific papers and patents' full texts by Russian authors.

The big-data-augmented foresight exercise allows appraising the benefits of the proposed approach for evidence-based STI policy and corporate strategic planning. Although the productivity growth of analysts and experts' work due to technology identification on big-data sources cannot be measured directly, because this type of information processing, clearly, could not be performed manually altogether, the inner circle tools of mapping (aimed at preliminary filtering of data and at technology classification suggestions) help analysts, in our experience, to rise work productivity by 3–5 times. It takes analysts much less time to review hundreds instead of thousands of terms due to filtering, choose most important technologies and categorize them with the help of machine-generated suggestions on places of technologies in sectoral ontology. Furthermore, productivity is improved due to the "single window" effects of automatic attachment of expansions (of abbreviations), definitions and translations of technology-signifying terms. These options decrease the negative consequences of cognitive flow disruptions produced by switching between data sources in search of information, which is dominant in traditional analysts' working settings.

The outer circle mapping tools (aimed at helping to present the results easily and effectively), such as semantic map and trend map shown above, can be used by both the internal analysts of a company conducting a foresight study and the invited experts. These tools provide very useful information products for structuring expert interviews and panel discussions or for complex topic briefings. Therefore, there is a capacity for an automation of some aspects of the production of evidence on emerging technologies for supporting decision-making.

Complete mechanization of the foresight process is not a desired outcome, as the continuous dialogue among stakeholders and expert community is of vital importance for the sustainable governance. However, the demonstrated tools of big-data analysis give decision-makers the leverage in the negotiation with the sectoral groups, scientists and consultants. The latter will have to address practical questions about the technologies they advertise for government or corporate support based on objectively collected data on the parameters of these technologies in the current S&T discourse.

The demonstrated text-mining approach will not replace the domain experts as analysts in the foreseeable future but will transcend the foresight exercises from local ontology building to the high-level ontology interpretation. Many foresight studies have been restricted to collective mapping of trends in some domain in hope that the mapping is precise and full. If mapping is automated, resources are available for what foresight is really about: building an integrated and balanced vision of the future based on an intense interpersonal communication of domain experts.

Acknowledgements The book chapter was prepared within the framework of the Basic Research Program at the National Research University Higher School of Economics and supported within the framework of the subsidy by the Russian Academic Excellence Project '5-100'.

Annex: Prospective Technological Areas for the Innovative Scenario of Russian A&F Sector Development Between 2020 and 2030

Advanced precision agriculture technologies
– Big data and Internet of Things technologies for agriculture – New electronic technologies, wireless networks and microsensors – Advanced robotic technologies based on artificial intelligence, swarm intelligence and machine learning – Application of unmanned aerial vehicles in agriculture – Application of nano- and pico-satellites in agriculture – Fully automated operation of agricultural machinery – Technologies for deep processing of agricultural materials at the place of their harvesting (mobile semiautomated mini-plants)
Urban agriculture technologies
– Urban hothouse complexes including robotic hothouses, cassette planting of vegetable seedlings, etc. – Vertical farms and skyscraper farms – Home aeroponics and hydroponic installations for city apartments – Recirculation aquaculture and aquaponics technologies – Aeroponics technologies for food self-sufficiency of manned spacecraft, ships, submarines and other autonomous objects
LEISA technologies (organic agriculture, integrated pest control systems, water- and soil-saving agriculture, restoring fertility of degraded lands, sustainable fishing and fish farming)
– No-till farming and mulching technologies – Tractor-less plant growing technologies (agricultural bridges—stationary installations capable of working round the clock implementing a specified programme, without human interference) – Soil erosion prevention technologies, antierosion farming, soil crumbling and field smoothing technologies – Salination-preventing irrigation techniques – Techniques for dealing with desertification of agricultural lands and dangerous hardening of soil – Technologies for preventing nutrients' washing out from soil and water reservoirs' eutrophication – Substituting chemical-based weed killers (herbicides) with agrotechnical approaches (pre-emergence and post-emergence harrowing of crops, etc.) – Technologies for rehabilitation of soils damaged by excessive agricultural use, including depleted, salinated, desertified, hardened, polluted with toxic substances and radionuclides – Technologies for rehabilitation of lands used for dumping and disposal of household and industrial waste and rehabilitation of former industrial enterprises' grounds and territories

(continued)

– Nanostructured geotextile coatings for soil protection and country road building – Technologies to produce nanostructured protective coatings for seed material – Nanoremediation and nanobioremediation systems for soils and groundwaters – Technologies for preventing solid waste from garbage dumps and garbage filtrate getting on agricultural lands, into soil and groundwater used for agricultural production – Organic, biodynamic agriculture technologies, managing organic products' supply chains and automated remote control of complying with organic agriculture certification requirements – Land melioration technologies (improving hydrological regime, acidity parameters, soil fertility), with minimum interference into ecosystemic processes – "Blue" technologies (sustainable fishing and fish farming) – Technologies for minimizing by-catch, non-destructive trawling and other fishing techniques – Ships with zero emissions and discharge into the environment and low-level noise pollution – Technologies for biodegrading plastic-based garbage in oceans – Technologies for cleaning water reservoirs from non-organic and organic pollutants
Technologies for full on-the-spot utilization and recycling of agricultural, fishery and food industry waste, including production of new valuable chemical and pharmaceutical products
– Technologies for safe application, in environmental, sanitary and epidemiologic terms, of organic animal farming sewage for irrigation, in a partially closed water cycle – Technologies for on-the-spot production and compact, environmentally safe, and long-term storage of high-quality animal and poultry forage, without losing its nutritious properties – Low-waste and wasteless production technologies for animal farming, including biotechnologies to accelerate bioconservation and integrated recycling of animal farms' waste.
Smart bioenergy convergent technologies (local smart grids and biofuels from agricultural waste for rural settlements' energy self-sufficiency)
– Technologies for production of second-generation biofuel (bio-petrol and bio-diesel, high-molecular spirits from cellulose-containing plant growing and forestry waste) – Technologies for manure and dung processing in biogas systems – Technologies for disposal of food industry's primary processing, storage and production waste, to supply energy to agricultural companies and rural areas – Technologies for on-the-spot processing of agricultural and rural settlements' waste, to produce electricity, heat and construction materials – Efficient and environmentally friendly gas-generator plants to power agricultural machinery
System integration technologies for managing agricultural sector's logistics, based on supercomputers, big data and machine learning, robotic storage facilities and transport systems
– Decision-making support systems for the agricultural sector – Technologies for tracking supply chains and remote automated control of agricultural products' origins – Information technologies for agricultural insurance and trading – Technologies for automated product flow management and dynamic optimisation, real-time stock control and distribution – Highly accurate short- and medium-term weather forecasting techniques – Technologies for predicting and preventing anthropogenic disasters due to water-based economic activities
Technologies for production of next-generation personalized and functional foods
– Technologies to produce nanodispersed and nanostructured foods with new absorbency properties (high, selective, delayed) – Functional albumin pastes – Technologies to produce enriched dairy products with bifidogenic properties and immunostimulating effects – Technologies for making edible meat product packaging, including flavoured collagen films – Technologies to make composite materials using grain shell fibres

(continued)

– Technologies for 3D scanning of meat products for total screening to detect alien insertions and check their hygienic suitability – Technologies for long-term energy-efficient storage of food materials and products, with minimal loss of their valuable components, production of aseptic semi-finished products, hydrothermal treatment, preservation, microfiltration, shock freezing, low-temperature vacuum drying, cold processing, production of ready-to-eat foods in protective packaging with integrated biocide effect, technologies for long-term room-temperature storage of products traditionally considered as perishables
Technologies for production of synthetic foods
– Technologies for growing meat tissues in artificial mediums – Technologies for production of synthetic meat and synthetic eggs from protein materials, with taste indistinguishable from natural ones – Technologies for producing milk and dairy products based on yeast culture bioreactors – Technologies for production of forage and extraction of valuable proteins from food and forage waste – Technologies for directly synthesizing nutritious substances from chemical and mineral materials – Technologies for production of foods indistinguishable from traditional ones from new unconventional raw material sources (e.g. algae, insects, etc.) – Food printing technologies
Advanced technologies for the fisheries sector
– Technologies for complete processing of by-catch – Recirculation aquaculture and aquaponics technologies – Phytobiotechnologies to produce forage, fertilizers and biofuel from algae and microalgae – Phyto- and micro-biotechnologies for protection and cleaning of water and sewage pipes from biofouling – Technologies for genetic modification of microalgae to produce new bioproducts and pharmaceutical substances – Ships with zero emission and discharge into the environment and low-level noise pollution – Technologies for cleaning water reservoirs from non-organic and organic pollutants – Technologies for biodegrading plastic-based garbage in oceans

Source: HSE

References

Aarkrog A (2003) Input of anthropogenic radionuclides into the World Ocean. Deep-Sea Res II Top Stud Oceanogr 50(17):2597–2606

Aghion P, Griffith R (2008) Competition and growth: reconciling theory and evidence. MIT Press, Cambridge, MA

Bakhtin P, Saritas O (2016) Tech mining for emerging STI trends through dynamic term clustering and semantic analysis: the case of photonics. In: Daim TU, Chiavetta D, Porter AL, Saritas O (eds) Anticipating future innovation pathways through large data analysis. Springer International Publishing

Bakhtin P, Saritas O, Chulok A, Kuzminov I, Timofeev A (2017) Trend monitoring for linking science and strategy. Scientometrics 111(3):2059–2075

Bildosola I, Rio-Belver RM, Garechana G, Cilleruelo E (2017) TeknoRoadmap, an approach for depicting emerging technologies. Technol Forecast Soc Change 117:25–37

FAO (2009) The state of food insecurity in the world. FAO, Rome

Gokhberg L, Kuzminov I (2017) Technological future of the agriculture and food sector in Russia. In: The Global Innovation Index 2017: Innovation Feeding the World, Cornell University, INSEAD, WIPO. Geneva, New Delhi, pp 135–141. http://www.wipo.int/edocs/pubdocs/en/wipo_pub_gii_2017.pdf. Accessed 11 Jul 2017

Gokhberg L, Fursov K, Miles I, Perani G (2013) Developing and using indicators of emerging and enabling technologies. In: Gault F (ed) Handbook of innovation indicators and measurement. Elgar, Cheltenham

Gokhberg L, Kuzminov I, Chulok A, Thurner T (2017) The future of Russia's agriculture and food industry between global opportunities and technological restrictions. Int J Agric Sustain 15 (4):457–466

Halaweh M (2013) Emerging technology: what is it. J Technol Manag Innov 8(3):108–115

Hamza MA, Anderson WK (2005) Soil compaction in cropping systems: a review of the nature, causes and possible solutions. Soil Till Res 82(2):121–145

Hidalgo C (2015) Why information grows. The evolution of order, from atoms to economies. Basic Books, New York

Hoegh-Guldberg O, Bruno JF (2010) The impact of climate change on the world's marine ecosystems. Science 328(5985):1523–1528

HSE (2017) Science and technology foresight for agriculture and food sector in the Russian Federation until 2030. National Research University Higher School of Economics, Moscow

Huang Y, Zhang Y, Ma J, Porter A, Wang X (2015) Tracing technology evolution pathways by combining tech mining and patent citation analysis. Portland international conference on management of engineering & technology

Kallhauge AC, Sjöstedt G, Corell E (2005) Global challenges: furthering the multilateral process for sustainable development. Greenleaf Publishing, Sheffield

Keenan M, Cutler P, Marks J, Meylan R, Smith C, Koivisto E (2012) Orienting international science cooperation to meet global 'grand challenges. Sci Public Policy 39(2):166–177

Madani F (2015) 'Technology Mining' bibliometrics analysis: applying network analysis and cluster analysis. Scientometrics 105(1):323–335

Mani S (2004) Government, innovation and technology policy: an international comparative analysis. Int J Technol Global 1(1):29–44

Momeni A, Rost K (2016) Identification and monitoring of possible disruptive technologies by patent-development paths and topic modeling. Technol Forecast Soc Change 104:16–29

Montgomery DR (2007) Soil erosion and agricultural sustainability. Proc Natl Acad Sci 104 (33):13268–13272

O'Mara FP (2011) The significance of livestock as a contributor to global greenhouse gas emissions today and in the near future. Anim Feed Sci Technol 166:7–15

Omenn GS (2006) Grand challenges and great opportunities in science, technology, and public policy. Science-NY 314(5806):1696

Pimentel D (2006) Soil erosion: a food and environmental threat. Environ Dev Sustain 8 (1):119–137

Porter A, Cunningham S (2004) Tech mining: exploiting new technologies for competitive advantage. Wiley, Hoboken, NJ

Rotolo D, Hicks D, Martin BR (2015) What is an emerging technology? Res Policy 44 (10):1827–1843

Saritas O, Kuzminov I (2017) Global challenges and trends in agriculture: impacts on Russia and possible strategies for adaptation. Foresight 19(2):218–250

Smith K, Haux T (2017.) Evidence-based policy-making (EBPM)) In: Greve B (ed) Handbook of social policy evaluation. Edward Elgar, Cheltenham, pp 141–160

Socher R, Bauer J, Manning CD, Ng AY (2013) Parsing with compositional vector grammars. Proceedings of ACL 2013

Spitsberg I, Brahmandam S, Verti MJ, Coulston GW (2013) Technology landscape mapping: at the heart of open innovation. Res-Technol Manag 56(4):27–35

Srinivasan UT, Cheung WW, Watson R, Sumaila UR (2010) Food security implications of global marine catch losses due to overfishing. J Bioecon 12(3):183–200
Thurner T, Zaichenko S (2015) The feeding of the nine billion – a case for technology transfer in agriculture. Int J Innov Manag 19(2):1550026
Tubiello FN, Salvatore M, Rossi S, Ferrara A, Fitton N, Smith P (2013) The FAOSTAT database of greenhouse gas emissions from agriculture. Environ Res Lett 8(1):015009
United Nations, Department of Economic and Social Affairs, Population Division (2015) World population prospects: the 2015 revision, data booklet. ST/ESA/SER.A/377
Winkler J, Moser R (2016) Biases in future-oriented Delphi studies: a cognitive perspective. Technol Forecast Soc Change 105:63–76
World Bank (2007) World Development Report 2008: agriculture for development. World Bank, Washington

Leonid Gokhberg is First Vice-Rector of the HSE and also Director of HSE ISSEK. His area of expertise is statistics and indicators on STI as well as foresight and policy studies in this area. He has authored over 400 publications in Russian and international peer-reviewed journals, monographs, and university textbooks. Prof. Gokhberg has coordinated dozens of national and international projects funded by public agencies, businesses, and international organizations. He has served as a consultant of the OECD, Eurostat, UNESCO, and other international and national agencies. Leonid is also a member of the Global Innovation Index Advisory Board, the OECD Government Foresight Network, and OECD and Eurostat working groups and task forces on indicators for S&T and, as well as steering committees of various prestigious international and national initiatives. Prof. Gokhberg is Editor-in-Chief of the Scopus-indexed scientific journal Foresight and STI Governance and editor of the Springer academic book series Science, Technology, and Innovation Studies, and participates on the editorial boards of several other influential journals. He holds PhD and Dr. of Sc. degrees in Economics.

Ilya Kuzminov is Head of the Information and Analytical Systems Unit at HSE ISSEK. He is also Senior Research Fellow at the Research Laboratory for Science and Technology Studies, and the HSE International Research and Educational Foresight Center. Dr. Kuzminov is responsible for the coordination of research activities in the fields of future-oriented studies of environmental technologies, sustainable development, mining, agriculture, and forestry sectors. His scientific interests include theory, methodology, and practices of research in global challenges and grand responses in STI, priority setting, scenario designing, roadmapping, foresight evaluation and implementation into policy making, and environmental management, as well as big data and machine learning approaches to STI data analysis (including text mining, semantic analysis, and knowledge discovery). Dr. Kuzminov holds a PhD in Economic and Social Geography.

Pavel Bakhtin is Research Fellow at the Information and Analytical Systems Unit, Institute for Statistical Studies and Economics of Knowledge, National Research University Higher School of Economics (HSE). He is an expert in the field of data mining, text mining and machine learning for S&T foresight. Pavel plays a key role in the development of the HSE based big data intelligent foresight analytics system iFORA. Mr. Bakhtin owns software patents and over a dozen publications about text data mining for strategic planning and foresight, including publications in high impact international journals. He holds a technical background in the sphere of business informatics, master of management in the area of STI governance.

Anton Timofeev is Laboratory Assistant at the HSE ISSEK Information and Analytical Systems Unit. He is a software architect and specialist in the field of natural language processing, data visualization, knowledge management systems and semantic analysis. He leads the development and design of architecture of big data analytics systems iFORA. Mr. Timofeev participates in the development of software solutions and technical support for Foresight methodology, sectoral strategic analytics, horizon scanning, technology mining and technology landscape mapping. As a developer he was involved in a number of foresight and S&T monitoring activities. Mr. Timofeev holds a software license in the field of text mining and contributed to several publications.

Elena Khabirova is Expert at HSE ISSEK Information and Analytical Systems Unit. Ms. Khabirova graduated from HSE with a Master's degree in Sociology. She participates in sectoral research projects with the application of big data foresight analytics system iFORA particularly in the areas of agriculture and environment.

Technology Assessment for Container Closure Integrity Testing Technology for Biotech Industry

10

Qin Guo, Michael Clark, Mitsutaka Shirasaki, and Tugrul Daim

10.1 Background

The majority of patients place a high level of trust that their healthcare providers are administering drug products that have met US Food and Drug Administration (FDA) product safety standards (Healtfacts 1997). Patients, healthcare providers, and regulatory agencies (FDA, et al.) have an interest in ensuring a high level of patient safety. Pharmaceutical companies meet this expectation by following FDA guidelines for drug product sterility (US Department of Health and Human Services 2004). In 2008, the FDA issued a guideline that sterility testing is required but that container closure integrity (CCI) testing is preferred (US Department of Health and Human Services 2008). CCI testing's advantages are conservation of samples, faster results, and sensitivity that can pinpoint leaks (Brasten et al. 2014).

CCI is the ability of a container closure system to maintain the sterility and quality of drug products throughout their shelf life (Ewan et al. 2015). The FDA defines a container closure system as "the sum of packaging components that together contain and protect the dosage form. This includes primary packaging components and secondary packaging components, if the latter are intended to provide additional protection to the drug product. A packaging system is equivalent to a container closure system" (US Department of Health and Human Services 1999). Typical container closure systems include IV bag, ampoule, serum vial, cartridge, and syringes (White 2012).

Q. Guo · M. Clark · T. Daim (✉)
Portland State University, Portland, OR, USA
e-mail: guoqin@pdx.edu; mrclark@pdx.edu; tugrul.u.daim@pdx.edu

M. Shirasaki
Genentech, Beaverton, OR, USA
e-mail: mshira2@pdx.edu

D. Meissner et al. (eds.), *Emerging Technologies for Economic Development*, Science, Technology and Innovation Studies,
https://doi.org/10.1007/978-3-030-04370-4_10

Based on a review of the literature, there is not a preferred CCI testing technology within the pharmaceutical industry (Brasten et al. 2014; Ewan et al. 2015; White 2012). The primary two low-tech options are dye ingress testing and manual visual inspection. Dye ingress testing involves examining a sample base by a quality control technician visually inspecting for dye contamination of the product, a time-consuming process that destroys the samples reviewed. Manual visual inspection requires trained inspectors visually examining all of the products, a time-consuming process that can be prone to human error. The four high-tech options are CCD camera visual inspection, high-voltage leak detection, vacuum decay/pressure lead detection, and laser headspace analysis. The industry expert panel advised assessing high-voltage leak detection, vacuum decay/pressure lead detection, and laser headspace analysis. Automated solutions are preferred due to time, quality, and cost concerns. Additional information about these options is discussed in the technology candidates' section.

10.2 Technology Candidate

10.2.1 Dye Ingress Testing

The dye ingress leak testing is probably the most common leak test method used in pharmaceutical/biotech industry to ensure the container closure integrity is intact in various container/package types. In the simplest terms, the methodology works in three phases: the container is submerged in the bath of dye solution (typically methylene blue), and vacuum is drawn; the container is placed in ambient atmospheric pressure; a technician inspects the ingress of blue dye into the packaging. In this study, only tubing vials are considered as the target for primary container type.

Variables

- Vacuum level and dwell time
- Atmospheric pressure and dwell time
- Concentration of dye surfactant
- Inspector capability

Limitations

- This methodology can only be applied as sampling method.
- Sensitivity of the testing is depended on the differential between the vacuum versus the atmospheric pressure times the dwell time.
- Accuracy of the testing is depended on the inspector capability to detect the dye within the container.
- Some leak within the seal of the vial is not visible to the inspector.

As this methodology can only be applied as sampling based, this technology is excluded from further considerations.

10.2.2 Helium Leak

The helium leak testing is said to be the most accurate method of detecting leaks in any container types. This method uses a vacuum to draw out the helium inside the container and uses a mass spectrometer to detect the rate of helium leaking from the container. Because it only detects the helium molecules, it is not affected by the other gas molecules in the air.

Variables

- Vacuum level and dwell time

Limitations

- This method needs vials to be prefilled with helium gas.
- It needs certain amount of headspace.

As this methodology can only be applied to vials prefilled with helium in the headspace, the methodology is not generic enough to be considered in this study. This technology is excluded from further considerations.

10.2.3 Manual Visual Inspection

The manual visual inspection methodology is the most common 100% inspection method in the pharmaceutical/biotech industries. This methodology provides flexibility to inspect for all type of visually detectable defects in the container such as particulates in solution, bruises, scratches, and dented seals including container closure defects such as cracks, deformed or split stopper, un-crimped seal, etc. An inspector takes one or more vials (depending on the vial sizes) and places it in front of black and white background under preset lighting condition. Entire circumference of the vial is visually inspected including the solution.

Variables

- Inspector capability
- Lighting condition
- Container sizes and types

Limitations

- This method is highly depended on the inspector capability for accuracy and repeatability.
- Lack of focus from inspector due to all other defects being inspected at the same time.
- Typical manual visual inspection takes 20 seconds per vial depending on the vial sizes.
- This methodology is a highly labor-intensive work, and inspectors need to take a lot of "micro breaks" to keep effectiveness of inspection.
- Defect must be "visible" to the inspector. For example, a crack under seal of vial cannot be detected by the inspector unless it is grossly leaking.

This methodology is often used as the benchmarking method for high-tech alternatives such as high-voltage leak detection or vacuum decay methods introduced below. This method is typically used in conjunction with CCI testing as CCI testing alone does not offer inspection of all the defect types. Therefore, this technology is excluded from further evaluations.

10.2.4 CCD Camera Visual Inspection

The charge-coupled device (CCD) camera visual inspection is most commonly used in fully automated visual inspection machines. Several CCD camera stations are used to inspect full circumference of the container for various defect types including container closure integrity defect, such as cracks, un-crimped seal, deformed stopper, etc. The container is placed in spin stations and presented in front of inspection stations with different lighting techniques for detection of visually detectable defects. The CCD camera technology is versatile in detecting different types of defects as they appear in the images taken by the CCD cameras, and proprietary image processing software is programmed to detect various types of defects. The technology has evolved in speed and complexity in logic in recent years, and most defects that can be detected by manual visual inspection now can be detected by automated visual inspection.

Variables

- Size and type of container
- Lighting condition

Limitations

- The detection capability is highly dependent on the size of defect.

- The defect must be "visible" to the vision system. For example, crack under the seal of the vial will not be "visible" unless it is grossly leaking.

This technology is typically used in conjunction with CCI testing as CCI testing alone does not offer inspection of all the defect types. Therefore, this technology is excluded from further evaluations. This technology can be used in place of or in conjunction with the manual visual inspection.

10.2.5 High-Voltage Leak Detection

High-voltage leak detection (HVLD) is a nondestructive test method that applies high-voltage potential to liquid-filled non-conductive container (glass vial in this study). This methodology has been successfully validated and in use since the early 1970s in the pharmaceutical industry for detection of pinhole in glass ampoules for fully automated 100% inspection. The high-voltage potential is applied at multiple locations of the container covering the entire circumference. If there is a crack or a pinhole, electrical current is conducted through the liquid, and potential difference can be detected compared to the "good" vial. The method is fast and highly accurate and requires no sample preparation. This method is generally safe in all types of liquid product. However, there is a reported case of ozone generation in the headspace of liquid container affecting the efficacy of the protein-based liquid drug product.

Variables

- Conductivity of liquid solution.
- Fill volume (must be able to cover the entire length of vial when placed side way).

Limitations

- This method can only be used for liquid product with conductivity of $\sigma \geq 1$ μS/cm.
- Depending on the headspace content, there is a risk of ozone generation within the headspace of liquid-filled container.

10.2.6 Vacuum Decay or Pressure Decay Leak Detection

The vacuum decay or pressure decay leak detection system is a nondestructive test that can be used on liquid or solid product in rigid or flexible and conductive or non-conductive containers. The container is placed within the chamber that isolates the container from atmospheric pressure. Depending on the system, either vacuum or high pressure is applied to the container, and pressure change within the chamber is measured. Although the measurement time is short, the sensitivity of this measurement is depended on the dwell time and the amount of headspace present within the container. The vacuum decay is generally preferred over pressure decay because the vacuum decay technology can also offer measurement of moisture content drawn into the chamber.

Variables

- Amount and level of pressure of headspace

Limitations

- This method requires individual chamber for different container sizes and shapes.
- The accuracy of the measurement is depended on the dwell time. So the speed of the equipment is limited based on the accuracy required.
- The pressure change must be measured after vacuum or pressure level is established within the chamber. Therefore, missing a large part of container may not be detected as there is no change in the pressure.

The principle of vacuum and pressure decay technologies is the same except for the positive or negative pressure applied within the chamber. Therefore, only vacuum decay technology is considered within this study.

10.2.7 Laser Headspace Analysis

The laser headspace analysis is a nondestructive test method which uses frequency modulation spectroscopy. This method can be used on liquid or solid product in transparent container. Although the technology can be used for liquid-filled container, it requires a clear path for the frequency-modulated light to pass through the container and the headspace. For this reason, the liquid product is generally avoided for the application of this technology. The frequency of the laser can be modulated for different types of molecules of interest such as O_2 or water contents. The laser passes through the headspace, and the amount of light absorbed by the headspace is proportionate to the concentration of molecules in interest. This method requires a wait period once the container is closed depending on the level of existing vacuum in the headspace. The atmospheric air and moisture must ingress to the headspace for it to be detected.

Variables

- Amount and contents of headspace
- Vacuum level of headspace

Limitations

- The contents of the headspace must be different from the atmospheric air.
- The liquid product may produce droplets on the wall of the container that interferes with the light.
- Depending on the level of vacuum in the headspace, a longer wait period may be required for the accuracy desired.

10.3 GAP Analysis

Prior to select the best alternative for the current technologies, there is a need to identify the current situation and the desired outcome. Gap analysis is a project management tool that helps identify the gap between the current and future state, as well as the tasks that need to be done to bridge the gap (Mind Tools Ltd 2015). This project management tool is useful in the initial stages of a project. The processes of conducting a gap analysis are as follows (Mind Tools Ltd 2015):

- State descriptions.

  A gap analysis starts with analyzing the current state of the technology and then identifying the objectives that need to be achieved. The analysis can be quantitative, qualitative, or both. The key thing is to be specific and factual with an emphasis on identifying weaknesses for the current situation and improvements for the future state.
- Bridge the gap.

  This can be done through identifying whether a gap exists between the current and future state and what is needed to bridge the gap.

In this project, the TOP (technical, organizational, and personal) methodology is adopted to help understand the current and future states of container closure integrity testing technologies. This methodology breaks down the overall system and performs a detailed analysis from three perspectives: technical, organizational, and personal (Linstone 1999). Technical perspective considers rational factors. Personal perspective considers factors that emerge from decision-makers' or any stakeholders' own motivations. Organizational perspective is a collection of all personal perspectives and can also be similar to inter- or intraorganizational politics.

As shown in Table 10.1, the gap analysis is conducted to identify the gap between four candidates: manual visual inspection, high-voltage leak detection (HLVD),

Table 10.1 Gap analysis results

Gap analysis	Requirement	Capabilities	Gap
Technical	• 100% product inspection • Reliability • Accuracy • Ease of use • Fast • Low maintenance • Nonlabor intensive	Low tech • High reliability • Slow • No maintenance	Low tech • Need to improve accuracy • Need to reduce labor
		High tech • Fully automated • High-moderate accuracy • Easy to use • Fast • Nonlabor intensive	High tech • Need to minimize maintenance
Organizational	• 100% product inspection • Patient safety • Accuracy • Reliability • Efficacy • Minimized cost • Maintenance • Labor • Production capacity	Low tech • Low initial cost • Low to moderate accuracy • Low production speed • High labor cost	Low tech • Need to improve patient safety level • Need to improve inspector capability • Need to increase production speed
		High tech • High initial cost/low maintenance cost • High accuracy • High production speed	High tech • Need to minimize capital investment costs • Need to reduce maintenance costs
Personal	• Patient/ personal safety • Efficacy • Reliability • Easy to operate • Easy to maintain • Easy to qualify	Low tech • Low to moderate accuracy • Low initial cost	Low tech • Need to reduce manual labor processes • Need to increase degree of confidence for patient safety
		High tech • Easy to operate • Easy to maintain • Difficult to qualify	High tech • Need for ongoing training for engineers

Source: Authors

vacuum decay/pressure leak detection, and laser headspace analysis. Capabilities are the current situation for the four candidates of CCI testing technology. Requirements refer to the future state of the four candidates. The gaps are identified after defining the current situations and the future states. For example, in the technical aspect, the speed of the current technology, which is the manual visual inspection, is drastically slow. Meanwhile, the accuracy of the results varies from people to people. Thus, in the future situation, the speed of inspection should be raised, and the accuracy needs to be improved.

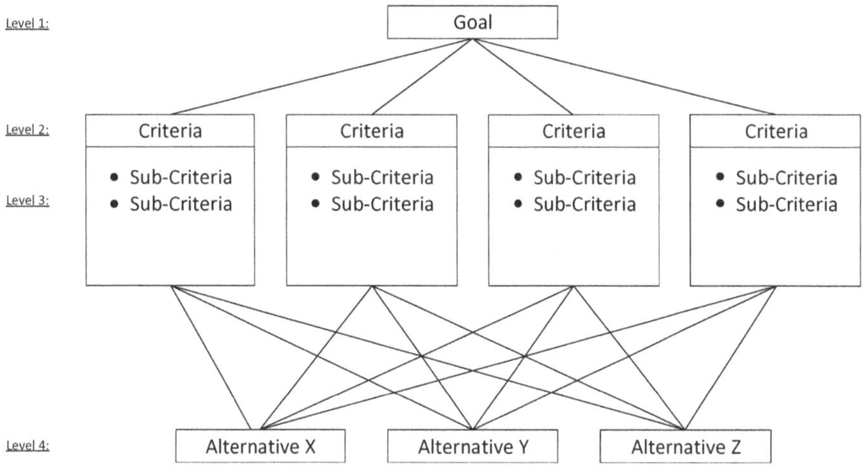

Fig. 10.1 HDM structure. Source: Authors

10.4 Hierarchical Decision Model Background/Methodology

In this project, a hierarchical decision model (HDM) is applied to evaluate the CCI testing technologies for biotech industry. The HDM is an effective tool for dealing with complex decision-making. By decomposing multi-criteria decisions to a series of pairwise comparisons, and then synthesizing the results, the HDM helps to capture both subjective and objective aspects of a decision (Lee et al. 2010). Additionally, the HDM incorporated a useful technique for measuring the consistency of the decision-maker's evaluations, thus reducing the bias in the decision-making process.

The HDM consists of a goal, a set of evaluation criteria and sub-criteria, and multiple alternatives among which the best decision is to be made (Turan et al. 2009). The HDM develops a set of weights for the evaluation criteria regarding to the decision-maker's pairwise comparisons. The higher the weight, the more important the corresponding criterion (Triantaphyllou and Mann 1995). Then, a score is assigned to each alternative, which is usually done by experts. The higher the score, the better the performance of the alternative regarding to the considered criterion. The third step is combining the weights and the scores to calculate an overall score for each alternative, and the one with the highest score is the best decision.

The HDM is typically structured into a hierarchal tree. The goal of the decision is represented at the top level of the hierarchal tree. The next level is the criteria and all possible sub-criteria. All the decision alternatives are placed at the lowest level of the hierarchal tree. Figure 10.1 below is a general representation of the HDM.

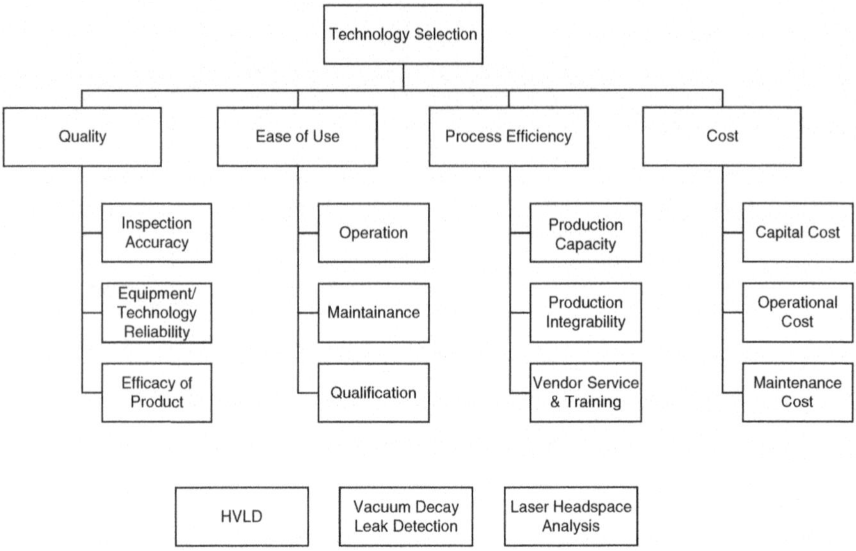

Fig. 10.2 HDM model used. Source: Authors

10.4.1 HDM Model

Based on the literature review and discussions with our expert panel, there is not a preferred CCI testing technology within the pharmaceutical industry. The two low-tech manual visual inspection technologies (dye ingress testing and manual visual inspection) are not preferred options due to their need to improve accuracy and production speeds while decreasing labor-intensive processes. Therefore, an HDM model was created to evaluate the three high-tech automated inspection technologies (high-voltage leak detection, vacuum decay/pressure leak detection, and laser headspace analysis).

The HDM model (Fig. 10.2) criteria and sub-criteria were identified from the literature and validated by experts. An online version of the model was sent to experts to complete the criteria pairwise comparison process. HDM Software was used for the analyses (Portland State University, Engineering & Technology Management Department (2015).

Quality

- Inspection accuracy
- Equipment/technology reliability
- Efficacy of product

From the technical, organizational, and personal perspectives, the technology should ensure an inspection process that yields quality drug products. One hundred percent of drug products should be inspected with a high degree of accuracy. The equipment/technology should perform the inspections with a high degree of reliability. As a result, patients can be assured of the efficacy of the drug products.

Ease of Use

- Operation
- Maintenance
- Qualification

From the technical and personal perspectives, the technology should be easy to operate, maintain, and qualify. The learning curve and ongoing training time should be minimal for operators, maintenance staff, and engineers.

Process Efficiency

- Production capacity
- Production integrability
- Vendor service and training

From the technical and organizational perspectives, the technology should improve the efficiency of the drug product inspection process. Faster inspection results due to automated processes should expand production capacity. The technology should integrate seamlessly with existing production processes and potentially improve workflows. The technology vendor should provide reliable, efficient service and training that avert any process slowdowns and inefficiencies.

Cost

- Capital
- Operational
- Maintenance

From the organizational perspective (pharmaceutical company), the acquisition and ongoing cost of a technology should be minimized without materially impacting other criteria. Costs include the initial capital investment (equipment, implementation, and training), operational (equipment and labor), and maintenance (equipment service and repair).

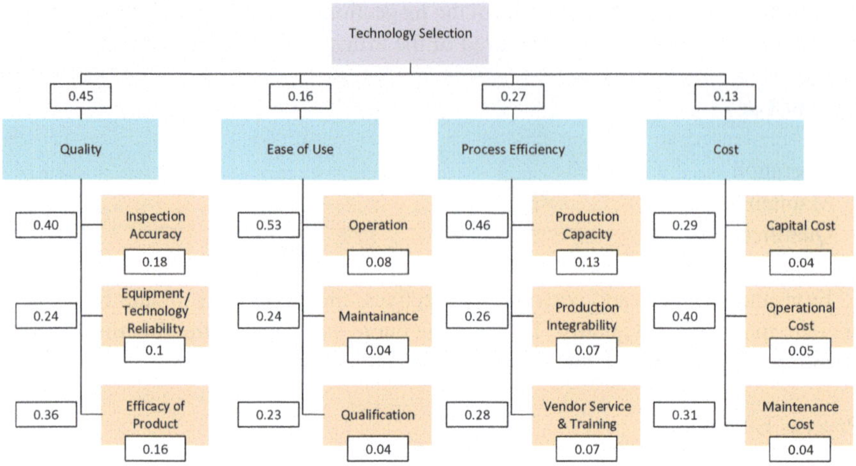

Fig. 10.3 Expert feedback results. Source: Authors

10.5 Result/Analysis

10.5.1 Expert Feedback

The result of pairwise comparison by the subject matter experts is shown below in HDM model. Sixteen experts provided feedback, including 13 experts in the USA, 2 in Canada, and 1 in Ireland with average experience of 18 years (Figs. 10.3 and 10.4).

There is a clear preference in "Quality" criteria having weight of 0.45, followed by "Process Efficiency" with weight of 0.27. The "Ease of Use" and "Cost" weighted the least with 0.16 and 0.13, respectively. This result shows the "Cost" does not have significant impact in the decision-making in the selection of CCI testing technology.

The disagreement among the expert was quite small with score of 0.05 as seen in the Appendix.

10.5.2 Final Scores

The final scores for technology selection were made based partially on user requirement specifications (URS) which had 162 line items. The URS was specific to the application for fully automated CCI testing equipment for liquid-filled glass vials. Each line item was categorized in the sub-criteria listed above and scored either "met" or "not met" and score "1" or "0." The sub-criteria of "Equipment/technology reliability," "Ease of use for operation," "Ease of use for maintenance," and "Production capacity" were scored solely on the URS line items "met" or "not met" scores. The other sub-criteria were scored based on our expert understanding of the technology and the equipment (Table 10.2).

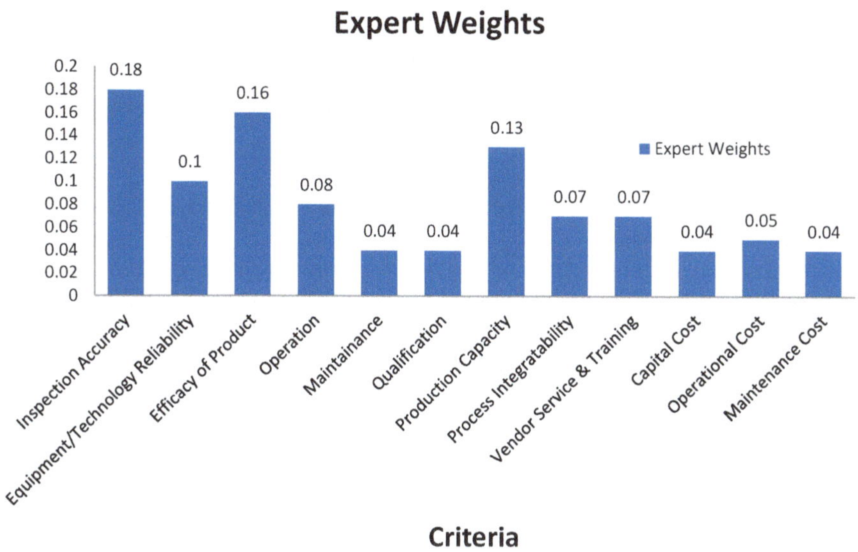

Fig. 10.4 Sub-criteria weights. Source: Authors

Table 10.2 Final score of each alternative

Sub-criteria	Weight	Alternatives		
		HVLD	Vacuum decay	Laser headspace analysis
Inspection accuracy	0.18	100	90	80
Equipment/technology reliability	0.1	100	100	100
Efficacy of product	0.16	80	100	95
Operation	0.08	100	100	100
Maintenance	0.04	100	100	100
Qualification	0.04	100	100	100
Production capacity	0.13	100	100	100
Process integrability	0.07	100	100	50
Vendor service and training	0.07	100	80	80
Capital cost	0.04	100	90	90
Operational cost	0.05	100	100	80
Maintenance cost	0.04	100	90	90
		96.8	96.0	88.9

Source: HSE

The research model presented is an initial technology assessment tool for the pharmaceutical industry. The model could be improved in the following areas:

- Safety should be added as a criterion distinct from quality/efficacy of product (patient safety), quality/inspection accuracy (product safety), and ease of use/operational (operational safety).
- The three technology alternatives could be combined with CCD camera visual inspection technology.
- Future research could forecast what new CCI technologies may develop.
- The expert panel could be expanded to represent the views of quality control experts.
- The experts' judgments were based on their experiences using the technology alternatives. Only 58% had experience with laser headspace analysis, a newer technology.
- Although cost was a low-weighted criterion, vendor cost estimates for each of the technology alternatives were not available.

10.6 Discussion and Conclusion

High-voltage leak detection equipment scored the highest of 96.8. HVLD scored slightly higher than the vacuum decay technology based on inspection accuracy, vendor service and training, capital cost, and maintenance cost. For the inspection accuracy, the HVLD can detect pinholes down to 2 μm pinholes. The HVLD technology has been in use in US pharmaceutical/biotech industry since the 1970s. The equipment vendor has a strong US presence and has better service and training. The HVLD equipment generally costs less than the other alternatives because the principle of the technology is simpler and has less moving parts. The HVLD equipment costs less to maintain for the same reason. On the other hand, the HVLD scored less for the efficacy of product as there is a reported issue where high potential voltage applied to the ambient air-filled headspace created ozone within the headspace, and in turn, the ozone oxidized the protein within the liquid solution degrading the active drug product. This issue can be avoided by placing the high-voltage prong on the bottom side where there is no air. The disadvantage of the HVLD is that it can only be used for liquid-filled container.

Vacuum decay leak detection equipment scored the close second highest of 96.0. The vacuum decay technology has a disadvantage in the inspection accuracy as the protein can clog the small pinholes or crack. The vacuum decay has no impact to the quality of the product and has an advantage over the HVLD. However, the vacuum decay technology has disadvantage under cost as the equipment is much more complicated than the HVLD alternative with individual chamber for different size containers, a need to create vacuum within that chamber and measure pressure change and moisture content. The equipment vendor is based in Switzerland, and the technology is relatively new in the industry. The company does not yet provide strong US local customer support with quick response and knowledgeable service engineer. The advantage of the vacuum decay is that the technology can offer a solution to both liquid-filled and lyophilized products. In this case, it did not translate to a higher score as the score is based solely on the liquid-filled product.

Laser headspace analysis scored the third highest based on disadvantage in the inspection accuracy, impact to the efficacy of product, process integrability, and vendor service and training. The inspection accuracy for the laser headspace analysis is impacted by the potential blockage by the droplet on the wall, which absorbs the laser and can create inaccurate result. This condition can also cause potential impact to the efficacy of the product. Although the laser used is low-energy laser, the laser has potential to alter the protein structure if the liquid solution is in the path of the laser. This technology has disadvantage in the process integrability as well as due to the wait time required after filling is completed. As the ambient air must permeate through the crack into the headspace of the container, depending on the size of crack that is desired to be detected, a longer wait period is required. This technology also does not work for the ambient air-filled vials as there is no change in the contents of the headspace. The technology is the newest of the three alternatives, and as with the vacuum decay technology, the representation in the USA is still weak and scores low on vendor service and training.

This chapter demonstrated the use of hierarchical decision model and multiple perspectives to assess technologies in the biotechnology sector through a case container closures. The model's strength is its ability to consider all perspectives that may impact the decision. The weakness of the approach is its reliance on experts and can be overcome by carefully selection of the right experts.

Appendix

Selection	Inspection Accuracy	Equipment/Technology Reliability	Efficacy of Product	Operation	Maintainance	Qualification	Production Capacity	Process Integrability	Vendor Service & Training	Capital Cost	Operational Cost	Maintenance Cost	Inconsistency	
ER	0.22	0.16	0.06	0.04	0.03	0.03	0.09	0.08	0.23	0.01	0.01	0.04	0.03	
ED	0.19	0.06	0.19	0.11	0.05	0.02	0.11	0.12	0.06	0.02	0.06	0.02	0.01	
GV	0.16	0.16	0.05	0.05	0.02	0.1	0.09	0.24	0.04	0.02	0.05	0.01	0.03	
JL	0.11	0.11	0.21	0.09	0.05	0.06	0.07	0.07	0.13	0.01	0.04	0.04	0.02	
JN	0.13	0.07	0.16	0.02	0.02	0.01	0.25	0.14	0.12	0.01	0.05	0.02	0.04	
JC	0.16	0.11	0.27	0.06	0.02	0.03	0.1	0.07	0.03	0.06	0.06	0.03	0	
JO	0.36	0.12	0.04	0.1	0.04	0.02	0.16	0.03	0.06	0.01	0.04	0.02	0.03	
LB	0.04	0	0.26	0.06	0.02	0.01	0.36	0.05	0.07	0.07	0.02	0.03	0.04	
MK	0.32	0.05	0.03	0.09	0.02	0.01	0.01	0.03	0.05	0.08	0.18	0.12	0.01	
MS	0.29	0.13	0.29	0.04	0.01	0.01	0.08	0.04	0.05	0.01	0.02	0.02	0.01	
PM	0.15	0.13	0.11	0.08	0.04	0.09	0.09	0.07	0.05	0.05	0.08	0.06	0.01	
RD	0.31	0.13	0.04	0.08	0.02	0.01	0.15	0.03	0.03	0.14	0.01	0.04	0.09	
RM	0.08	0.01	0.44	0.12	0.07	0.13	0.09	0.02	0.01	0	0.01	0.01	0.05	
RI	0.12	0.15	0.09	0.1	0.1	0.03	0.12	0.06	0.12	0.02	0.04	0.03	0.01	
RM	0.13	0.13	0.13	0.12	0.04	0.04	0.09	0.03	0.09	0.07	0.07	0.07	0	
TD	0.13	0.11	0.2	0.09	0.04	0.05	0.14	0.05	0.05	0.07	0.03	0.03	0.01	
Mean	**0.18**	**0.1**	**0.16**	**0.08**	**0.04**	**0.04**	**0.13**	**0.07**	**0.07**	**0.04**	**0.05**	**0.04**		1
Minimum	**0.04**	**0.01**	**0.03**	**0.02**	**0.01**	**0.01**	**0.01**	**0.02**	**0.01**	**0.02**	**0.01**	**0.01**		
Maximum	**0.36**	**0.16**	**0.44**	**0.12**	**0.1**	**0.13**	**0.36**	**0.24**	**0.23**	**0.14**	**0.18**	**0.12**		
Std. Deviation	**0.09**	**0.05**	**0.11**	**0.03**	**0.02**	**0.04**	**0.08**	**0.05**	**0.05**	**0.04**	**0.04**	**0.03**		
Disagreement													**0.05**	

References

Brasten L, Jacobs B, Brydzinski A (2014) Container closure integrity testing. Controlled Environment. p 18, 24 March 2014

Ewan S, Jian M, Stevenson C, Henderson O, Klohr S, Wessel M, Mehta P, Adler S, Lake C, Walsh J, Polomene T, Everaert V (2015) White Paper: Container closure integrity control versus integrity testing during routine manufacturing. PDA J Pharmaceut Sci Technol 69:461–465

Healtfacts (1997) Public trusts FDA. HealthFacts 22(6):2

Lee HB, Cho WN, Hong YS (2010) A hierarchical end-of-life decision model for determining the economic levels of remanufacturing and disassembly under environmental regulations. J Clean Prod 18(13):1276–1283

Linstone HA (1999) Decision making for technology executives: using multiple perspectives to improved performance, 2nd edn. Artech House, Norwood, MA, pp 560–570

Mind Tools Ltd (2015) Mind tools. Mind Tools Ltd, 1996–2015. https://www.mindtools.com/pages/article/gap-analysis.htm. Accessed 01 Dec 2015

Portland State University, Engineering & Technology Management Department (2015) HDM Hierarchical Decision Model. http://research1.etm.pdx.edu/hdm2/. Accessed 02 Dec 2015

Triantaphyllou E, Mann SH (1995) Using the analytical hierarchy process for decision making in engineering applications: some challenges. Int J Indus Eng Appl Pract 2(1):35–44

Turan T, Amer M, Tibbot P, Almasri M, Al Fayez F, Graham S (2009) Use of Hierarchal Decision Modeling (HDM) for selection of Graduate School for Master of Science Degree Program in Engineering in PICMET. Portland

U.S. Department of Health and Human Services (1999) Guidance for industry – container closure systems for packaging human drugs and biologics. Chemistry, Manufacturing, and Controls Documentation, May 1999

U.S. Department of Health and Human Services (2004) Guidance for industry. Sterile drug products produced by aseptic processing – current good manufacturing practice, September 2004

U.S. Department of Health and Human Services (2008) Guidance for industry. Container and closure system integrity testing in lieu of sterility testing as a component of the stability protocol for sterile products, February 2008

White EK (2012) The aseptic core: container-closure integrity. J Valid Technol Spring:10–16

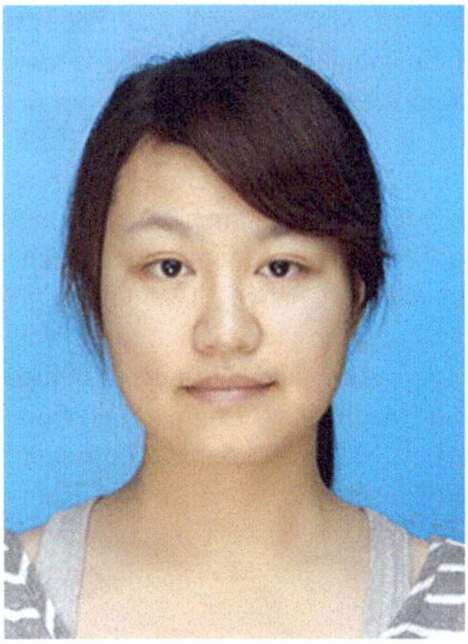

Qin Guo received Master of Science degree in Engineering and Technology Management at the Department of Engineering and Technology Management of Portland State University. She also has a Bachelor's Degree in Telecommunications Engineering from Shanghai University in China. Qin had worked at China Mobile as a network supervision coordinator. She is also a member of Omega Rho International Honor Society of Operations Research and Management Science.

Michael Clark is Adjunct Professor at the College of Engineering, Technology and Management at Oregon Institute of Technology. He is pursuing a PhD at the Department of Engineering and Technology Management at Portland State University. Mr. Clark has over 15 years of experience with research and grant administration in government and academic organizations. He received his BS in Management Information Systems: Management Accounting from Oregon Institute of Technology, MIB in International Business from Pepperdine University in California, MA in Business Administration from Instituto Tecnológico y de Estudios Superiores de Monterrey in Mexico, and MLS in Library Science from Southern Connecticut State University.

Mitsutaka Shirasaki is an inspection vision systems engineer at Genentech, a leading biotech company that discovers, develops, manufactures and commercializes medicines to treat patients with serious or life-threatening medical conditions. Previously he had worked at Eisai Machinery as a technical services director. Mr. Shirasaki recently completed his Master of Science degree in Engineering and Technology Management at the Department of Engineering and Technology Management of Portland State University. He also has Bachelor of Science Degree in Engineering in Electrical Emphasis and Mathematics from Texas Christian University. Mr. Shirasaki is also a member of Omega Rho International Honor Society of Operations Research and Management Science.

Tugrul Daim is Professor and Director of the Technology Management Doctoral Program in the Maseeh College of Engineering and Computer Science at Portland State University. Prof. Daim is also Director of the Research Group on Infrastructure and Technology Management and Faculty Fellow at the Institute for Sustainable Solutions. He is also a Leading Research Fellow at the Higher School of Economics in Moscow Russia. He is the Editor-in-Chief of IEEE Transactions on Engineering Management. He has been at various editorial roles in journals including International Journal of Innovation and Technology Management, Technological Forecasting and Social Change, Technology in Society, Foresight, Journal of Knowledge Economy and International Journal of Innovation and Entrepreneurship. Tugrul received his BS in Mechanical Engineering from Bogazici University in Turkey, MS in Mechanical Engineering from Lehigh University in Pennsylvania, MS in Engineering Management from Portland State University, and PhD in Systems Science: Engineering Management from Portland State University.

Emerging Technologies, Trends and Wild Cards in Human Enhancement

11

Ozcan Saritas

11.1 Background

The future role and place of human in the world is a widely debated issue. New technologies are developed continuously and are presented to markets momentarily. The key question is whether human will be at the centre of change as it has historically been or will be put aside by smarter machines. An ongoing concern on how the middle class is hollowed by technology indicates the global socio-economic extent of the issue (Saritas 2015). It has been already over 10 years since Kurzweil mentioned 'singularity' as a fateful moment when our technology becomes smarter than us and able to learn faster than we can (Kurzweil 2006). Machines then will become the principal creator of new technologies and race far ahead of humanity. As a result, humans may effectively fall out of the loop and can be eliminated. For instance, new software has already been developed, which is capable of collecting data about anything like statistics, financial reports, sports and so on and turns it into articles like this one (AP 2016).

While the machines are advancing, emerging technologies also offer new possibilities for human to remain competitive against machines through the enhancement of physical and mental capacities. The development and use of, such as, neuroscience, silicon chips and smart technologies offer new opportunities. A desirable future is that there would be an ecosystem of human and machine, where both complement each other rather than competing with each other. Machines would continue to support humanity's well-being and quality of life and offer new possibilities for advanced 'human-machine', 'human-human', 'human-work' and human-environment interactions.

O. Saritas (✉)
National Research University, Higher School of Economics, Moscow, Russia
e-mail: osaritas@hse.ru

D. Meissner et al. (eds.), *Emerging Technologies for Economic Development*,
Science, Technology and Innovation Studies,
https://doi.org/10.1007/978-3-030-04370-4_11

However, the future may bring some wild cards too. Machines may go ahead of human and take the control in a human-excluded world; a new social divide may occur between enhanced and non-enhanced human; or with the contribution of machines, human may move life to a virtual level to dominate the real life. With these ideas in mind, this article will first look at some technologies for human enhancement, which have uses starting from foetus and then the entire life of human. Some of these technologies might be biological through medical intervention or with the use of drugs; others maybe implantable through the integration of human body and electronic chips or externally through wearable technologies or other smart devices.

Following the exploration of technologies, this chapter will review and discuss the implications of human enhancement on the future of work, where the socio-economic impacts of emerging technologies can be well-observed. These technologies are expected to make revolutionary changes in working environment as people will be able to work harder, longer and smarter. While the opportunities are vast, a number of scientists and ethicists have already raised serious concerns about the trend of technologically augmented humans, such as the identity-affecting changes mentioned by Delaney (2016). The chapter will also address some of those ethical issues associated to human enhancement technologies. Then, overall conclusions will be drawn based on the developments presented in the paper with a few examples of wild cards which may cause future surprises and shocks.

11.2 Technologies for Human Enhancement

Human enhancement technologies involve broad areas of scientific and technological augmentation, which are fundamentally different from each other in terms of the ways and timings of use, as well as their impacts. Therefore, it is useful to categorise them and present the emerging technologies under each category.

Three broad areas can be mentioned where human enhancement will be observed:

1. Medicine/drugs/diet
2. External/wearable technologies
3. Internal/implanted technologies

Below, technologies will be presented in each category and examples will be given for each.

11.2.1 Human Enhancement Based on Medicine, Drugs and Diet

Technologies for human enhancement based on medicine, drugs and diet can be applied through the entire life of human and even earlier. These technologies will be examined under three categories:

11.2.1.1 Technologies Related to Fertilisation

The earliest phase of human enhancement can be undertaken at the fertilisation phase. A new technological procedure called 'preimplantation genetic diagnosis (PGD)' is a screening test used to detect genetic or chromosomal disorders within embryos created through in vitro fertilisation (IVF) with the aim of preventing certain diseases or disorders before passing on to the child (APA 2016). The PGD process begins with an embryo screening. This is done before embryos are located to the uterus. In this process, couples are able to make decisions about the subsequent phase in the in vitro fertilisation. Embryos without genetic or chromosomal disorder are selected and transferred to the uterus. PGD process is recommended in the cases of:

- There is a case of heritable genetic disorder in one or both couples.
- One or both of couples have a chromosomal abnormality.
- When mother is beyond the maternal age.
- The mother had previous miscarriages (Penn Medicine 2016).

A large number of clinical preimplantation genetic diagnosis cycles have been achieved across the world resulting with a number of healthy babies.

11.2.1.2 Body Enhancement Technologies

So far, physical enhancement has largely focused on restoration. In the next few years to come, this will be extended to cover also healthy individuals (The Academy of Medical Science 2012). A wide variety of enhancement applications can be mentioned under this category including technologies for enhancement of physical functions of human in various ways such as via transfer of organs and tissues; stem cells; plastic surgery and cosmetic modifications; and drug and diet-based enhancement.

Transplantation of Organs, Tissues and Limbs

Transplantation or organs dates back to the earlier twentieth century, when a series of experimental studies were undertaken. In 1912, Alexis Carrel was awarded the Nobel Prize for his pioneering scientific and clinical work. It was after the World War II, when surgical transplantation of human organs from dead and living donors started. Since then, the transplantation of human organs, tissues and cells has advanced dramatically both in terms of quality and quantity across the world and helped to save thousands of lives. Recent years have witnessed greater progress in understanding organ and tissue rejection, more demand for organs and tissues as well as higher number of donations (WHO 2010).

Considerable work is undertaken on tissues, particularly to replace degenerating tissues through tissue engineering and regenerative medicine. Applications in simpler tissues, e.g. bone and joint tissues, have entered. Progress has also been made on tissues with higher complexity, such as tracheas. Although organ production is still difficult, treating specific parts of an organ is possible. Using stem cell together with cardiac pumps emerges as a new research area. This may be extended to cover other organs in human body (Aidil Bin Ahmad 2016).

Beyond organs and tissues, technologies are being developed to improve limb functions. Devices like bionic limbs and exoskeletons, mimic or surpass the functionality of human limbs are promising to overcome the traditional challenges of user control, energy efficiency and usability as they become more automated and easy-to-use. While helping disabled and elderly people to gain their physical capabilities back by restoring mobility (Galle et al. 2017), future applications in limb technologies are also expected to help healthy individuals, whose jobs involve manual labour (The Academy of Medical Science 2012).

Hearing and Sight Enhancements

Ageing population is a fact in today's world. As people live and work longer, or operate under more and more extreme conditions such as in construction, army or space work, there is a greater need to maintain and restore the sensory functions of the body (The Academy of Medical Science 2012). A number of technologies and products to aid hearing and sight are already available. Advanced hearing aids will support prolonged auditory abilities. Technologies like retinal implants, gene transfer and replacing photoreceptors in the eye are among the new ways to restore and maintain vision. Night vision for human is considered to be near (Aidil Bin Ahmad 2016).

Stem Cells

Different from the other cells in human body, stem cells are 'unspecialised' and have the potential to become any type of cell in the body. Moreover, they are able to divide and multiply themselves for long period of time (NIH 2016). It is possible to transform stem cells into cells of the blood, bones, skin, muscles, heart and even brain. Sources of stem cells can be different. However, all stem cells may be transformed into different types of cells (Health News 2013).

Laboratory studies are underway to understand the properties of cells and what makes them different from other specialised cells. Stem cells have already been used to screen new drugs and detect the causes of defects in birth. Stem cell research is rapidly advancing. Recent work focuses on the development processes of an organism from one cell and the ways of replacing damaged cells with the healthy ones in organisms (NIH 2016). With a number of potential further advancements, stem cell research appears to be one of the most fascinating areas of research as they are expected to become the future of medicine.

Plastic Surgery and Cosmetic Modifications

Plastic surgery and cosmetic enhancement are in continuous and increasing demand. Younger appearance is believed to have implications for individuals' social and employment relations. Humans have been enhancing their appearance since we worked out how to use tools. Archaeological evidence suggests cosmetics date at least 6000 years and cosmetic surgery over 4000 years. At present, cosmetic surgery can be considered as one of the most common forms of human enhancement.

Enhancement of appearance goes beyond the physical body to extend into digital world (HEB 2010). As people's virtual second life is overtaking their real physical life, it will be more common to see avatars, which will represent not real humans but an enhanced or a new form of how they wish to look like.

11.2.1.3 Mental Enhancement Technologies

Besides physical enhancement, technologies enhancing memory, concentration and motivation will be covered under the category of mental enhancement technologies. Some of the most noteworthy technologies in this category are mentioned below.

Neuroscience and Neurotechnology

Neuroscience is concerned with the scientific study of the nervous system. Today neuroscience goes beyond being a branch of biotechnology to become an interdisciplinary science at the intersection of disciplines like chemistry, computer science, engineering, linguistics, mathematics and medicine as well as philosophy, physics and psychology (Bear 2017). Based on the scientific background, neurotechnologies under development provide a number of opportunities for the enhancement of cognitive functions while opening new ways of communication between human brains and brain and machines. For instance, mind mapping for understanding brain's structural and functional connections may help to understand how these change in the cases of Alzheimer's and schizophrenia and may provide solutions for treatment. Brain-like computers may reason, predict and react like human neocortex. Devices connected to brain are promising both for improving lives of patients with physical and neurological conditions and opening vast amount of opportunities for applications in automotive, education, gaming and security industries (Doraiswamy 2015). Moreover, brain-to-brain communication allows the direct linkage of human brains. Direct brain-to-brain interfaces (BBIs) are technologies that combine neuroimaging and neurostimulation methods to exchange information between brains directly in neural code (Chantel et al. 2015). Transmitting thoughts from one brain to another without speech or text represent an important first step in exploring the feasibility of complementing or bypassing traditional language-based or motor-based communication (IFLscience 2016).

Nootropics (Smart Drugs)

Nootropics include smart drugs, neuroenhancers, memory enhancers, intelligence enhancers and cognitive enhancers. The main functions of nootropics include improving mental functions such as memory, cognition, intelligence, attention, concentration and motivation. They are delivered in the forms of drugs, nutraceuticals, functional foods and supplements. Cognitive enhancement drugs aim to make people more productive at work, allowing them to perform better and enjoy their tasks more. Some of the most commonly used smart drugs are:

- Piracetam was the first smart drug on the market. It has been used 'to enhance, elevate, and stimulate the functions of the (ACh) Acetylcholine receptors implicated in memory processes and developments'. Because the drug enhances cognitive functions related to the central nervous system, it is used not only by individuals with neurodegenerative and central nervous system but also by healthy individuals (Memolition 2014).

- Modafinil was developed as a treatment for narcolepsy to improve attention and make tasks more enjoyable, compared with placebo. Lower rate of accidents among shift workers was observed following the use of the drug (Jha 2012).
- Ritalin is a central nervous system stimulant, which is commonly used to improve focus. Although legitimately prescribed for the treatment of ADHD, Ritalin has been misused by individuals for a stimulant effect and to increase academic or athletic performance (Kroutil et al. 2006).
- Aniracetam aims at improving memory recall and immune function, increasing intellectual clarity as well as giving well-being and healthy feelings. In essence, 'Aniracetam is an ampakine and nootropic of the racetam chemical class purported to be considerably more potent than piracetam. It is lipid-soluble and has possible cognition-enhancing effects' (Wikipedia 2017).
- Pramiracetam enhances long-term memory functionality and helps choline transformation into neurons (SDFT 2016).

Drugs developed for brain conditions such as Alzheimer's and schizophrenia are increasingly used by healthy individuals to boost their cognitive skills—though the access to these drugs for healthy individuals, potentials to create addiction and abuse, and their side-effects on healthy brains are still in question (Keane 2008).

Functional Food

There is an increasing tendency towards healthy eating. This trend brings discussions on the potential benefits of certain foods and ingredients. Although there are a lot of hypes, scientific evidence also supports that some of these food and ingredients may have positive impact on health and well-being. Recently, foods have been enhanced to increase their benefits for health and reduce diseases. They are called 'functional foods', which include healthy additives as well as vitamins. Some of the food enhancements have been made to meet legal regulations, for instance, through addition of iodine into salt or vitamin D into milk. This sort of 'invisible fortification' is not considered as functional food (OMICS 2014). On the contrary, functional food helps to improve physical and mental state by improving:

- cognitive performance
- mood and vitality
- reaction to stress
- short-term memory
- vigilance and attention
- changes in memory and other mental processes during ageing. (EU 2010)

For instance, glucose may improve mental performance, including memory and time to make decisions. Caffeine may lead to improvement in cognitive performance with effects on reaction time, vigilance, memory and psychomotor performance.

Sucrose may reduce pain perception. Vitamin B can be used to improve cognitive performance and maintenance of mental health in older people. The amino acid tryptophan can reduce the time taken to fall asleep. Tyrosine and tryptophan may help to recover from jet lag. Ingredients like S-adenosylmethione (SAMe) and folic and n-3 fatty acids may help to improve depression (EU 2010).

Overall, research on nutrition is increasingly concerned with new technologies such as nutrigenomics, imaging techniques as well as other converging technologies. A number of benefits can be mentioned such as food development for a certain group of population suffering from diseases or risk factors such as obesity, diabetes, allergy or cardiovascular disease. Similarly personalised diets and foods can be developed in line with individual genetic information to support optimal health for population.

Longevity

A combined consequence of the technologies presented above would be the longevity of human life. In this respect a number of trends can be mentioned in this field. These include (Canton 2006):

1. Within ten years, humans routinely living beyond one hundred will be an accepted reality
2. Longevity Medicine will postpone aging and promote health, enabling people to be more active, more productive, and enjoy longer lives
3. Health-enhancement rights, fueled by the wealth of aging baby boomers and the fusion of nano, bio, IT and neuro innovations, will become fierce social issue
4. Mapping personal DNA profiles, and linking that knowledge to prevent illness, will radically change medicine, making it boldly predictive
5. Health enhancement via biotech, stem cells, and genomic drugs will enhance human intelligence
6. Supercomputers, artificial intelligence, and advanced medical information technology will usher in a new era that will empower doctors to extend the quality of life
7. Personalized DNA diets will greatly enable longevity as people learn which foods enhance their health and prevent illness
8. Life-extension treatments, from genetic vaccines and designer DNA "surgery" to smart drugs and neuro-medical devices, will augment health, improving intelligence, and maximizing beauty
9. Cognitive brain-science breakthroughs will protect the aging mind, refreshing vital memories, improving physical agility, and promoting human performance enhancement
10. The evolutionary transformation of human beings, via emerging breakthroughs in Longevity Medicine, will provide vast new choices of an astounding and alarming nature for individuals and society".

11.2.2 Human Enhancement Technologies Based on Internal and Implanted Technologies: Towards Transhumanism

Human ICT implants such as cochlear implants and cardiac pacemakers have been in common clinical use for many years, which helped to form close links between technology and the body. Recent years have witnessed advancements in implanted medical devices in their functionality and uses. Some of these are able to modify behaviour by directly interacting with the human brain and to restore functionality (Gasson 2012).

Implantable devices can be categorised as 'medical' or 'non-medical' devices. These may be passive or active devices. The passive implants are typically those structural devices such as artificial joints, artificial valves and vascular grafts. Active implantable medical devices are integrated into human body either medically or surgically. Implantable non-medical devices are usually in the form of electronic chips. An example of a passive device is the radio frequency identification (RFID) device. Active devices may use electrical impulses to interact with the human nervous system (EGE 2004).

Bionic eye can be given as an example to such implantable technologies (Monash 2016). A research group at Monash University, Australia, has been developing a direct-to-brain bionic eye system. This allows blind people to see by wearing a pair of glasses with a digital camera that captures low-resolution black-and-white images like a retina. First tests have been undertaken with people who lost their sight in traumas and then with the others who have been blind since birth. But it is a precursor of developments that will change the way we deal with disabilities in society (Kraft 2012).

One of the most striking recent advancements in human enhancement is the use of DNA as a storage device. As the building block of life, DNA may also be the natural repository for digital data of all sorts. Scientists have already successfully converted 739 kilobytes of hard drive data in genetic code and then retrieved the content with 100% of accuracy. The only obstacle for the practical application of the technology is its cost. Sequencing, and especially synthesising the DNA, is a costly process. Like other new technologies, it is expected that costs will be reduced to affordable levels in the near future. DNA data storage could be feasible at a large scale by 2023 (Draxler 2013).

A more and intensive use of human enhancement technologies will enhance the progress towards transhumanism (H+ or h+). Transhumanism is concerned with fundamentally transforming the human condition through the use of technologies. The aims are to enhance human physical, psychological and intellectual capacities as well as to eliminate ageing. There are an increasing number of studies on the benefits and risks of transhumanism to study their moral and ethical impacts. The extreme use of technologies would transform humans into a 'post-human' state (Peragine 2013). During the Global Future 2045 International Congress in Moscow in 2012, the following developments were foreseen for the movement towards transhumanism across a timeline (GF2045 2013):

- 2012–2013: Emergence of new transhumanist movements and parties amid the ongoing socio-economic crisis.
- 2014: New centres for cybernetic technologies to radically extend life, where the 'race for immortality' starts.
- 2015–2020: Creation of robotic human copy (avatar) and robots to replace human tasks as well as smart machines like flying cars.
- 2025: Creation of brain-to-brain communication systems, brain transplantation into avatar bodies, and life extension.
- 2030–2035: Emergence of 're-brain', mapping human brain and a step closer to 'understanding the principles of consciousness'.
- 2035: The first successful transplantation of personality—'epoch of cybernetic immortality begins'.
- 2040–2050: The arrival of bodies 'made of nano-robots' that can take any shape and 'hologram bodies'.
- 2045–2050: Drastic changes to the social structure and sci-tech development with 'spiritual self-improvement' takes precedent.

11.2.3 Human Technologies Based on External and Wearable Technologies

11.2.3.1 Exocortex

Exocortex is 'a theoretical artificial external information processing system that would augment a brain's biological high-level cognitive processes. An individual's exocortex would be composed of external memory modules, processors, IQ devices and software systems that would interact with, and augment, a person's biological brain'. The interaction between human and exocortex is undertaken through a direct brain-computer interface. This interaction makes the extensions functionally a part of the individual's mind, which make them a step closer to cyborgs or transhumans (Synthetic Telepathy 2012).

A gadget developed by Thalmic Labs, namely, Myo, measures the electrical impulses produced by muscles using the 'electromyography' technique. Myo's sensors are sensitive enough to detect gestures, which are then translated into a digital command for a computer or a mobile device. 'When people go to move a hand, they are using muscles in their forearm which, when they contract and activate, produce just a few microvolts of electrical activity. The sensors on the surface of the skin amplify that activity by thousands of times and plug it into a processor in the band, which is running machine learning algorithms. Similar technology is found in high-tech arm and hand prosthetics, as well as the Necomimi Brainwave Controlled Cat Ears'. The continuous use of Myo over time provides the system to learn and develop its accuracy (Mitroff 2013).

11.2.3.2 Wearable Computers

Wearable computing devices are projected to increase in popularity over the next years as a wave of new gadgets continuously hit the consumer market. ABI Research forecasts the wearable computing device market will grow to 485 million annual device shipments by 2018 (ABI 2013). Currently, the largest share of wearable technologies consists of sports and activity trackers. These have recently been complemented by a number of other wearable technologies including cameras, smart glasses, smart watches and 3D motion sensors.

11.2.3.3 Robotics and Smart Devices

A robot is a device or a system, which is capable of undertaking a specific task such as welding, surgical operations, drones and exploratory vehicles among the others. Different than robots, smart devices are equipped with embedded computing capabilities. Today there are billions of mobile phones in the world as well as sensors and other computing devices with increasing interconnections and networks of them (WEF 2012).

Recent advancements in robot technologies mean that it can now move from a mere focus on technical innovation and applications towards a focus on integration with other devices in an ecosystem of smart devices, humans, as well as wider society. Robots and smart devices have traditionally been controlled by humans for certain functions like production, exploration and defence purposes. With the introduction of the Internet of Things (IoT) systems, an ecosystem will be generated with intelligent machines, devices and robots. Whether being autonomous, fixed or mobile, specialised for a specific task within a specific environment or generalised, these devices will be capable of performing a number of general or specialised services. In all instances, they will need to operate and co-operate with humans and other robots within a smart ecosystem. Humans, robots and other entities need to learn, adapt and modify their behaviours to maintain a synergy within the ecosystem to obtain useful and optimal services. This will lead to an ecosystem of the Internet of Everything (IoE) to include human at centre of the network system. Thus, the following areas can be considered as research priorities in this domain:

- Smart device ecology connectivity
- Event and transaction models for the smart device ecology
- Smart device ecology service models, especially adaptive and emergent paradigms
- Data mining contextual information from live video feeds within an intelligent environment, with an emphasis on retrieving human behavioural information to aid the environment to provide better service responses
- Biologically-inspired approaches to smart devices ecologies and robots
- All aspects of robot interaction within the ecology, but especially with humans
- Behavioural paradigms and architectures for intelligent robots, covering traditional controllers (IQ based) and socially orientated controllers (EQ based) i.e. exploring various psychometric traits such as the irrational, rational, ego centric and so on
- Psychological and practical impacts of intelligent robots working alongside humans, particularly in the envisioned domestic environment, exploring the

adaptations of space and human behaviour with case studies and other experimental evidences. (IntEnv 2009)

In the future humans will be able to have the night vision capability and will be able to hear sounds in a wider frequency spectrum. These technologies will open new possibilities for new communication channels, networks and technologies. Unavoidably, all these transformations will have implications for socio-economic activities of humans. Major changes are expected in social relations, culture, work environments, skills and employment patterns, among the other spheres of life. The future may be prone to advancements and opportunities, as well as challenges and tensions because of these changes. Below, a discussion will be undertaken in relation to the impacts of human enhancement and the future of work. Further multidisciplinary research needs to be undertaken in this domain for broader implications.

11.3 Implications for Human Enhancement on Work, Society and Ethics

It is expected that the automation of knowledge work will affect US$5–7 trillion of the world's economic activity (Saritas 2015). Advancing technologies are already confronting existing working population. Computers and machines have already started replacing human in routine jobs. For instance, Amazon is planning to deliver packages to customers within 30 min with the use of drones (Amazon 2014). Improvements in productivity, costs and efficiency are considered to be the main drivers for the use of machines. This trend is pushing the employees to take jobs at the lower or higher ends of occupational skills while affecting the middle class jobs at a greater extent.

Although human appears to be in a disadvantaged position when it comes to employment, the enhancement technologies presented above are expected to open new horizons for the future of work and employment. Enhancement technologies will help to overcome some of the challenges like ageing workforce and changing, and more challenging, working practices.

Widespread use of enhancements might influence an individual's ability to learn or perform tasks and might open new opportunities for entering new professions; influencing motivation; enabling people to work in more extreme conditions or, into old age, reduce work-related illness; or facilitating earlier return to work after illness (The Academy of Medical Sciences 2012). Increased efficiency through enhancement would lead to a better work-life balance. Occupational accidents and illnesses may be reduced with better and more secure working places and practices. Moreover, there might be a new human enhancement sector, which would lead to multidisciplinary innovations, new markets, high-skilled employment as well as economic gains for economies.

The availability and access to these technologies will play a major role in their widespread implementation. High costs, at least at the beginning, would affect the equality of accessing them. Even if the costs are low, all products on the market may not be equally beneficial. They need to be regulated, and their benefits and risks should be well explained to the potential users. This is especially crucial for technologies such as pharmacological cognitive enhancers.

The use and benefits of enhancements will vary with the context of application. By maximising benefits from enhancement technologies, the emerging new occupations and working practices should be understood and involve individuals in the design and integration of technologies and in the design of any regulatory frameworks. Further effects like the consequences of delayed workforce retirement on youth employment may also need to be taken under consideration. Therefore, continuous monitoring to inform the reassessment of any policy or regulatory decisions is considered to be crucial.

With the widespread applications of human enhancement technologies, serious concerns have been raised. Besides opportunities, a number of challenges are brought in terms of health and safety as well as ethical, social and political considerations. All the impacts of using enhancement technologies need to be considered through deliberative dialogue with users as well as wider society including working and nonworking populations, elderly, employers, trade unions, policy-makers as well as scientific researchers and producers. Scientists including social scientists, philosophers, ethicists, policy-makers and the public need to be in a continuous dialogue to discuss the ethical and moral consequences of enhancement and thus maximise benefits and minimise any harmful effects.

For a more responsible development of human enhancement technologies, a set of social, technological, economic, environmental, political and value/cultural (STEEPV) aspects (Miles et al. 2016) need to be considered. These are (WEF 2012):

1. Social dimension to ensure equal access to these technologies by providing necessary subsidies for the ones, who are in utmost need, such as disabled and elderly. This requires not only physical but also psychological and sociological understanding of people's needs, desires, capabilities and traits and a neurocognitive understanding of the physical and behavioural dimensions.
2. Technological dimension involves the responsible innovation in the field. Besides engaging users and broader stakeholders in a deliberative process of technology development, process and products need to address expectations related to physical and psychological health of the potential users. Technologies would aim to achieve not only functional but also intelligent devices and applications, which may interact with humans.
3. Economic dimension is a crucial one as the technologies should be made available to a broader range of users to ensure that they are accessible not only for wealthy ones but also for the middle- and low-income users.
4. Environmental dimension is concerned about the ecological impacts of the products developed for human enhancement. These include not also physical waste but also chemical and biological ones, which need to be able to be treated appropriately with no or less environmental impact.
5. Political and legal dimension is concerned with the development of regulatory structures for machines, humans and human-machine interactions. A set of new

issues is likely to arise at the interfaces, which need to be addressed with a more proactive and predictive assessment.

6. Value/cultural: This dimension also includes developing an ethical understanding of what is being offered with human enhancement technologies, preventing any doubts about what it would mean about being human and possible clashes between humans and machines as well as enhanced and un-enhanced people due to the fear of losing their jobs, relationships and uniqueness. There is a need to ensure that technologies function globally, in different cultural contexts by considering different beliefs, genders, ethnicities, generations as well as social classes.

11.4 Weak Signals and Wild Cards in Human Enhancement

Human enhancement technologies presented above will certainly introduce fundamental changes to human life and society. During this process some intended and unintended consequences will arise. To conclude this article on these grounds, a list of weak signals and wild cards are presented below to address potential uncertainties in this domain:

- Machines may go ahead of human and take the control in a human-excluded world.
- New social divide: A new social divide may occur between enhanced and non-enhanced human or with the contribution of machines.
- People will have two or three personalities: Move life to a virtual level to dominate the real life.
- Future humans will share highly interactive virtual environments with robots and with advanced IT. Humans may experience states of consciousness that current human brains cannot access.
- If brain can be provided with unlimited memory, unlimited calculation ability and instant wireless communication ability, it will be possible to produce a human with unsurpassable intelligence. This kind of link between brain and machine is expected to be demonstrated in the future (Baker 2008).
- A neural prosthesis will be implanted in the brains of great apes which may endow them with 'synthesised speech'. This would be an unprecedented breakthrough in interspecies communication (Saniotis 2009)
- As scientists develop new opportunities for self-improvement through technology, society is likely to be faced with choices about accepting, rejecting or regulating these new kinds of possibilities, which may have implications on the meaning of democracy and social responsibility.
- As humanity finds ourselves able to use enhancement technologies to change more and more of what was formerly understood as 'given', in Dworkin's words, we experience the disruption of 'the boundary between chance and choice' (Honnefelder 2008).

- Human and cognitive enhancement may challenge the long-standing assumptions about personal identity and human responsibility as well as self-understanding (Caldera 2008).
- A complete ban may be seen on the use of biotechnology for enhancement purposes. These arguments revolve around the claim that enhancement technologies will change the basis of human nature, leading to differences in equality and legal protection between enhanced and non-enhanced persons.

What is clear is that further scientific, economic, social and ethical investigations will be required to better assess the advantages and disadvantages of such technologies and to decide to what extent to use them.

Acknowledgements An earlier version of this work was published in Saritas, O. (2013).

The book chapter was prepared within the framework of the Basic Research Program at the National Research University Higher School of Economics and supported within the framework of the subsidy by the Russian Academic Excellence Project '5-100'.

References

ABI (2013) Wearable computing devices. https://www.abiresearch.com/press/wearable-computing-devices-like-apples-iwatch-will/. Accessed 31 Jan 2017

Aidil Bin Ahmad MA (2016) Human enhancement: physical enhancement. http://afifaidildec5d.blogspot.ru/p/physical-enhancement.html. Accessed 31 Jan 2017

Amazon (2014) Amazon prime air. http://www.amazon.com/b?ref__tsm_1_tw_s_amzn_mx3. Accessed 31 Jan 2017

AP – Associated Press (2016) AP expands minor league baseball coverage. Press Release. https://www.ap.org/press-releases/2016/ap-expands-minor-league-baseball-coverage. Accessed 31 Jan 2017

APA – American Pregnancy Association (2016) Preimplantation genetic diagnosis (PGD) benefits & concerns. http://americanpregnancy.org/infertility/preimplantation-genetic-diagnosis/. Accessed 31 Jan 2017

Baker S (2008) The rise of the cyborgs. http://discovermagazine.com/2008/oct/26-rise-of-the-cyborgs. Accessed 31 Jan 2017

Bear MF (2017) Neuroscience: exploring the brain. Content Technologies

Caldera EO (2008) Cognitive enhancement and theories of justice: contemplating the malleability of nature and self. J Evol Technol 18(1):116–123

Canton J (2006) The extreme future: the top trends that will reshape the world for the next 5, 20, and 20 years. Penguin Group, New York

Chantel PS, Losey DM, Cronin JA, Wu J, Abernethy JA, Rao RPN (2015) Playing 20 questions with the mind: collaborative problem solving by humans using a brain-to-brain interface. PLoS ONE 10(9):e0137303. https://doi.org/10.1371/journal.pone.0137303

Delaney JJ (2016) Human enhancement and identity-affecting changes: the problem of controversial cases. Ethics Med Publ Health 2:499–506

Doraiswamy M (2015) 5 brain technologies that will shape our future. WEF – World Economic Forum. https://www.weforum.org/agenda/2015/08/5-brain-technologies-future/. Accessed 31 Jan 2017

Draxler B (2013) Data storage in DNA becomes a reality. http://blogs.discovermagazine.com/80beats/category/technology/page/2/#.WJDA7FOLSUl. Accessed 31 Jan 2017

European Group on Ethics (2004) The ethical aspects of ICT implants in the human body. In: Proceedings of the Roundtable Debate, Secretariat of the EGE. http://bookshop.europa.eu/sk/the-ethical-aspects-of-ict-implants-in-the-human-body-pbKA6505876/downloads/KA-65-05-876-EN-C/KA6505876ENC_002.pdf;pgid=y8dIS7GUWMdSR0EAlMEUUsWb0000k7wBFdvu;sid=IRlyh_ueZzhyh6i_rJDsIJm7Jsp0VPxNJ8o=?FileName=KA6505876ENC_002.pdf&SKU=KA6505876ENC_PDF&CatalogueNumber=KA-65-05-876-EN-C. Accessed 31 Jan 2017

European Union (2010) Functional foods. Directorate-General for Research. http://www.eurosfaire.prd.fr/7pc/documents/1276590504_functional_foods_en_publi_ce.pdf. Accessed 31 Jan 2017

Galle S, Derave W, Bossuyt F, Calders P, Malcolm P, De Clerq D (2017) Exoskeleton plantarflexion assistance for elderly. Gait Posture 52:183–188

Gasson MN (2012) Human ICT implants: from restorative application to human enhancement. Inf Technol Law Ser 23. https://doi.org/10.1007/978-90-6704-870-5_2

GF2045 (2013) Towards a new strategy for human evolution. In: Congress proceedings and transcripts of global future 2045: towards a new strategy for human evolution. Lincoln Center, New York City, June 2013

Health News (2013) What are stem cells? Medical Life Sciences News. http://www.news-medical.net/life-sciences/What-are-Stem-Cells.aspx. Accessed 31 Jan 2017

Honnefelder L (2008) Bioethics and the normative concept of human selfhood. In: Diiwell M, Rehmann-Sutter C, Mieth D (eds) The contingent nature of life. Springer. https://archive.org/stream/springer_10.1007-978-1-4020-6764-8/10.1007-978-1-4020-6764-8_djvu.txt. Accessed 31 Jan 2017

Human Enhancement and Biopolitics (2010) Make realistic the unrealistic ideal body image. https://hplusbiopolitics.wordpress.com/2010/05/04/make-realistic-th-unrealistic-ideal-body-image/. Accessed 31 Jan 2017

IFLscience (2016) Direct brain-to-brain communication used in humans. http://www.iflscience.com/brain/direct-brain-brain-communication-used-humans/. Accessed 31 Jan 2017

IntEnv (2009) The 5th international conference on intelligent environments. Special session on domestic robotics and smart device ecologies, 20–21 July 2009. Technical University of Catalonia, Castelldefels, Barcelona, Spain. http://www.intenv.org/sites/default/files/IE09%20-%20Special%20Session%20on%20Domestic%20Robotics%20and%20Smart%20Device%20Ecologies.pdf/. Accessed 31 Jan 2017

Jha A (2012) Report raises ethical concerns about human enhancement technologies. The Guardian. https://www.theguardian.com/science/2012/nov/07/report-ethical-concerns-human-enhancement-technologies. Accessed 31 Jan 2017

Keane H (2008) Pleasure and discipline in the uses of Ritalin. Int J Drug Policy 19:401–409

Kraft A (2012) Bionic eye to help the blind see. http://www.zdnet.com/article/bionic-eye-to-help-the-blind-see/. Accessed 31 Jan 2017

Kroutil LA, Van Brunt DL, Herman-Stahl MA, Heller DC, Bray RM, Penne MA (2006) Nonmedical use of prescription stimulants in the United States. Drug Alcohol Depend 84:135–143

Kurzweil R (2006) The singularity is near, when humans transcend biology. Penguin Group, New York

Memolition (2014) Nootropics: smart drugs. http://memolition.com/2014/04/19/nootropics-smart-drugs/. Accessed 19 Apr 2014

Miles I, Saritas O, Sokolov A (2016) Foresight for science, technology and innovation. Springer, Berlin

Mitroff S (2013) Your next computer will live on your arm. Wired. https://www.wired.com/2013/02/thalmic-labs/. Accessed 31 Jan 2017

Monash (2016) Bionic eye. http://www.monash.edu/bioniceye. Accessed 31 Jan 2017

National Institutes of Health (2016) NIH stem cell information home page. In: Stem cell information. National Institutes of Health, U.S. Department of Health and Human Services, Bethesda, MD. https://stemcells.nih.gov/info/basics/2.htm. Accessed 31 Jan 2017

OMICS (2014) Functional food. OMICS International. http://research.omicsgroup.org/index.php/Functional_food. Accessed 31 Jan 2017

Penn Medicine (2016) Penn preimplantation genetic testing program. https://www.pennmedicine.org/for-patients-and-visitors/find-a-program-or-service/penn-fertility-care/embryo-screening. Accessed 31 Jan 2017

Peragine M (2013) The universal mind: the evolution of machine intelligence and human psychology. Xiphias Press, San Diego, CA

Saniotis A (2009) Present and future developments in cognitive enhancement technologies. J Fut Stud 14(1):27–38

Saritas O (2013) Emerging technologies in human enhancement. Foresight Russia 7(1):6–13 (In Russian: Саритас О. (2013). Технологии совершенствования человека: перспективы и вызовы, Форсайт 7(1):6–13)

Saritas O (2015) SPIEF2014: science, technology, energy & Russia's future development agenda. Foresight 17(3):233–239

Smart Drugs for Thought (2016) Pramiracetam: what is Pramiracetam? https://www.smartdrugsforthought.com/what-is-pramiracetam/. Accessed 31 Jan 2017

Synthetic Telepathy (2012) Exocortex. http://www.synthetictelepathy.net/human-computer-interaction/exocortex-your-brain-on-internet/. Accessed 31 Jan 2017

The Academy of Medical Sciences (2012) Human enhancement and the future of work. Workshop report. https://www.acmedsci.ac.uk/viewFile/publicationDownloads/135228646747.pdf. Accessed 31 Jan 2017

Wikipedia (2017) Aniracetam. https://en.wikipedia.org/wiki/Aniracetam. Accessed 31 Jan 2017

World Economic Forum (2012) Global Agenda Councils: robotics and smart devices. http://reports.weforum.org/global-agenda-council-2012/councils/robotics-smart-devices/. Accessed 31 Jan 2017

World Health Organization (2010) WHO guiding principles on human cell, tissue and organ transplantation. http://www.who.int/transplantation/Guiding_PrinciplesTransplantation_WHA63.22en.pdf. Accessed 31 Jan 2017

Ozcan Saritas is Professor of Innovation and Strategy at the HSE and editor-in-chief of "Foresight"—the journal of future studies, strategic thinking and policy. He worked as Senior Research Fellow at the Manchester Institute of Innovation Research, The University of Manchester. His research focuses on innovation and policy research with particular emphasis on socio-economic and technological Foresight. With a PhD from the "Foresight and Prospective Studies Program," he introduced the "Systemic Foresight Methodology", and has produced a number of publications on the topic. Dr. Saritas has extensive work experience with the international organizations, including the United Nations, OECD, and the European Commission. He has been involved in large scale national, multinational and corporate research and consultancy projects on sectors including energy, climate change, agriculture, food, water, transportation, and ICT among others; published a number of articles in respected journals; and have delivered keynote speeches in more than 50 countries across the world. Besides his research and publication activities, he designs and delivers academic and executive education courses on Foresight and Strategic Planning. He has recently co-authored a book, entitled "Foresight for Science, Technology and Innovation" published by Springer, which has become one of the key readings in the field.

Part IV

Security

Illuminating the "Dark Side" of Emerging Technologies

12

Aharon Hauptman

12.1 Introduction

Foresight studies are important in order to anticipate and assess potential impacts—sometimes quite unexpected and surprising—of new technologies on society. Although the future is practically unpredictable and inherently uncertain, especially if human whims, desires, and fears are involved, foresight studies can help us to at least contemplate about alternative futures and so, hopefully, to increase the prospects of informed decisions leading to a desired future.

There is probably no need to convince the reader that almost every new technology, even if developed for the benefit of society, can have a "dark side". This potential wicked face of technology can be manifested, for instance, by harmful effects on the environment or on human health or by other unexpected unintentional—or intentional—consequences. This situation reflects a pressing need for foresight studies which are invaluable for continuous analysis of the unfolding technology landscape for potential threats, opportunities, and other interesting implications.

In this chapter we focus the attention on two important categories of consequences—security threats and implications on privacy. In the security context, we concentrate on potential abuse of new technologies by negative actors like terrorists or criminals. In the privacy context, we not only deal with the increased threat of excessive privacy intrusions but also with potential changes in people's perception of privacy—induced by emerging technologies. In some cases, the same technologies may have implications both on security and privacy. Special attention is paid to new, even surprising opportunities opened by the convergence of various technologies.

A. Hauptman (✉)
Tel Aviv University, Tel Aviv, Israel

D. Meissner et al. (eds.), *Emerging Technologies for Economic Development*, Science, Technology and Innovation Studies,
https://doi.org/10.1007/978-3-030-04370-4_12

12.2 Emerging Technologies and Security Threats

A foresight study on the potential dark side of emerging technologies in terms of their security threats was carried out within the European project FESTOS[1] (conducted in 2009–2011 under the "Security" theme of the European Commission's Seventh Framework Programme, FP7). The goals of this project were to assess potential abuse threats posed by selected emerging technologies, to cope with the dilemma of knowledge control versus the freedom of research and to propose policy guidelines to reduce the likelihood of abuse.

12.2.1 FESTOS Horizon Scanning

The horizon scanning effort in FESTOS focused on five fields: information and communication technologies (ICT), nanotechnologies, biotechnology, robotics, new materials, and converging technologies (nano-bio-info-cogno). In the first stage, 80 technologies were identified and briefly described, including related threat indications. Three broad categories of threats were observed:

- *Disruption* of certain applications, e.g. cyberattack on intelligent transportation systems. This conspicuous category is increasingly important with our growing dependence on technologies.
- *Increased accessibility* to technologies that once were confined to the military sector or to unique laboratories and were prohibitively expensive, e.g. commercial off-the-shelf components that may generate damaging electromagnetic pulses (EMP).
- *Surprising malicious uses* of new technologies that are being developed for beneficial purposes, for instance, employing advanced toy robots for terror attacks or using synthetic biology to engineer bacteria that consume fuel instead of producing it.

All the three categories are important in addressing emerging threats. FESTOS concentrated mainly on the third, where we may find the most unexpected potential threats, signals to "wild cards" (low-probability high-impact events or developments). "Wild cards" are increasingly recognized in recent years as an important foresight method, in light of apparent growing frequency of "strategic surprises" and the lack of preparedness (Lee and Preston 2012) or even denial of decision-makers (Schwartz and Randall 2007), and their elicitation can inspire the envisioning of interesting surprising scenarios.

[1]FESTOS: **F**oresight of **E**volving **S**ecurity **T**hreats p**O**sed by emerging technologie**S**. The project partners were Interdisciplinary Center for Technology Analysis and Forecasting (ICTAF), Tel Aviv University (Israel), Technical University of Berlin (TUB) (Germany), Finland Futures Research Centre, University of Turku, Foundation for European Scientific Cooperation (FEWN) (Poland), and EFP Consulting (UK).

12.2.2 FESTOS Expert Survey

About 280 experts participated in the online survey run by FESTOS, in which their opinions were elicited concerning the threat potential of 33 selected technologies. For each technology the experts were asked to assess:

- When will this technology be sufficiently mature[2] to be used in practice?
- How easy will it be to use it for malicious purposes[3] (1 = not easy at all, 5 = very easy)?
- How severe is the potential security threat posed by this technology (1 = very low severity, 5 = very high severity)?
- The likelihood that it will actually come to pose a security threat, in different time frames (1 = very unlikely, 5 = very likely).
- To which societal spheres it will pose a security threat.

One striking general finding was that majority of experts agree that the public is rather badly informed about dangers of new technologies and that governments tend to underestimate the potential threats. Main technology-oriented results of the survey are presented below. We show here only the results for 17 selected technologies which may be regarded as converging technologies in the broad sense, including certain nano-/biotechnologies and cognition/brain technologies (for full results, see Hauptman and Sharan 2013).

12.2.2.1 Maturity Timeframes

Table 12.1 presents the distribution of the estimated time of "sufficient maturity to be used in practice" for each technology. Naturally, for many new technologies, the uncertainty regarding their time of maturity is high, and therefore disagreement between experts is not surprising.

12.2.2.2 Abuse Potential: Severity of Threat and Easiness of Abuse

Table 12.2 presents the experts' assessments (averages) of the severity of threats posed by each technology and the easiness of its malicious use. A useful way to prioritize the technologies is by simply multiplying the easiness of malicious use by the severity of threat, which can be interpreted as the ***potential of abuse***.

12.2.2.3 Likelihood to Actually Pose a Security Threat in the Future

It should be noted that a technology can be maliciously used even before it is sufficiently mature for regular purposes, e.g. if it is in a phase of prototype testing

[2]"Sufficiently mature" means that the technology was at least demonstrated and validated outside the laboratory, through testing of prototypes. (This is similar to TRL-5 or higher, on the "technology readiness scale" used in many technology assessments).

[3]"Easy" means that the technology is easily available/affordable/adaptable or "disruptable". "Malicious" refers to terrorism and crime.

Table 12.1 Technologies descriptions and maturity timeframes

Technology	Potential threats (examples)	Till 2020 (%)	2021–2035 (%)	Later or never (%)	Median
Advanced artificial intelligence Artificial general intelligence (AGI) is aiming at general purpose systems with intelligence comparable to the human mind. According to some forecasts, desktops may have the processing power of human brains by 2029, and by 2045 AI could be able to "improve itself" at a rate that far exceeds anything conceivable in the past	New opportunities for malicious use, e.g. by interpreting facial expressions and human intentions, design of "smart" malware for cyberattacks, or malicious use of autonomous robots. More speculatively, could it lead to machines that have (malicious) intentions themselves?	49	34	17.0	2018
AI-based robot-human interaction and coexistence Japan and S. Korea are preparing for "human-robot coexistence society", foreseen to emerge before 2030. A striking feature of this development is "social robots" with AI, with which people will have emotional and even intimate interactions	New opportunities for malicious use of robots that have close intimate interactions with people who trust them?	36.9	47.4	15.8	2023
Swarm robotics Coordination of large numbers of robots, inspired by the observation of insects. Large number of simple individuals can interact to create collectively intelligent systems. Tiny (millimetre-sized) robots could be mass-produced in swarms and programmed for applications such as surveillance, micro-manufacturing, medicine, cleaning, and more	Self-adaptation and self-reprogramming could be employed for malicious behaviour of the swarm. The ability to easily mass-produce tiny robots for swarms makes the threat even more concrete	22.2	77.8	0.0	2030

(continued)

Table 12.1 (continued)

Technology	Potential threats (examples)	Till 2020 (%)	2021–2035 (%)	Later or never (%)	Median
Molecular manufacturing Assembling various products "bottom-up" and molecule by molecule, possibly in small "nanofactories"	Creation of new hazardous materials or new types of weapons	30	50	20.0	2023
Self-replicating nanoassemblers Nanoassemblers, envisioned as tools for molecular manufacturing, could self-replicate exponentially. Uncontrolled "runaway replication" has been described in speculative scenarios of futuristic nanotechnology	Even if uncontrolled "runaway replication" is unlikely, one can't rule out an *intentional* design for malicious purposes	16.7	41.6	41.7	2030
Medical nanorobots Nanoscale robots inserted in the body are envisioned to conquer diseases. A possible "shortcut" to bionanorobotics could be engineered viruses or bacteria to create bio-devices	Nanorobots instructed to harm humans and/or to remotely control human actions	18.2	54.5	27.3	2030
Molecular nanosensors Sensors with molecular precision will enable advanced nano-diagnostics and will detect where a person has been by sampling environmental clues	Could make people "molecularly naked", with personal information abused by criminals.	84.6	15.4	0.0	2018
Nano-enabled brain implants Neural interface implants in the central nervous system to treat motor disorders or to control prosthetic limbs or external devices. Future implants could enhance brain functions.	Thought/behaviour control of people or equipping future perpetrators with "super mental power"	13.3	60	26.7	2030

(continued)

Table 12.1 (continued)

Technology	Potential threats (examples)	Till 2020 (%)	2021–2035 (%)	Later or never (%)	Median
According to one report, 2035 chips wired directly to the user's brain would make information be accessible through cognition and might include synthetic sensory perception beamed direct to the user's senses					
Brain-to-brain communication Advances in brain-computer interface (BCI) may lead to "radiotelepathy" enabled by direct conversion of neural signals into radio signals and vice versa	Control of people's behaviour and actions	11.1	66.7	22.2	2030
Cyborg insects Insects controlled through implanted electrical stimulators. Researchers envision living communication networks (for sensing, surveillance, etc.) by implanting electronics in insects. Advanced capabilities could be offered by micro/nano technologies	Could be used by perpetrators for harming people or agriculture, for spying, or for other malicious activities	46.2	38.4	15.4	2023
Brain-computer interface; "mind-reading" gadgets Systems that enable people with disabilities to operate gadgets "by thought" have been demonstrated, as well as systems for the gaming industry. Further progress is envisioned	Malicious distortion of communication between users and gadgets. Could hacking such a device enable influencing user's actions?	50	41.7	8.3	2023
Human enhancement based on NBIC convergence Converging technologies offer unprecedented	Malicious use of specific capabilities by perpetrators and hacking of the	23.5	64.7	11.8	2030

(continued)

Table 12.1 (continued)

Technology	Potential threats (examples)	Till 2020 (%)	2021–2035 (%)	Later or never (%)	Median
enhancement of human performance: augmentation of physical and mental abilities and new modes of interaction. Some experts envision convergence of human and machine intelligence	involved (implanted) devices				
Metamaterials and optical cloaking "Cloaks" made of metamaterials with negative refraction index can hide objects from sight in certain wavelengths (possibly in all wavelengths in the future) or make them appear as other objects	"Invisibility cloaking" or perfect camouflage for malicious purposes	36	12	52.0	2030
Programmable matter Materials that can be programmed to self-assemble, alter their shape and physical properties to perform a desired function, and then disassemble—in response to user input or autonomous sensing	Reconfigurable shape-changing tools with perfect performance, including weapons (that pass security checks), adaptable to changing requirements	24	28	48.0	2030
3-D printing Future low-cost printers could be able to self-copy and to use a wide variety of materials for a wide range of products	"Home-made" (undetectable?) weapons or cheap manufacturing of fake products	50	23	27.0	2018
Synthetic biology Building of biological agents from interchangeable bio-components. Could enable synthetic genome and then using it to control a recipient cell. Could lead to engineering of living systems to perform new functions not found in nature	Accessible tools to build novel bioweapons, virulent pathogens, dangerous organisms, etc.	50	34.6	15.4	2018

(continued)

Table 12.1 (continued)

Technology	Potential threats (examples)	Till 2020 (%)	2021–2035 (%)	Later or never (%)	Median
Induced pluripotent stem cells (iPS cells) Scientists used viruses to flip genetic switches in the DNA of skin cells from adult mice to turn them into iPS cells that are functionally equivalent to embryonic stem cells	May enable genetically engineered traits into cells to create "designer embryos" or camouflage at the cell level? Reproductive cloning by "rogue organizations"?	29.4	41.2	29.4	2023

Source: Author

and doesn't not comply yet with safety standards. The (average) likelihoods in different future timeframes, as assessed by the expert survey participants, are shown in Table 12.3. For most technologies, the likelihood rises with time, but interestingly in some cases (certain IT technologies not presented in this chapter), the likelihood declines in later stages, presumably because the experts envision that preventive means will be effectively applied.

12.2.2.4 The Impact on Society

The experts were asked to assess to which of the following societal spheres each technology will pose a security threat (multiple choice): people, infrastructures, economy, environment, political systems, and values. The percentages of respondents who opted for each societal sphere are shown in Table 12.4. Due to multiple choices, the sum of percentages across each technology can vary between 0 and 600. This sum (the right column) is interpreted as the so-called overall intensity of threat.

Evidently, some technologies potentially threaten several societal spheres, while others affect fewer spheres. Broadly speaking, most technologies pose threats to *people*, with significant threats to infrastructures and the environment as well. Interestingly, several technologies are perceived as potentially threatening the political systems and values—especially converging technologies related to the human brain and human body in general.

Some free-text comments submitted by respondents underscored the insufficient awareness (even by experts!) of potential security threats and hence the importance of foresight studies like FESTOS. A notable remark of one expert: "*Going over your questions* I suddenly *realized how smilingly innocent technologies can pose severe security threats in the near future. Being a relatively experienced individual, yet unaware of the above, I would say now that the public is not informed about such threats…*".

Table 12.2 Ranking of selected technologies by their abuse potential

Technology	A: Easiness of malicious use	B: Severity of threat	Abuse potential (A × B)
Advanced AI	3.21	3.43	11.01
Synthetic biology	3.16	3.40	10.74
Cyborg insects	3.33	3.08	10.26
AI-based robot-human interaction	3.00	2.94	8.82
Swarm robotics	2.89	3.00	8.67
Brain implants	2.73	3.07	8.38
Human enhancement	2.63	3.13	8.23
Nanoassemblers	2.75	2.92	8.03
3-D printing	2.89	2.71	7.83
Metamaterials and cloaking	2.50	2.95	7.37
Programmable matter	2.29	2.79	6.39
Molecular manufacturing	2.50	2.50	6.25
Medical nanorobots	2.27	2.73	6.20
Brain-to-brain communication	2.25	2.56	5.76
Brain-computer interface	2.33	2.42	5.64
Molecular nanosensors	2.08	1.85	3.85
iPS cells	1.44	1.87	2.68

Source: Author

Table 12.3 Likelihood of posing a threat in different time intervals

Technology	2016–2020	2021–2025	2026–2035	After 2035
Advanced AI	2.56	3.13	3.43	3.71
AI robots	2	2.25	2.65	2.94
Swarm robotics	1.44	1.89	2.56	2.89
Molecular manufacturing	1.88	2.44	3	3
Nanoassemblers	1.1	1.67	2.11	2.73
Medical nanorobots	1.25	1.78	2.44	3.11
Molecular nanosensors	1.88	2.44	3	3
Brain implants	1.64	2.07	2.46	2.85
Brain-to-brain communication	1.11	1.44	2	2.56
Cyborg insects	2.38	2.92	3.08	3.17
Brain-computer interface	1.9	2.11	2.2	2.22
Human enhancement	1.69	2.07	2.67	3.13
Optical cloaking	1.99	2.56	3.33	3.82
Programmable matter	1.95	2.41	3.14	3.40
3-D printing	2.68	3.18	3.53	3.56
Synthetic biology	2.59	3.14	3.64	4.15
iPS cells	1.31	1.69	1.86	2.31

Source: Author

Table 12.4 Intensity of threats

Technology	Economy (%)	Environment (%)	Infrastructures (%)	People (%)	Political systems (%)	Values (%)	Overall intensity of threat
Advanced AI	67	38	76	82	51	31	345
Human enhancement	38	6	38	94	69	81	326
Swarm robotics	67	67	78	78	11	22	323
Cyborg insects	42	67	92	92	8	17	318
AI-based robot-human interaction	41	41	59	88	35	41	305
Programmable matter	45	64	73	73	18	18	291
Brain-to-brain communication	40	10	10	100	60	70	290
Molecular manufacturing	50	88	38	75	25	13	289
Self-replicating nanoassemblers	55	82	55	73	18	0	283
3-D printers	75	33	58	75	17	8	266
Metamaterials and optical cloaking	46	23	85	69	31	8	262
Synthetic biology	23	69	15	100	23	31	261
Nano-enabled brain implants	7	20	13	93	33	67	233
Brain-computer interface	8	8	17	100	33	50	216
Molecular sensors	20	30	20	100	40	0	210
Medical nanorobots	18	46	9	100	9	18	200
iPS Cells	9	18	0	82	0	27	136

Source: Author

Some experts asserted that although the technologies are new, the threats are not, because similar malicious uses are already possible by existing technologies. In this context, technologies like nanoassemblers and molecular manufacturing are regarded as "too speculative" by some respondents, but there is a clear disagreement about that: While one expert claims that molecular manufacturing is *"too speculative to deserve attention at the moment"*, another one says that it is *"no more dangerous than organic chemistry"*, and a third one thinks that the same technology *"could be dangerous by scaling down the resources and facilities needed to manufacture other risky technologies [...] enabling creative malefactors to invent entirely new categories of threats there is no general preparedness or recognition of"*. Many experts stressed the well-known fact that *all* technologies are inherently prone to potential misuse and that the necessary regulations usually lag behind the technologies. Moreover, risk policy is problematic because, as one expert put it, *"Society tends to misestimate risks, and be very over-confident about its estimates. Certain minor risks are exaggerated, while others are regarded as silly"*.

12.2.2.5 "Wild Card" Threat Scenarios

The selected technologies with their threat indications can be regarded as "weak signals", some of them hinting at potential "wild cards": surprising low-probability high-impact events (Petersen and Steinmüller 2009). In recent years there is a growing recognition that the elicitation of potential wild cards as part of a foresight study is not just an interesting intellectual exercise in imaginative thinking but may prove as an essential means for preparedness to critical future surprises. As an occurrence of a wild card has a very high impact on specific systems/stakeholders, organizations are usually especially vulnerable to such events, and paying special attention to wild cards in foresight studies undertaken by these organizations could alleviate this vulnerability (Hauptman and Steinmüller 2019).

In a special scenario workshop organized by FESTOS, experts were invited to share ideas about wild cards inspired by selected emerging technologies. These ideas were subsequently developed into four full-fledged "wild" narrative scenarios (some would call them science fiction stories or outlines for such stories), vividly displaying the dark side of several technologies. Special attention was given to potential combinations or convergence of technology trends.

As a case in point, the Internet of Things (IoT) could, in combination with programmable matter and/or molecular manufacturing, give rise not only to a revolution in manufacturing but also in the use of "intelligent", "nano-enabled" everyday gadgets. Such sophisticated networked future objects (which may also emerge as outcomes of recent developments in so-called 4-D printing) could be capable of self-healing, self-reconfiguration, self-upgrading, or automated recycling—an interesting development in itself. But what if a malware or a malicious remote signal transforms self-healing into self-destruction? This wild card, "disassembling of nano-enabled products by remote signal", was the basic idea that evolved into one of the four narrative scenarios, entitled "At the flea market". Why flea market? The title hints at a market that flourishes in a world scared of self-destructing gadgets, with high demand to old "pre-nano" things that did not fall victim to the virus or malicious signals. Here is an excerpt from this narrative scenario:

> "Now, it has gotten into the hairdryer!" Sandra cursed loudly. Yesterday, it was sitting on the shelf and today there was just an unsightly heap. The device had started to ooze like a block of Camembert. A couple of metallic parts protruded from the heap. A small sign that read "Made by NanoTrust, Inc. China" was visible. Sandra cleaned up the mess with a hand broom and dustpan. Yesterday, she had been hoping that the hairdryer wouldn't be infected. The disintegration, however, crept into everything, especially the new ones. "Never again nano!" She swore that to herself weeks ago, when her television stopped working, closely followed by her espresso machine, her washing machine, her new living room lamp and her smartphone. At that time, the washing machine company had been accommodating, offering a life-long guarantee and providing a new machine. Then the plague started to afflict the entire city. One after the other, dealers started going bankrupt, manufacturers were no longer reachable and the military patrolled the streets to prevent looting. . . .

Due to space limitation, we only briefly mention here the other three narrative scenarios fully developed in the above-mentioned workshop[4]:

- "Cyborg-insects attack!": Swarms of cyborg insects (insects with implanted electronics) attack people, animals, and agriculture crops.
- "The Genetic Blackmailers": DNA of human individuals is misused for extortion.
- "We'll change your mind. . .": A terrorist group uses a virus to change the behaviour of a portion of the population for a certain period of time.

For the full narrative scenarios, the reader is referred to the FESTOS reports (Dienel and Peperhove 2011).

12.3 Emerging Technologies and Privacy

The impact of technology on privacy, including technology-driven changes in privacy perceptions, is not a new phenomenon, as can be exemplified by the history of photography: After its invention in the nineteenth century, it was mostly used to make private portrait photographs of wealthy people in special studios with cumbersome static cameras. The photographed person had full control of the result, and it was unheard of that a private photo portrait is reproduced (not to mention distributed in public) without permission. After a few years, technology advances led to smaller hand-held cameras, by which a picture could be taken without its subject even knowing that a camera was present. The rapid spreading of journalistic snapshot photography, practiced in public (and often in private) locations, heralded in a way "the end of privacy", at least for celebrities. No wonder that the first (and still famous) publication advocating the right to privacy in the USA was written by

[4]The Scenarios Workshop was moderated by the Futurist and Science Fiction writer Dr. Karlheinz Steinmüller, who also subsequently wrote the narrative scenarios based on the workshop.

Warren and Brandeis[5] in 1890 largely in response to this new trend. Prophetically, it referred not only to photography but also to other "modern devices":

> the existing law affords a principle from which may be invoked to protect the privacy of the individual from invasion either by the too enterprising press, the photographer, or the possessor of any other modern device for rewording or reproducing scenes or sounds.

Fast-forward to the twenty-first century: The social network phenomenon, with its photo-sharing and other features, is of course one present-day example of changing sensitivity to privacy, especially among young people. Incorporation of face recognition features in photo-tagging and in smartphones has stirred debates. But this is only the tip of the iceberg—if we think about new technologies that may be part of our life in the future, such as nano-enabled personalized medicine, ubiquitous ultrasensitive sensors, or "synthetic telepathy".

The famous physicist Freeman Dyson speculated about a system of such implantable chips and its potential to become a powerful instrument of social change, for good or for evil purposes (Dyson 2009). Even present-day brain scanning technologies, still far from "real" synthetic telepathy, already raise non-trivial questions related to privacy. Consider real-time functional MRI, already in use for medical research and diagnosis. Although "reading thoughts" by fMRI is impossible, it is possible to "read" emotional states and to detect lies in a reliability much higher than the old-fashioned polygraphs. Such advancing capabilities made researchers think whether this may lead to a new frontier—mental privacy—with interesting potential implications on what we perceive as "private":

> The foreseeable ability to read the state of a person's brain and thereby to know aspects of their private mental experience, as well as the potential to incorrectly interpret brain signals and draw spurious conclusions, raises important new questions for the nascent field of neurotics. One can imagine the scenario of someone having their private thoughts read against their will, or having the contents of their mind used in ways of which they do not approve (deCharms 2008)

Consider some other examples of technologies that may become ubiquitous in the next decades: Anthropomorphic household robots are people's companions. Machines resembling humans, speaking with humans, and behaving almost like humans are part of daily life. Besides the obvious privacy concerns related to the surveillance capabilities of robots (the ability to sense, process, and record their surroundings), there are less obvious social and psychological implications. Would you feel comfortable to undress in front of a friendly smiling robot? Or would you speak about your intimate secrets, while the robot is listening? Why not, don't you do it in front of your cat? The answer is not trivial.

Nanotechnology advances may significantly improve our quality of life, but they are also likely to enable unprecedented capabilities of surveillance and information

[5]Warren and Brandeis, "The Right to Privacy", Harvard Law Review, December 15, 1890 http://groups.csail.mit.edu/mac/classes/6.805/articles/privacy/Privacy_brand_warr2.html

gathering—with obvious privacy implications. Indeed, surveys have shown that "losing personal privacy" is the public's biggest fear about nanotechnology. One of the emerging nanotechnology developments is ultrasensitive nanosensors, which will have a dramatic impact on medicine and security—and on privacy. Molecular nanosensors can detect a single molecule and distinguish between different molecules. It is expected that such sensors will be able, for example, to detect drugs from saliva samples of people or to infer where a person has been by quickly sampling minute environmental clues on clothes. No wonder that researchers have expressed concerns about so-called nano-panopticism. In an article published in the prestigious journal *Nature*, one of the leading scientists, active in this field, called the emerging molecular nanosensors "plenty of eyes at the bottom"[6] and expressed his worries about a "molecularly naked patient" whose insurance company knows more about his body than himself (Toumey 2007):

> With so many huge databases of personal information, plenty of eyes at the bottom, molecularly naked patients and more, it is hard to imagine how multiple developments in nanotechnology will not intrude further into our privacy.

Bearing in mind the above-mentioned reflections, special effort was intentionally made in the European project PRACTIS to look forward beyond the information and communication technologies (the "usual suspect" in privacy-oriented debates) and to discuss the privacy aspects (sometimes unexpected and surprising) of new technologies emerging from other fields, such as nanotechnology, robotics, or cognition, including technologies enabled by the convergence of certain fields. The goals of PRACTIS[7] (conducted in 2010–2013 under the "Science in Society" theme of FP7) were to assess impacts of emerging technologies on privacy, to propose means to cope with potential risks to privacy while maximizing the benefits of these new technologies, and to formulate a framework for thinking about ethical and legal issues related to privacy in the future when these technologies prevail. In both projects special attention was paid to new, even surprising opportunities opened by the convergence of technologies.

The project referred to three main types of impacts: *threats* to privacy, *enhancement* of privacy (better protection), and *changing our perceptions* of privacy. The first type of impact is the most straightforward and includes for instance technologies that make it easier for information to be collected about people. The second type refers mainly to innovations that could enable new privacy-enhancing technologies

[6]Paraphrasing Richard Feynman's famous visionary lecture "Plenty of Room at the Bottom" from 1959.

[7]PRACTIS: **PR**ivacy-**A**ppraising **C**hallenges to **T**echnologies and Eth**I**c**S**. The project partners were the Interdisciplinary Center for Technology Analysis and Forecasting (ICTAF); Tel Aviv University (Israel); University of Lodz (Poland); Research Centre in Informatics and Law of the University Faculties of Notre-Dame de la Paix (FUNDP) (Belgium); Interdisciplinary Centre for Comparative Research in the Social Sciences (ICCR) (Austria); Nexus, Berlin (Germany); and Finland Futures Research Centre, University of Turku.

(PETs), such as methods for anonymization of information. Certain emerging technologies have both a potential to pose new threats to privacy and to enhance privacy (sometimes indirectly), depending on the specific application.

The third kind of impact, namely, the change of perception, is the most complex and is hard to attribute to any one technology by itself. We already know that perceptions of privacy (or more precisely the sensitivity to what is perceived as privacy intrusion) may change over time. Part of the explanation for such change lies in technologies or, more specifically, in the way that people may get used to intrusive technologies such that they no longer consider them a threat to their privacy. People may be willing to accept certain privacy violations and even not to perceive it at all as a threat to privacy, if the benefits in terms of better health service or improved security are perceived as worthwhile. In other words, there is a trade-off between the (perceived) privacy and the (perceived) benefits: in certain conditions people may be willing to "sacrifice" some of their privacy for concrete benefits such as improved security, lower insurance costs, better health services, and the like. Privacy seems to be a sociocultural construct, depending on dominant values of a society, its socio-cultural heritage, and contemporary technological developments. According to this understanding, the perceptions of privacy can change in time with technology being an important driver in shaping our concept of privacy, as it directly influences our daily lives and our values (for more details on these issues and other aspects of project PRACTIS, see Hauptman et al. 2011).

12.3.1 PRACTIS Expert Survey

Two hundred sixty-six experts responded to the PRACTIS survey questions and submitted their opinions on 39 emerging technologies. For each technology the respondents were requested to assess the foreseen time frame of its widespread use, its threat to privacy (on a scale 0–5: 0 = no threat; 5 = very high threat), its influence on changing people's sensitivity about their privacy (on a scale 0–5: 1 = becoming less sensitive; 3 = no change; 4 and 5 = more sensitive), and, where applicable, its possible ability also to contribute to privacy enhancement.

In this chapter only some results are presented for selected technologies closely related to technologies' convergence, and we omit technologies with obvious privacy implications such as facial recognition or mobile phone tracking (these were indeed identified as near-term threats at the time of the survey—2011).

Highest threat levels are attributed to the following technologies (the years in brackets indicate the likely time frames of widespread use): cyborg insects (2023), brain-to-brain communication (2028), mind-reading commercial gadgets (2023), and reality mining (2018). With regard to change of perception, most technologies are likely to make people more sensitive about their privacy.

In general, the experts tend to associate higher threats to privacy with larger increases in sensitivity, especially for technologies associated with the human body. Top increases in sensitivity are attributed to cyborg insects (2023), invisibility cloaking (2028), nano-based surveillance (2023), portable full genome sequencing

(2023), brain-to-brain communication (2028), and advanced artificial intelligence (2023). Important impacts (threats as well as sensitivity changes) are attributed to several technologies in the domains of cognition, biology, nanotechnology, and robotics (Table 12.5).

Some comments submitted by respondents of the PRACTIS expert survey reflected the opinion that the combination or convergence of several technologies may have much higher impact on privacy than single technologies, for example, the combination of advanced sensors with wireless networks and sophisticated software. In a vivid portrayal of one expert:

> Cyborg implants, memory chips, augmented reality, wireless networking, cloud computing, etc., in conjunction with brain-computer interfaces (BCI) offer amazing post-human capabilities... the potential is there to hack...create false memories, provide false behavior-changing information, perhaps even record and make public individuals thoughts, emotions and past experiences.

Another conspicuous opinion was that in a modern society, absolute prevention of privacy intrusion is impossible due to a trade-off between benefits and some privacy sacrifice. Indeed, an important change of perception of privacy, supported by several experts in the survey, relates to individuals "getting used" to technologies and their willingness to accept certain privacy violations (or even not to perceive it at all as such), if the benefits (e.g. better health service or improved security) are perceived as worthwhile. In other words, there is a trade-off between the (perceived) privacy and the (perceived) benefits.

An interesting opinion of one expert (probably not shared by many) was that a future with less privacy should be viewed as a rather positive trend, as less privacy means more decent behaviour and self-control:

> it may be better to get over our recent modern obsession with privacy and accept that any and every action and thought that we have may be open to investigation and to public exposure... The abolition of privacy (or perhaps a return to a very limited degree of privacy, similar to the levels in which human social behaviour and morality developed) might make for a better, fairer world.

One of the observations that stemmed from the PRACTIS study was that many of the emerging technologies have the potential either to pose a threat to privacy or to enhance it and in several cases—both at once. In other words, at this point in time, the technologies have not yet acquired a social meaning, which can develop in either way (Ahituv et al. 2013). Moreover, many of the emerging technologies and their operation will be complex and difficult to understand: while the average data subject will be able to notice some of the technologies in operation (such as visible robots), other technologies, such as nanotechnologies, are not visible. The complexity affects privacy in the sense that the data subjects will find it more difficult to figure out how these technologies process their data and whether the data is aggregated, matched with other subjects' data, transferred to third parties, etc. The technological

Table 12.5 Threats and enhancements of selected technologies

Technology	Potential threat/enhancement	Threat level	Sensitivity change
Nano-enabled personalized medicine By providing better diagnostics and genetic information, nano-enabled devices may lead to personalized medicine, based on comparison of diagnostic information about one's body with similar information about other people	Threat: Implies storing vast amounts of personal medical information in large centralized systems. People may become "molecularly naked" in front of insurance companies and other interested parties	2.90	3.78
Nano-based surveillance Nanotech-enabled miniature highly sensitive surveillance devices, with increased computing power and higher storage capacity. For example, cheap video cameras with the size and aerodynamic characteristics of a mosquito	Threat: Proliferation of devices which make it much easier and cheaper to carry out surveillance could imply a growing threat to privacy Enhancement: Could provide also anti-surveillance monitoring	3.80	4.24
Molecular nanosensors Sensors with molecular precision that can detect any number of molecules and distinguish what type of molecule is being detected. For example, such sensors would extract more information from fingertips by quickly detecting drug metabolites present in sweat	Threat: Deriving "lifestyle intelligence" from fingertips, knowing where a person has visited by sampling environmental clues on clothes, etc. Enhancement: In police use, thanks to high accuracy and zero false alarms, such sensors could help to focus on "real" suspects and reduce privacy intrusion to most people	2.70	3.60
Portable full genome sequencing (FGS) Complete and rapid DNA sequencing of a person, by sampling minute amounts of biological material, such as a hair follicle or a drop of saliva. In the future this technology might be simple enough even for a layperson to use	Threat: Combined with micro sampling devices, this could enable easy and clandestine access to personal biological samples, making genetic data available to interested bodies Enhancement: Individuals may be able to ascertain their own health status and avoid having legally proscribed disclosure from being applied. It could provide a unique personal identification	3.89	4.24
Intelligent medical implants A combination of miniaturized electronics, biomaterials, and signal processing techniques applied in medical implants such as sub-retinal implants, auditory	Threat: Patients' private medical information could be extracted and their implants reprogrammed without the patients' authorization or knowledge. Enhancement: With better coded	2.58	3.42

(continued)

Table 12.5 (continued)

Technology	Potential threat/enhancement	Threat level	Sensitivity change
brainstem implants, cardiac defibrillators, etc.	implants. Privacy might be better protected in relation to present data systems		
Medical nanorobots Nanorobots assimilated into the human body, or residing on the skin, could revolutionize medical diagnostics and treatment, to conquer disease, ill health, and ageing. Future advances could include controlled swarms of molecular nanorobots, reacting faster than neurons	Threat: If nanorobots can "intrude" the human body by food or drinks, it may pose new risks for privacy intrusion, including a risks of being used against someone's will Enhancement: Could protect medical privacy—medications and procedures that are intrusive could become more private	2.36	2.33
Advanced artificial intelligence	Threat: AI-enabled autonomous robots could be used for target-oriented surveillance and "spying" Enhancement: If appropriate "computer ethics" is applied in the intelligent networks to which all machines are connected. Intelligent enforcement of rules defined by the user	2.94	4.07
Robots as social actors "Anthropomorphized" robots designed to interact more socially. Research has shown that people tend to interact with sufficiently human-like machines as if they are real humans.	Threat: "Chilling effect"—impact on our sense of being alone and able to act with freedom. The mere presence or a "humanoid" robot could create the feeling of being observed	3.30	3.50
"Cyborg insects" Future miniature flying robots may be insect-like or even based on real insects with implanted devices ("cyborg insects")	Threat: Unprecedented capabilities for large-scale surveillance. Could covertly enter offices or houses Enhancement: Could also be used as protection from privacy intrusions of other robots	4.64	4.50
Invisibility cloaking Specially engineered materials with tailored refraction index could enable hiding objects from sight (in certain wavelengths and in the future possibly in all wavelengths) or make them appear as other objects	Threat: A perfect camouflage could enable spying on people without being detected Enhancement: Perhaps could be used also as a countermeasure against surveillance	3.13	4.36
Brain-to-brain communication Advances in brain-computer interface may lead to	Threat: Ultimate privacy intrusion, when even one's	4.73	4.10

(continued)

Table 12.5 (continued)

Technology	Potential threat/enhancement	Threat level	Sensitivity change
"radiotelepathy" enabled by direct conversion of neural signals into radio signals and vice versa	thoughts would not be secret anymore		
Reality mining This term has been coined by MIT researchers as a new paradigm in data mining based on the collection of machine-sensed data from mobile phones, cellular tower identifiers, GPS signals, etc., to discover patterns in daily user activity. Such large-scale data mining allows the modelling of large communities of individuals to provide insight into the dynamics of individual and group behaviour	Threat: Massive collection of data pertaining to human social behaviour obviously raises privacy questions, especially if the anonymization of data is difficult Enhancement: Innovative solutions for privacy-preserving data mining have been proposed, possibly handing control back to the users	4.08	3.74

Source: Author

complexity affects the transparency, awareness, and, ultimately, the data subjects' ability to control their personal data (Ahituv et al. 2013).

12.3.2 PRACTIS Privacy Scenarios

Based on the PRACTIS expert survey, interviews with experts, and related analysis, five alternative future scenarios of privacy were constructed, reflecting different versions of changing privacy perceptions influenced by emerging technologies.

The five scenarios range from utopia to dystopia, reflecting different combinations of attitudes to privacy protection and to new technologies. Their brief summaries are presented below (Auffermann and Luoto 2011).

12.3.2.1 Scenario 1: "Privacy Has Faded Away"

The majority of people's perception of privacy has changed along with the "new rules" posed by emerging technologies. Emerging technologies are readily embraced and widely accepted. The border between private and public space has vanished since people have given up their privacy voluntarily. Various technologies have become more vital than privacy, which is still a value, but people have started to sell their "moral" for money. People make **trade-offs** in favour of goods and services (e.g. health care and safety provided by new technologies) and have given up privacy **voluntarily**.

12.3.2.2 Scenario 2: "People Want to Maintain as Much Privacy as Possible"

People value their privacy, want to protect it, and do not accept new technologies as easily as before because of former experiences of privacy intrusions. The effective use of privacy-enhancing technologies (PETs) is established and ensured by appropriate legislation. Most people are worried about the merge of information into one database covering different areas of life. In order to avoid the creation of such a database, the state controls the usage of information. The popularity of social media has decreased, since the majority of people prefer real-world interaction. Once public exposure had gone to extremities, most people started to value their privacy and private information once again.

12.3.2.3 Scenario 3: "People Have Lost Their Control of Privacy"

Dependence on new technologies has led to a situation where people are permanently monitored by the state or the private sector both at home and in public places. Information is gathered constantly about everybody. Citizens have equal access to others' information. Without noticing, people have been "sleepwalking" into a no-privacy world, and suddenly they have no choice but to live with it. People have no control over their own information (which is actually not their own anymore) as they have become dependent on new technologies, so they simply have to accept the public exposure of their private lives in order to use them.

12.3.2.4 Scenario 4: "Segmented Privacy"

Emerging technologies and privacy are perceived differently depending on which "class" a person belongs to. Privacy has become **a market value.** Technology providers have made two versions of their technology applications—one with high privacy settings and a higher price and another with low privacy settings and a lower price. Only wealthy people can afford to buy technologies where privacy settings are considered and highly valued. Instead of a trade-off, privacy and security seem to go hand in hand. Most wealthy people have moved into gated communities where they feel safer than in their former residential areas and with their privacy better protected when they live in these well-demarcated areas outside the rest of the society. Past developments have led to a situation in which income inequality has grown extremely high and the society is very prone to conflicts. This has increased the tension between different social classes.

12.3.2.5 Scenario 5: "Tailor-Made Privacy"

Privacy is understood differently by each individual. Privacy-enhancing technologies (PETs) are available to everyone. Technology applications allow people to choose privacy settings which are tailored to their individual needs. Through education, people have become more aware of the possibilities and disadvantages of emerging technologies and know how the information gathered about them is used and for what purposes. Sufficient funding has been directed at data protection. Private companies and schools have hired supervisors in charge of data protection, and civil society actors are also active. The majority of people remain open towards new technologies, since they are perceived to offer major benefits and people do not have to worry too much about their privacy risks.

The PRACTIS scenarios were analysed in depth in order to recognize the potential changes in privacy climates and explore their impacts on ethical principles. The results of the scenario analysis have provided valuable inputs for further discussion on ethical frameworks and legal considerations for the interface between privacy and technology (not elaborated in this chapter).

12.4 Conclusions

The foresight processes which helped to assess evolving security threats and privacy implications of new technologies reflect a pressing need for continuous analysis of the unfolding technology landscape for potential threats, opportunities, and other interesting implications such as the changes in public perception as exemplified in the case of privacy. The technologies covered in projects such as FESTOS and PRACTIS described in this chapter are only the tip of the iceberg; every emerging technology may embody its potential dark side as well as its bright side. Emerging technologies, especially in their early stages of development, can be viewed as conveying signals of change. Such signals may be hardly perceptible at present but are likely to constitute a strong trend, in some cases even a surprising wild card, in the future.

In addition to exploring the technology developments, both projects exemplify how foresight studies can deal with important societal issues. FESTOS put on the table the dilemma between knowledge control—limiting the access to sensitive knowledge—and the freedom of scientific research. The dark side of technology needs activities from the side of actors and decision-makers balancing between the security needs of our open democratic societies and the freedom of knowledge and scientific research. Probably more promising than top-down measures are bottom-up approaches: Instead of legislation and coercive measures with rather questionable outcomes, FESTOS proposed to develop "soft" and optional measures that base first of all on self-regulation, self-control, and education of engineers and scientists. Codes of conduct, ethical guidelines, and educational measures may be established on substate levels but develop in the future towards national, European-wide, and global regimes (Auffermann and Hauptman 2012). These should be complemented with dedicated R&D efforts which have been proposed in the context of each potentially threatening technology aiming at adoption of a "security-by-design" approach.

Similarly, privacy intrusions can be minimized or prevented (if this is indeed desired by the society) not only by top-down regulatory means (although these are probably unavoidable) but by increased user awareness, ethical considerations, and responsible use of specific technologies. These considerations should be applied in the early stages of technology development, implying the high importance of the notion of "privacy by design" (PbD). Implementation of privacy by design can reduce the risk of privacy intrusion and enable the data subject clear and simple options as to the collection of personal data, including the option to turn off the collection of data altogether. PbD can be supported by regulations, which can either require it or encourage it. Perhaps most importantly, PbD might also be driven by a market demand—depending on the appropriate awareness of users.

At any rate, in both privacy and security realms, alarmism is not recommended, but continuous assessment of the potential dark side of new technologies is always needed. It is an essential prerequisite to the implementation of the developing notion of responsible research and innovation (RRI)—an absolute necessity if we wish that after all the bright side of new technologies prevails.

References

Ahituv N et al (2013) Privacy implications of emerging technologies for stakeholders: research summary and policy recommendations. PRACTIS Deliverable D6.3

Auffermann B, Hauptman A (2012) EFP Foresight Brief No. 225. http://www.foresight-platform.eu/wp-content/uploads/2012/10/EFP-Brief-No.-225-FESTOS.pdf

Auffermann B, Luoto L (2011) Integrated security threats report. FESTOS Deliverable D3.3, with contribution from all partners

deCharms RC (2008) Applications of real-time fMRI. Nat Rev Neurosci 9(9):721

Dienel HL, Peperhove R (2011) Final scenario and indicators report. FESTOS Deliverable D4.3

Dyson F (2009) “Radiotelepathy”, the direct communication of feelings and thought from brain to brain. Edge 2009. www.edge.org/q2009/q09_3.html

Hauptman A, Sharan Y (2013) Foresight of evolving security threats posed by emerging technologies. Foresight J Future Stud Strateg Think Policy 15(5):375–391

Hauptman A, Steinmüller K (2019) Surprising scenarios: imagination as a dimension of foresight. In: Peperhove R et al (eds) Envisioning uncertain futures. Springer, Berlin (forthcoming)

Hauptman A, Sharan Y, Soffer T (2011) Privacy perception in the ICT era and beyond. In: von Shomberg R (ed) Towards responsible research and innovation in the information and communication technologies and security technologies fields. European Commission. http://ec.europa.eu/research/science-society/document_library/pdf_06/mep-rapport-2011_en.pdf

Lee B, Preston F (2012) Preparing for high-impact, low-probability events – lessons from Eyjafjallajokull. A Chatham House report

Petersen JL, Steinmüller K (2009) Wild cards. In: Glenn JC, Gordon TJ (eds) Futures research methodology – V3.0. The millennium project

Schwartz P, Randall D (2007) Anticipating strategic surprises. In: Fukuyama F (ed) Blindside – how to anticipate forcing events and wild cards in global politics. Brookings Institution Press, Washington, DC

Toumey C (2007) Plenty of eyes at the bottom. Nat Nanotechnol 2:192–193. https://doi.org/10.1038/nnano.2007.93

Aharon Hauptman is Senior Researcher at the Unit for Technology and Society Foresight at Tel Aviv University and Fellow at Zvi Meitar Institute for Legal Implications of Emerging Technologies, IDC Herzliya. His research deals with Technology Foresight, its relations with R&D policy and the evaluation of trends and impacts of emerging technologies. Dr. Hauptman has been involved in foresight in many areas, including Delphi surveys and "wild card" analysis, for example in EU projects such as FESTOS, PRACTIS, iKNOW, RACE2050 and FORCE. He holds a PhD in Engineering from Tel Aviv University and BSc from the Technion (Israel Institute of Technology).

Defence and Security: New Issues and Impacts

13

Andrew James

13.1 Introduction

In a previous paper (James and Teichler 2014), we observed that public domain foresight studies in Europe were characterised by a shift away from a focus on state-centric military threats to a much broader view of security risks. In the intervening years, much has changed. Many of those changes (the global financial crisis, the growth of authoritarianism and the rise of an increasingly assertive Russia using hybrid warfare including cyber) were barely mentioned in these foresight studies or (infamously in the case of the global financial crisis) not at all. The focus of today has returned to state-centric hybrid threats. Events have overtaken foresight studies.

This chapter provides a meta-analysis of some of the main themes emerging from public domain defence and security foresight studies conducted since the 9/11 attacks on the United States. There are also a number of studies undertaken that—because of the sensitivity of the subject area—have either not been published or have only been published in an abridged form. For instance, the *UK Science and Technology Strategy for Countering International Terrorism* makes reference to a scenario study conducted by the government's defence laboratory DSTL on future technological threats to the United Kingdom (see HM Government 2009). The chapter focuses mainly on studies undertaken in Europe and argues that these foresight studies reflect a shift in security thinking away from a focus on state-centric threats towards a much broader view of security risks. This expanded perspective includes risks presented by the vulnerability of European society to the failure of critical infrastructure, to pandemics, environmental change and resource-based conflicts. The chapter places a particular emphasis on the treatment of technological change in these defence and security foresight studies and argues

A. James (✉)
Manchester Institute of Innovation Research, The University of Manchester, Manchester, UK
e-mail: Andrew.James@manchester.ac.uk

D. Meissner et al. (eds.), *Emerging Technologies for Economic Development*,
Science, Technology and Innovation Studies,
https://doi.org/10.1007/978-3-030-04370-4_13

that the growing importance of dual-use technologies is likely to mean that defence will play a declining role as a sponsor and lead user of advanced technologies in the future.

The chapter is structured as follows: Sect. 13.2 presents some recent defence and security foresight exercises, highlighting who has sponsored the studies, their objectives and participation. The next five sections identify some of the main themes identified from a meta-analysis of these studies. Section 13.3 stresses the core theme that these foresight studies reflect a wider shift in security thinking away from a focus on state-centric threats towards a much broader view of security risks. Section 13.4 considers their treatment of the issue of resource-based conflicts. Section 13.5 notes the importance of the topic of technological vulnerability and the potential misuse of science. Section 13.6 notes how these foresight studies emphasise the changing nature of knowledge production and use in the future, and Sect. 13.7 emphasises how in the future defence is likely to be a technological follower rather than leader in most fields. Section 13.8 goes beyond James and Teichler (2014) to consider the meaning of "emerging technologies" in a military context and what they mean for the future of war. Section 13.9 provides some conclusions.

13.2 Recent Foresight Exercises

A number of foresight exercises on defence and security matters have been conducted in recent years and placed in the public domain. These exercises have been sponsored by agencies of national governments, international organisations as well as the European Commission.

At the national level, the UK Ministry of Defence's Defence Concepts and Doctrine Centre leads an ongoing Global Strategic Trends Programme. In France, the Ministry of Defence's Délégation Aux Affaires Stratégiques undertakes an ongoing programme of foresight studies. In Finland, the *FinnSight 2015* programme on the outlook for science, technology and society undertaken by the Academy of Finland and the Tekes innovation and technology agency considered security issues. In the United States, the National Intelligence Council (an institution supporting the Director of National Intelligence within the US intelligence community) has conducted a series of *Global Trends* studies.

Similarly, international organisations and foundations have engaged in foresight studies contributing to the discussion of future defence challenges. The United Nations and the Bertelsmann Foundation have supported studies dealing with the future of a globalised world, which also address security and defence issues (UNO 2004). Finally, NATO has more particularly addressed the defence challenges that the Atlantic community is likely to face in the coming decades (NATO 2006).

In addition, there have been a series of studies undertaken in Europe that have been sponsored directly or indirectly by the European Commission. The Seventh Framework Programme included for the first time a security research theme and within that there was a funding stream on foresight, scenarios and security as an evolving concept. This aimed at research in broad societal foresight to capture new and emerging threats as well as other aspects of security as an evolving concept (e.g. ethical and economic aspects). The Commission has funded projects such as FORESEC on *Europe's evolving security: drivers, trends and scenarios* and FESTOS which has as its goal "to identify and assess evolving security threats posed by abuse or inadequate use of emerging technologies and new S&T knowledge, and to propose means to reduce their likelihood". Within the FP7 Social Sciences and Humanities funding stream, the Commission has also funded foresight studies. For instance, the theme *Blue Sky Research on Emerging Issues Affecting European S&T* funded Project SANDERA which focused on the future relationship between the EU science and technology policy strategy to move towards the European Research Area and those EU policies focused on the security of the European citizen in the world both through EU defence policies and EU security policies. This chapter draws on some of the insights from SANDERA which brought together academics and think tanks from eight European countries.

The Commission was also heavily involved in ESRIF (the European Security Research and Innovation Forum). ESRIF describes itself as "a European strategy group in the civil security research domain". ESRIF was established in September 2007 by the EU Member States and the European Commission with the objective to develop a mid- and long-term strategy for civil security research and innovation in Europe. ESRIF included a working group on foresight and scenarios, and its final report—delivered in 2009—includes scenarios.

In addition the European Union Council has sponsored studies discussing future defence challenges such as the European Defence Agency's *Long-Term Vision for Defence* and the EU Institute's for Security Studies *The New Global Puzzle* (European Defence Agency 2006; Gnessotto and Grevi 2006).

The purpose of these foresight studies has been to inform policy-makers, provide a basis for policy priorities and raise awareness amongst key stakeholders. Their main published outputs have been lists of drivers of change, and some studies have also developed scenarios (e.g. ESRIF and SANDERA).

In comparison to the studies sponsored by national governments, the EU reports have addressed the topics from a security rather than defence perspective. Hence, their principal focus has been on the identification of "risks" or "hazards" to security at either the national or European level. Military threats are addressed only in a strictly limited way, relating to the Petersberg Tasks and the fight against terrorism.[1] Whilst these reports, like other foresight studies, have aimed to simulate public debate, they have sought to do so in a different manner by involving stakeholders across the EU and claiming the topic of security as a responsibility of the Commission.

These defence and security foresight exercises have found it difficult to engage with civil society. The FORESEC study and final conference raised the problem that

[1]This reflects the status of defence matters in the European Union which, prior to the Lisbon treaty, were strictly intergovernmental and from which the European commission was largely excluded.

European security foresight exercises rely almost exclusively on a community of security "experts". By omission or commission, the broader European scientific community and civil society have been effectively excluded from such exercises.

The remaining sections of this chapter report a meta-analysis of these foresight exercises undertaken as part of Project SANDERA, and from that we have identified five broad themes:

- Defence and security foresight studies reflect a wider shift in security thinking away from a focus on state-centric threats towards a much broader view of security risks.
- Defence and security foresight studies emphasise that resource-based conflicts are likely to grow in importance in the future.
- Defence and security foresight studies tend to emphasise the technological vulnerability of European society and the potential for the misuse of science
- Defence and security foresight studies emphasise the changing nature of knowledge production and use in the future and how that might increase security risks to Europe.
- Defence and security foresight studies emphasise how in the future defence is likely to be a technological follower rather than a technological leader in most fields.

13.3 From "Defence" to "Security": The Emergence of a New Security Paradigm

The first point from our meta-analysis is that the focus of the defence and security foresight studies that we have reviewed is less on state-centric threats and the potential for state-on-state conflict and more on the security risks posed by intentional and unintentional human actions, technological change and environmental factors.

In this way these foresight studies reflect a broader shift in academic and policy thinking from the traditional notion of national security, which puts the emphasis on the security of the state and military threats to its territorial integrity towards a thinking in broader terms to include new types of threats (e.g. ecological, economic) to new objects of security (the human beings or the citizen, society), calling for a new means to ensure security. Several concepts have been developed to capture these or parts of these notions such as "human security", "total defence", "societal security", "security of the citizen" or the "all-hazards" approach [for a discussion see, e.g. Giegerich and Pantucci (2008)].

Many of these studies assume a shift from traditional state-centric warfare towards a much broader view of security risks. This perspective is captured in the UK Ministry of Defence's Development Concepts and Doctrine Centre (DCDC) report which observes that "Greater interdependence and intensifying competition are likely to be a defining feature of the next 30 years. This tension is likely to heighten preoccupation with risk at every level, from the personal to the international" (DCDC 2007).

Almost every one of these foresight studies places an emphasis on the increasing interdependence of large parts of the world economy. In the future, the world is seen as one characterised by interdependencies through supply chains, financial flows as well as knowledge flows. These studies are agreed that the twenty-first century will be "the Asian century" (Délégation aux affaires stratégiques 2008; DCDC 2007). They also emphasise how almost ubiquitous communications technologies are likely to strengthen that sense of interdependence. What is striking is how the world has changed in ways not anticipated by these foresight studies. The emergence of authoritarianism and anti-globalisation, not least with the election of President Donald Trump, raises profound and worrying questions about these easy assumptions about growing interdependence that have been commonplace in many foresight studies (although for balance it ought to be noted that the DCDC future reports warn of the tensions arising from globalisation).

There is an emphasis on competition in this new globalised and multipolar world. This competition is seen not only in economic terms although this is recognised as important. Equally, attention is paid to the nature of political competition in a world where the United States is likely to face near-peer competitors such as China and India and in which Europe struggles to retain its position in the world. The competition is also seen in terms of a competition between the state and the emergence of new actors and the importance of non-state terrorist groups and disaffected individuals. Competition also takes the form of competition between competing identities, ideologies and religious world views.

The emphasis on risk also means that these foresight studies consider the consequences for European society of unintended dangers such as natural or man-made disasters. Interdependence means that Europe will become more vulnerable to pandemics no matter whether a virus is spread by tourists, terrorists or both. Equally, food security, energy security and so forth are matters of concern as well as increasingly frequent and violent weather events as a consequence of climate change.

13.4 The Rise of Resource-Based Conflicts

The second observation from our meta-analysis is that defence and security foresight studies emphasise that resource-based conflicts are likely to grow in importance in the future.

Climate change is seen as likely to exacerbate existing conflict situations by intensifying already stressed security situations, particularly in regions with weak institutions that are not able to mitigate or adapt to the changed climatic circumstances (Academy of Finland and Tekes 2006; NIC 2008). The role of climate change in spurring an increase in uncontrolled migration is also emphasised not least from Africa as the sub-Saharan region experiences increasingly prolonged droughts and famines. There is likely to be an increasing prevalence and frequency of human and animal pathogens as a consequence of climate change, international flows of people and sociocultural change, and this is likely to lead to an increasing number of global pandemics (Délégation aux affaires stratégiques 2008).

Some of these foresight studies also note that demographic developments may have security implications. Asia, Africa and Latin America will account for virtually all population growth over the next 20 years amounting to 1.2 billion more people by 2025 (UNO 2004). In combination with continued economic growth, this will significantly increase demand for energy, food and water resources and amplify the problem of climate change. In countries with significantly more young males than females ("youth bulge"), economic and social institutions need to develop in order to avoid that these countries (in particular Afghanistan, Nigeria, Pakistan and Yemen) continue to be prone to instability and internal conflict (DCDC 2007; National Intelligence Council 2008).

Geopolitical struggles over material and energy resources are also identified as a potential source of growing interstate tensions in the future (DCDC 2007; Délégation aux Affaires Stratégiques 2008). This may take three forms: first, states might use their control over energy resources for political coercion and influence. Second, terrorists and pirates might pose threats to transit routes calling for military protection of those routes, a situation we witness currently at the coast of Somalia. Finally, domestic instability and conflict within strategic energy-producing states could trigger intervention from outside (National Intelligence Council 2008).

13.5 Dangerous Knowledge: Technological Vulnerability and the Misuse of Science

Another theme that we have identified is that defence and security foresight studies tend to emphasise the vulnerability of European society due to its growing dependence on technology and the potential for the misuse of science and technology.

A common theme is that in the future, cyberspace is likely to be a key area for conflict and that security will need to be ensured in this domain. A number of security foresight studies note how critical infrastructures including energy supply, transport and water may be vulnerable to deliberate or accidental failure due to their dependence on networked information and communication technologies (Academy of Finland and Tekes 2006; NIC 2008).

Another theme is the potential for misuse of science and technology. A number of the foresight studies emphasise that there is likely to be accelerating convergence and interaction between major enabling technologies such as information technologies, biotechnologies, nanotechnologies and neuro- and cognitive sciences (Institute of the Future 2005; James et al. 2008). These converging technologies could have a profound transformative impact on European societies and are seen as presenting significant economic, social, cultural and ethical opportunities and challenges. Military application of nanotechnology is seen by some as having potentially destabilising effects, and there may be spillovers from the military use of converging technologies to crime and terrorism (Nordmann 2008). Thus, ESRIF's final report says:

> There is no doubt that rapid evolution in ICT/cyber security and its misuse will continue and even accelerate. Some technologies already identified as candidates for misuse are nanotechnology, artificial intelligence and 'synthetic biology' (i.e., the use of DNA technology to 'engineer' living organisms). These threats will have to be continually monitored and countered. (ESRIF 2009: 25)

13.6 Beyond the Military-Industrial-Scientific Complex: The Changing Nature of Knowledge Production and Use

A further theme emphasised by defence and security foresight studies is the changing nature of knowledge production and use in the future and how that might increase security risks to Europe. The globalisation process is likely to further weaken the ability of states to control research, production and exportation of sensitive technologies and goods, and even maintaining secrecy about sensitive technologies and systems will be extremely difficult (DCDC 2007). These studies suggest that technological innovation will continue to be predominantly commercially led, and commercial dynamics mean that companies will seek to exploit technologies in as many new applications and markets as possible (DCDC 2007). Global capital flows and cross-border ownership will mean that the ownership of companies will become increasingly complex (Délégation aux affaires stratégiques 2008). This may provoke tensions with the security community who are likely to regard the proliferation of some scientific and technological knowledge as a threat its potential to offer terrorists an increasing access to sensitive technologies (National Intelligence Council 2008).

They also emphasise that we are likely to see the emergence of new scientific superpowers. Science in the twenty-first century will be more like a network, with multiple, linked centres of excellence. The United States, Britain and other current leaders will still be important centres of research and innovation but will be joined by India and, probably, China. A host of small countries or regions, including South Korea, Taiwan, Israel and Brazil, will also develop world-class capabilities in strategic specialties or interdisciplinary areas, building targeted programmes that fuse global scientific knowledge with local technical, natural or even cultural resources (Institute for the Future 2005).

A further element in these studies is that we are likely to see new knowledge circulation patterns. The globalisation of scientific and technological activity is likely to continue. There will be increasing information exchange across borders, and we are likely to move "from brain drain to brain circulation (Institute for the Future 2005). This circulation is likely to be driven by the growth of research and entrepreneurial opportunities in emerging countries, the general lowering of global barriers to migration and the erosion of the standard career model in business and academia (Institute for the Future 2005).

These foresight studies also suggest that an increasing body of scientific and technological knowledge and technique is likely to be widely available as a consequence of the emergence of new scientific superpowers and the diffusion of knowledge. This is likely to be facilitated by modern means of communication (mainly, but not only, the Internet) (Délégation aux affaires stratégiques 2008). The Internet and information technologies have broadened access to scientific knowledge and are starting to lower the barriers to participation in scientific research. In the next decades, the spread of pervasive computing technologies, low-cost sensors, flexible electronics and desktop manufacturing tools, combined with commons-based, peer-reviewed scientific production systems, will broaden the range of opportunities for wider participation in science and technology (Institute for the Future 2005).

These trends and technologies, some of the studies suggest, will lower the barriers to participation in science for individuals, groups and emerging countries (Institute for the Future 2005). The result may be the growth of "amateur science", new scientific and technical centres of excellence in developing countries and a more global distribution of world-class scientists and technologists (DCDC 2007).

There is a concern in the defence and security communities that the pace of change in many science and technology domains may exceed the speed at which governments can respond and that the spread of scientific knowledge and technological capabilities may become able to outperform states who can be constrained by their decision-making processes, technological path dependencies and commercial and political factors (DCDC 2007).

This is an emerging trend, and such developments may be regarded as a threat to the security community since it has the potential to widen the tools available to both small terrorist groups and to emerging countries seeking to increase their ability to wage war by both old and new means (DCDC 2007). The diffusion of the knowledge underpinning weapons of mass destruction and of dual-use knowledge in the life sciences that has the potential to be applied to biological weapons is frequently noted and discussed (DCDC 2007; Délégation aux affaires stratégiques 2008). In addition, terrorists and/or criminals might abuse new emerging technologies for their purposes, in particular the output of robotics, nanotechnology in combination with medicine, cognitive science, sensors, networks and smart materials (Délégation aux affaires stratégiques 2008). This driver is supported also by a trend that innovation, research and development will originate from more international and diffuse sources and will proliferate widely, making regulation and control of novel technologies more challenging (DCDC 2007).

At the same time, such developments may also be seen as an opportunity by the defence and security communities, and they will increasingly seek to access the potential of "democratised innovation" as part of a move towards an open innovation model for defence. This may stimulate efforts on the part of the defence and security policy communities to build closer and cooperative relationships with the wider civil research community although how that civil research community may respond is an open question as we will see in the next section.

13.7 Defence as a Technological Follower Rather than a Technological Leader

Defence and security foresight studies emphasise how in the future defence is likely to be a technological follower rather than a technological leader in most fields.

Defence has played an important role as a sponsor and lead user of some advanced technologies at some stages in their development. The growing emphasis of defence science and technology policy on accessing commercial off-the-shelf technologies as well as a move towards an open innovation model that will depend upon accessing globally available technological knowledge suggests that defence may play a declining role as a sponsor of advanced technologies in the future. Equally, defence procurement has declined in absolute terms in Europe and also when compared to the size of other (civilian) markets (James et al. 2008). This is a continuation of past trends, and defence and security science and technology research may be of declining importance in the European science policy mix.

The rapid pace of nondefence origin technological change and the likelihood that absolute defence R&D spending in Europe will continue to decline (or at best remain stable) are likely to impact the way that governments and industry conduct defence R&D. The role of defence R&D is already shifting from the development of new technologies in large specialised defence research establishments to partnerships that can access and exploit technologies that are the product of commercially funded research. This trend is likely to continue (James et al. 2008).

Specific national government R&D investment in in-house expertise and technology development is likely to continue where there is no civilian equivalent and where there are particular concerns about the need to retain national operational sovereignty, for example, in some critical defence and security technologies, such as cryptography, nuclear, counterterrorism and chemical, biological, radiological and nuclear (CBRN) defence (DCDC 2007). In other fields, however, governments are already developing R&D programmes that draw on the increasingly diverse global science and technology base in industry, SMEs and universities and adapt and apply that knowledge for military use. This approach is likely to broaden and deepen.

The pace of this move towards an open innovation approach is likely to depend in the ability of the civil and military research communities to work together. On the other hand, it will require the willingness of those nontraditional sources of scientific and technological knowledge to engage with the defence sector. There are major cultural differences between universities and the defence sector. On the other, traditional defence firms will need to embrace new business models and marketing practices reflecting the different dynamics of value creation in civilian markets and requirements linked to commercial customers.

13.8 Emerging Technologies and the Future of War

So far this chapter has closely followed an earlier journal paper (James and Teichler 2014). This section goes beyond that earlier paper to consider the meaning of "emerging technologies" in a military context and what they mean for the future of war.

This question is important because radical technological change is back on the military's agenda. In his last major address as Defence Secretary, Chuck Hagel announced the Defence Innovation Initiative, aimed at fostering a third "game-changing" offset strategy. Eisenhower's "New Look" doctrine in the 1950s led to the development of new types of nuclear weapons, long-range delivery systems and active and passive defences to offset the Soviet Union's quantitative force advantage. The offset strategy of the 1970s and 1980s led to leap-ahead capabilities like standoff precision strike, stealth, wide-area surveillance and networked forces. The Defence Innovation Initiative calls for "a new Long-Range Research and Development Planning Program [that] will help identify, develop and field breakthroughs from the most cutting-edge technologies and systems, especially in robotics, autonomous systems, miniaturization, big data and advanced manufacturing, including 3-D printing".[2] At the time of writing, it was not possible to judge the Trump administration's view of the Defence Innovation Initiative although the President's 2017 budget request had included a proposed substantial increase in spending of defence research and development.

13.8.1 Emerging Technologies in the Military Context

This chapter has already shown that visions of the military future almost always have a strong technological element. Emerging technologies feature prominently in foresight studies. They identify a host of emerging technologies that may have implications for security, military capability and—in some cases—the conduct of future war. These include:

- **Autonomous systems and artificial intelligence**: self-thinking, deciding and organising partially sentient devices that mimic aspects of human intelligence and decision-making are being developed and may move/reduce the need for human input and reduce the manpower burden.
- **"Big data"** information analysis and exploitation may lead to better decision-making and enhanced intelligence analytic capability.
- **Developments in nanotechnology and microsystems** promise sensors of small size and improved performance.

[2]"Hagel Announces New Defence Innovation, Reform Efforts" http://www.defense.gov/news/newsarticle.aspx?id=123651 (last accessed 7 February 2015).

- **Human enhancement and augmentation**: a range of technologies offer considerable scope for enhancing and augmenting the physical and cognitive performance of humans, including prosthetics, drugs and genetic manipulation.
- **Synthetic biology**: the design and fabrication of biological components and systems that do not already exist in the natural world with the potential to produce to create novel threats in the form of "designer" bio-weapons.
- **Social and behavioural sciences**: developments are expected to provide additional insights into the intent and behaviour of individuals and groups leading to new opportunities to influence them.

However, before going any further, it is important to define what is—and what is not—meant by "emerging technologies". The UK's Defence Technology Plan defines emerging technologies as follows: "Emerging technologies can be characterised as: immature technologies in the early proof-of-principle stages; [and] more mature technologies but where a novel defence application has been identified". Whilst this definition appears clear and straightforward (and this chapter will use it), it is the case that a feature of much of the discussion of emerging technologies is a lack of clarity as to the subject of analysis.

"Emerging" is used variously to examine technologies that analysts regard as potentially emerging in the far future (e.g. the latest UK MOD DCDC programme report looks out to 2040 and consciously examines what technological developments *may* occur). In contrast, "emerging" is sometimes used to describe technologies that have reached a stage that we know that they *will* find application in a weapon system in the near future [e.g. many of the "emerging" IT technologies discussed by Bruce Berkowitz in his 2003 book are now in military service at least with the US military (Berkowitz 2003)]. Sometimes analysts conflate the far future and the soon to be fielded as "emerging technologies" giving the impression to the unwary that (true) emerging technologies on the technological far horizon are as certain to be fielded as those in late-stage development. This raises important questions about timing that are critical to discussions about emerging technologies. It also raises issues about uncertainty. Both issues will be discussed later in this chapter.

There is an important distinction here that is sometimes missed by military analysts of emerging technologies. The distinction is between technologies and the weapons, their delivery systems and the infrastructure that supports military capability. *Technologies* underpin *weapon systems* but are distinct from them. Thus, nanotechnologies may be important to the military, but only if they find application in weapon systems. Consequently, how emerging technologies and other factors are combined into military capability should be the critical consideration not the emerging technologies themselves.

Equally, new or improved classes of weapon rarely (if ever) comprise only new ("emerging") technologies but instead combine new technologies with mature technologies. Thinking influenced by the economist Joseph Schumpeter emphasises that innovation can be new combinations of existing technologies and stresses the potential significance of combining existing technologies in a new use. Innovation that produces modern weapon systems is increasingly based on this kind of dynamic recombination of generic technologies which are often information technologies (Hasik 2008). The DCDC Strategic Trends study identifies the rapid asymmetric insertion and exploitation of widely available commercial technologies—GPS, low-cost unmanned aerial vehicles, mobile telephones—as a significant concern. Indeed, the experience of Iraq and Afghanistan provides graphic illustrations of how such tactics can have devastating asymmetric effects. The contrast between the rate of combinatorial innovation of this kind and the pace of developments in the traditional defence acquisition has been striking (DCDC 2014).

13.8.2 Emerging Technologies and the Future of War

Most emerging technologies represent incremental improvements to what went before and enhance the competencies of the military along dimensions that they have traditionally valued. This kind of technological development presents relatively few challenges to the military, although their insertion into existing platforms can be difficult. In contrast, it is new technologies that are radical, competence destroying and create new sources of military advantage along dimensions not traditionally valued or poorly understood by the military that tend to be the focus of attention and concern.

Fundamentally, these types of new technologies can change the environment in which military forces operate. In *The Pursuit of Power*, William H McNeill (1982) charts the consequence of technological change on the balance of power. In *War and Power in the 21st Century: The State, Military Power and the International System*, Paul Hirst (2001) analysed how new military technologies change the way that wars are fought and how power relations change as a result.

A radical new technology can change the balance of power or create new forms of insecurity. The most dramatic illustration of the impact of new technology was the Allied development of the atomic and hydrogen bombs during the Second World War and the subsequent development of similar capability by the Soviet Union. In turn, the development of inertial navigation technologies added the prospect of accuracy to devastating lethality.

It is a commonplace that today's emerging technologies may lead to the proliferation of novel disruptive threats. Many—most—of the emerging technologies are not the preserve of the military and governments. Most are emerging out of work being conducted in universities, firms and garages across the world. Some require only modest resources. For example, synthetic biology is an area of S&T that has a growing and Internet-linked "DIY community".

New technologies may also influence the likelihood of conflict. The emergence of the hydrogen bomb arguably reduced the threat of conflict. The increased availability and capability of remotely operated vehicles and their increasingly autonomous successors may reduce the threshold for their use by reducing the political risk of military casualties, likewise cyberwarfare. "The anonymity that cyberspace can offer reduces the risk of retribution, so may increase the attractiveness of making an attack". In *The Future of War* (2004), Christopher Coker talks about the "re-enchantment of war in the twenty-first century". Coker argues that developed societies are likely to continue with war in the future because technological change—not least that associated with the information revolution—may make it more rational and precise than ever before. Indeed, he says: "if war is seen as merely one end on a spectrum of violence, death is not essential to it. Killing could be made redundant (though probably not optional), leaving physical coercion or the will to power by other means" (p. 141).

New technologies can redefine the way that warfare is conducted or create new types of warfare. Technology and military doctrine are closely coupled and interdependent (Alic 2007). Blitzkrieg, the AirLand Battle and Carrier Strike are but three examples of how new technologies combined with organisational and doctrinal change led to new ways of warfare (Williamson and Murray 1996). The Revolution in Military Affairs provides another example. The Internet and its widespread application have created the possibility of a new form of warfare—cyberwarfare—that was hardly imaginable 20 years ago.

Emerging technologies may also pose profound ethical and moral questions. Many areas of emerging technology will pose ethical challenges. Take the use of biotechnology, for instance. Christopher Coker (2004: 140) argues:

> by enhancing, modifying or altering our genes, we may be able to enhance the things we do well, and have always done well, as a species. One of the things we have done particularly well over the centuries has been war, and there is nothing to suggest that we will be going out of the war business—indeed quite the opposite.

In his latest book, *Warrior Geeks*, Coker (2013) warns that technological change is threatening to create a battlespace that has no place for human qualities such as courage, sacrifice or honour and even more fundamental categories such as subjectivity, agency and ethics.

13.9 Conclusion

This chapter has analysed some of the main themes emerging from public domain defence and security foresight studies conducted since the 9/11 attacks in the United States. We have emphasised how these foresight studies reflect a shift in security thinking away from a focus on state-centric threats towards a much broader view of security risks that includes risks presented by the vulnerability of European society to the failure of critical infrastructure, pandemics, environmental change and resource-based conflicts.

This chapter has also emphasised how emerging technologies may influence the future of war. This chapter has noted how the role of defence for the development of new technologies is likely to change dramatically in the future. In particular, defence and security foresight studies have emphasised that the growing importance of dual-use technologies is likely to mean that defence will play a declining role as a sponsor and lead user of advanced technologies in the future. This can be seen as the continuation of trend developments over the last two decades. At the same time, whilst the role of defence in the creation of knowledge has changed along these dimensions, there are also continuities. "Defence", i.e. the call to protect the security of a country, its population and assets against threats or from any harm, can be expected to remain an accepted justification for extraordinary political action and the channelling of political, financial and industrial resources. We have pointed to the "securitisation" of cyberspace and of critical infrastructure, areas that have formerly been considered outside the realm of defence, governments and private actors.

This chapter concludes with a final reflection. Rémi Barré has rightly observed that "The objective, themes and content of a foresight have specific meaning and intention".[3] Nowhere is more true than in the field of defence and security foresight. Defence and security represent distinct epistemic and policy communities. Risk is the lens through which these policy communities view the world and this is reflected in the often pessimistic character of the visions of the future that emerge from defence and security foresight exercises. Many of the foresight exercises are essentially closed activities that draw upon expertise from within the policy community (although in some exercises there have been attempts to "reach out" to broader expertise). Like all policy fields, there are strong vested interests, and the proper role of defence and security in Europe is controversial and contested. The meaning and intention of security and defence foresight activities are deserving of further academic scrutiny.

Acknowledgements I wish to acknowledge the contribution of Dr Thomas Teichler to the journal paper on which parts of this chapter are based. I also wish to thank the participants in the Clements and Strauss Centers Seminar "Emerging technologies and the future of war" at the University of Texas at Austin in February 2015. Section 13.8 draws on my comments to that seminar and the useful discussion that followed. I also wish to express my thanks to all the members of the SANDERA consortium for their contribution to our thinking on these matters and in particular the participants in the SANDERA workshops in Manchester and Valencia and the contributors to the SANDERA discussion papers. All errors and omissions are entirely my responsibility.

References

Academy of Finland and Tekes (2006) FinnSight 2015 - the outlook for science, technology and society. Academy of Finland, Helsinki

Alic JA (2007) Trillions for military technology: how the pentagon innovates and why it costs so much. Palgrave MacMillan, New York

Berkowitz B (2003) The new face of war: how war will be fought in the 21st century. The Free Press, New York

Coker C (2004) The future of war: the re-enchantment with war in the twenty-first century. Blackwell, Oxford

Coker C (2013) Warrior geeks: how twenty-first century technology is changing the way we think about war. Hurst & Company, London

[3]Intervention by Rémi Barré at the AUGUR project conference Sharing Visions on Europe in 2030: Lessons from Comparative Approaches of Recent Foresight Exercises 2 June 2010, Brussels.

DAS (2008) Geostrategic prospectives for the next thirty years. Report made under the direction of the Délégation aux affaires stratégiques. Délégation aux affaires stratégiques, Paris

DCDC (UK) (2007) The DCDC's global strategic trends programme 2007–2036. Report prepared by the UK Ministry of Defence Development, Concepts and Doctrine Centre, DCDC, Shrivenham

DCDC (UK) (2014) Global strategic trends: out to 2040, 4th edn. Development, Concepts and Doctrine Centre, Ministry of Defence, London

European Defence Agency (2006) An initial long-term vision for European defence capability and capacity needs. EDA, Brussels

European Security Research and Innovation Forum (2009) Security research: final ESRIF report. European Commission, Brussels

Giegerich B, Pantucci R (2008) FORESEC deliverable 2.4 synthesis report

Gnesotto N, Grevi G (2006) The new global puzzle. EU Institute for Security Studies, Paris

Hasik J (2008) Arms and innovation: entrepreneurship and alliances in the twenty first century defense industry. University of Chicago Press, Chicago

Hirst P (2001) War and power in the 21st century: the state, military power and the international system. Polity Press, Cambridge

HM Government (UK) (2009) UK science and technology strategy for countering international terrorism. Home Office, London

Institute for the Future (2005) 2005–2055 science and technology perspectives. Institute for the Future for UK Government Office of Science & Innovation

James AD, Teichler T (2014) Defence and security: new issues and impacts. Foresight 16(2): 165–175

James AD, Hartley K, Lazaric N, Gasparini G (2008) Study on how to measure the strengths and weaknesses of the DTIB in Europe. European Defence Agency, Brussels. http://www.eda.europa.eu/documents/09-11-24/Study_on_how_to_measure_Strengths_and_Weaknesses_of_the_DTIB_in_Europe

McNeill WH (1982) The pursuit of power: technology, armed force and society since AD 1000. University of Chicago Press, Chicago

NATO (2006) Future world scenarios. North Atlantic Treaty Organisation, Brussels

NIC (2008) Global trends 2025: a transformed world. National Intelligence Council, Washington, DC

Nordmann A (2008) Converging technologies - shaping the future of European societies. European Commission, Brussels. http://www.google.com/url?sa=t&source=web&cd=1&ved=0CBYQhgIwAA&url=http%3A%2F%2Fciteseerx.ist.psu.edu%2Fviewdoc%2Fdownload%3Fdoi%3D10.1.1.133.2322%26rep%3Drep1%26type%3Dpdf&rct=j&q=Converging%20Technologies%3A%20Shaping%20the%20Future%20of%20European%20SocietiesNordmann%2C%202004&ei=-4EgTp6CJ4WBhQeS7KTBAw&usg=AFQjCNFQrEo7g9Qtx4hRijIuSHXInMJF4A&cad=rja. Accessed 12 Dec 2008

UNO (2004) A more secured world: our shared responsibility. Report of the high-level panel on threats, challenges and change. United Nations, New York

Williamson W, Murray AR (eds) (1996) Military innovation in the inter-war period. Cambridge University Press, Cambridge

Andrew James is Professor of Innovation Management & Policy at Alliance Manchester Business School and a former Director of the Manchester Institute of Innovation Research. His research focuses on technology and innovation management and policy in the aerospace, security and defence sectors. Between 2009 and 2011, he was the Scientific Coordinator of SANDERA (The Future Relationship between Security and Defence Policies and the European Research Area) funded under the Seventh Framework Program Socio-Economic Sciences and Humanities theme Blue Sky Research on Emerging Issues Affecting European S&T.

Defense 4.0: Internet of Things in Military 14

Serhat Burmaoglu, Ozcan Saritas, and Haydar Yalcin

14.1 Introduction

Scientific and Technological (S&T) developments have been influencing military concepts and practice, particularly following the inception of the scientific revolution in the late sixteenth century. This interaction has not always been one way. Defense has traditionally been one of the key drivers of S&T advancements due to large amount of funding it received particularly by national governments. A number of technologies have been developed for defense, found their civilian applications, and vice versa. Wherever the boost for change comes from, the nature of warfare has changed radically both due to S&T advancements and changing socioeconomic and geopolitical contexts. Despite of the barriers due to strict organizational culture, armies have adapted themselves into changing characteristics of warfare through new concepts and instruments. Examples can be found from the earlier Revolutions in Military Affairs (Burmaoglu and Saritas 2017). A recent example is NATO's changing concepts to tackle with the recent hybrid wars, which is characterized by the involvement of non-state actors in warfare.

Among S&T developments, recent advancements in Information and Communication Technologies (ICTs) bring enormous opportunities as well as challenges for defense. One of the recent phenomena emerged with the rapid development of ICTs

S. Burmaoglu
Faculty of Economics and Administrative Sciences, Izmir Katip Celebi University, Izmir, Turkey

O. Saritas (✉)
National Research University, Higher School of Economics, Moscow, Russia
e-mail: osaritas@hse.ru

H. Yalcin
Faculty of Economics and Administrative Sciences, Izmir Katip Celebi University, Izmir, Turkey

Faculty of Humanities and Social Sciences, Izmir Katip Celebi University, Izmir, Turkey

D. Meissner et al. (eds.), *Emerging Technologies for Economic Development*,
Science, Technology and Innovation Studies,
https://doi.org/10.1007/978-3-030-04370-4_14

is the Internet of Things (IoTs), which affects every aspect of life with a growing number of devices communicating with each other. The number of connected devices is expected to reach 50 billion connected devices by 2020, and the potential market value is expected to be between $2.7 and $6.2 trillion per year by 2025 (Mantyika et al. 2013). While the possibilities introduced by the IoT have been providing immense benefits, the increasing number of connections makes the system ever more complex and vulnerable because of the difficulty of securing huge networks. If one of the main platforms for warfare is going to be the cyberspace and if the combatants of the future are going to be irregulars, then digitalization of warfare and cyberterrorism can be considered as the logical paradigm of future conflict (Rathmell 1997), which should be countered by authorities appropriately.

The aim of this chapter is to discuss how the IoTs will affect the military affairs and to propose future scenarios for exploring alternative trajectories (Miles et al. 2016) while contributing to the existing literature on the opportunities and threats brought by new and emerging technologies as well as changing nature of warfare. Thus, the outline of the chapter is as follows. In the second section, digitalization and IoT concepts are reviewed from multiple dimensions by considering their social, technological, economic, environmental, political, and value/cultural (STEEPV) aspects. Then in the third section, the relationships between defense and IoTs are investigated with emerging opportunities and threats. In the fourth section, two scenarios are presented. These predictable and possible scenarios portray alternative roles the digitalization and IoTs can play in the future warfare. Finally, findings are discussed, and future implications for research and policy are discussed in the fifth section.

14.2 Digitalization and Internet of Things

The Internet of Things (IoTs) can be considered as an important enabler of the information society by providing advanced services through interconnected physical and virtual things, based on interoperable information and communication technologies.[1] The term was first coined by Ashton (2009):

> the Internet has been almost completely dependent on people for its supply of information. But in the future, things will be able to input data themselves. It will be as though a net is laid over the physical world, linking up and processing the abundance of data generated by "smart" things and ubiquitous sensors. This is expected to reveal patterns and make everything from energy to logistics transparent and potentially open to real-time optimization.

It is a new term, but not a new process. The precursor operations of IoT were known as "pervasive computing," "ubiquitous computing" (ubicomp), and "ambient intelligence." Ashton (2009), in his presentation at Procter & Gamble, recommended

[1]http://handle.itu.int/11.1002/1000/11559-en?locatt=format:pdf&auth

the use of RFID technology in the company's supply chain application. This kind of operations could not make progress until the early 1990s because of the high investment costs of database storage technologies. Once the data storage expenditures became cheaper, a new data storage model, namely, "cloud," showed up. The cloud system used from the 2000s enables IoT, because it provides an infrastructure, which can replace a central server. It is still possible to access a file or piece of information, which would normally be hosted by the central server, since these bytes of information are "distributed" or "replicated" throughout the network. This kind of applications enables new services like "distributed computing," where every device on the network is used as a potential node for storing information. The IoTs is meant to move these technologies once step further to describe a network between many and different types of devices.[2]

Buildings, cars, consumer products, and people then become information spaces connecting with each other through "Radio Frequency Identification Tags" or sensors and transmitting all kinds of data through this tags. By connecting the things, the world has become an interface or a living organism that makes "real-time data workflow of the connected things." This organism gives a chance to make a smart ecology in our everyday life. In order to profit the advantages of the real-time information flows, we must learn how to make sense and use it. We must have an ability to read data as "data" or "information," not a noisy or unusable thing (Daim et al. 2016). In other words, with the gadgets, sensors, and machines that track our every move in the real world, it is possible to develop apps and infrastructures that may learn and predict our actions and emotions. Cloud-based apps are the key to using leveraged data. IoTs does not function without cloud-based applications to interpret and transmit the data coming from all the stakeholders. The cloud enables "turning information into action" via linked data. Cloud-based technologies and IoT focus more on the functionality and the data, not the devices. In other words, IoT is more about the data than the hardware that serves it. The hardware is just there to serve the data to the user's needs for all aspects of everyday life and thus enables cheaper IT choices that connect everything with each other. Connected devices could have their own connected channels. Dedicated channels could also serve as a backbone for things (devices, etc.) to communicate in case of emergencies. That way, one network can still stay up if the other one becomes overloaded or offline, for instance, millions of people streaming a popular video will not bypass an emergency call or alert.

Through the improvement of the information architecture, IoT technologies not only improve everyday life of humans but also transform some of the key industries. Examples are given in Table 14.1 with key changes and potential benefits provided for the users.

A number of standards and protocols are needed to regulate the IoT systems and ensure the secure operation of the IoT services. These are discussed in the next session.

[2]http://www.theinternetofthings.eu/what-is-the-internet-of-things

Table 14.1 The internet of things: a transformational force

Industry	Key change	Potential benefits
Automotive and transportation	Real-time driving behavior, traffic, and vehicle diagnostics	Improved customer experience, reduced pollution, increased safety, and additional revenue streams
Healthcare	Remote monitoring of staff and patients ability to locate and identify status of equipment	Improved employee productivity, resource usage, and outcomes that result in efficiency gains and cost savings
Manufacturing	Quick response to fluctuations in demand; maximized operational efficiency, safety, and reliability, using smart sensors and digital control systems	Enhanced agility and flexibility and reduced energy consumption and carbon footprint
Retail	Stock-out prevention through connected and intelligent supply chains	Ability to predict consumer behavior and trends, using data from video surveillance cameras, social media, Internet, and mobile device usage
Supply chain	Real-time tracking of parts and raw materials, which helps organizations preempt problems, address demand fluctuations, and efficiently manage all stages of manufacturing	Reduced working capital requirements, improved efficiencies, and avoidance of disruptions in manufacturing
Infrastructure	Smart lighting, water, power, fire, cooling, alarms, and structural health systems	Environmental benefits and significant cost savings with better utilization of resources and preventive maintenance of critical systems
Oil and gas	Smart components	Reduced operating costs and fuel consumption
Insurance	Innovative services such as pay-as-you-go insurance	Significant cost savings for both insurers and consumers
Utilities	Smart grids and meters	More responsive and reliable services; significant cost savings for both utilities and consumers resulting from demand-based and dynamic pricing features

Source: Ericsson, M2M Magazine 2013, Zebra Consulting/Forrester Research, IBM, McKinsey & Co., Data Informed, ZDNet

14.2.1 Standards, Protocols, and Applications for the IoTs

Besides the standards and protocols set by the Internet Engineering Task Force (IEFT), there are several other protocols which are also under discussion. For instance, Message Queue Telemetry Transport (MQTT) is a lightweight publish/subscribe messaging transport connectivity protocol. Developed by IBM, the protocol is integrated with the IBM WebSphere application server. Another IoT solution, ZigBee (or XBee), is a set of application profiles for creating low-rate wireless mesh networks which has been built upon the 802.15.4-2003 standard. DASH7 Alliance operated at the 433 MHz frequency range. Among other uses, the system enables tag-to-tag communications with a range up to 1 km. It is suitable, for instance, for setting up friendly fire warning applications for soldiers. BACnet is used as a communication protocol for Building, Automation and Control networks. BACnet is essentially used in HVAC systems (heating, ventilation, and air-conditioning), lighting control, and access control. It is also suitable for both in-house and outdoor smart headquarter applications (Bandyopadhyay et al. 2013).

With the developments on ICTs, Bluetooth has become a key technology in computing and product markets. It is a key for wearable products and plays an important role enabling IoT through smartphones, smartwatches, and other wearable technologies. There are also new developments about Bluetooth technology such as Bluetooth Low Energy (BLE) or Bluetooth Smart, which are designed not for file transfer but more for small and real-time dataflows. It has a major advantage certainly in a more personal device context over many competing technologies given its widespread integration in smartphones and many other mobile devices. It has a huge potential about smart applications and devices. ZigBee is another option for IoT applications. Profiles that produce via ZigBee use the IEEE802.15.4 protocol. This is an industry standard for wireless networking technology operating at 2.4 GHz. The technology targets applications with relatively infrequent data exchanges and at low data rates over a restricted area. This can be military headquarters and within a 100 m range or inside a building. ZigBee or RF4CE has some significant advantages. It consumes low power and provides better security, robustness, and scalability. With its high node counts, it takes the advantage of wireless control and sensor networks, in particular with M2M and IoT applications (Ding et al. 2009). Z-Wave is a low-power RF communication technology like BACnet . It is preferable for microenvironmental automation for products such as light controllers and sensors. It supports full mesh networks without the need for a coordinator node and is very scalable, enabling control of up to 232 devices. In other words, it makes an infrastructure for calm technology focusing on data which focus on a broad mix of information.

Sigfox is an alternative wide-range technology. Its range falls between Wi-Fi and cellular. The system uses the ISM bands to transmit data to and from connected objects over a very narrow spectrum. These are free of charge and do not require licensing. Sigfox uses an Ultra Narrow Band (UNB) technology. The system is designed to handle low data transfer speeds from 10 to 1000 bits per second. It can be used for real-time data gathering in areas which need small data transaction, such as military headquarters. Similarly, LoRaWAN targets wide area network (WAN) applications. It is designed for low-power WANs to support low-cost mobile secure bi-directional communication, which is needed by the IoT systems, M2M, as well as smart city and industrial applications.

Already deployed in tens of thousands of connected objects, the network offers a robust, power-efficient, and scalable system that is able to communicate with millions of battery-operated devices across areas of several square kilometers, making it suitable for various M2M applications such as smart meters, patient monitors, security devices, street/traffic lighting, and environmental sensors currently being rolled out in major cities especially across Europe. With these options there is a necessity for an information system architecture, which makes sense the data for making decisions. Hadoop is one of the powerful options for this kind of applications. Hadoop is "a framework that allows for the distributed processing of large data sets across clusters of computers using simple programming models. It is designed to scale up from single servers to thousands of machines, each offering local computation and storage. Rather than relying on hardware to deliver high-availability, the library itself is designed to detect and handle failures at the application layer, so delivering a highly-available service on top of a cluster of computers, each of which may be prone to failures."[3] It has a big potential in big data age such as Ambari, which is "a web-based tool for provisioning, managing, and monitoring Apache Hadoop clusters which includes support for Hadoop HDFS, Hadoop MapReduce, Hive, HCatalog, HBase, ZooKeeper, Oozie, Pig and Sqoop. Ambari also provides a dashboard for viewing cluster health such as heat maps and ability to view MapReduce, Pig and Hive applications visually along with features to diagnose their performance characteristics in a user-friendly manner."[4]

With IoTs, a number of new technologies have been deployed as solutions. These technologies can be seen in many areas, from the creation of data structures that enable the continuous use of linked data to the next generation of database types. Technologies made it possible to develop decision support mechanisms that enable real-time data mining through machine learning. Some of these technologies with examples for their customized use are as follows:

- Avro (a system for data serialization)
- Cassandra (a scalable multi-master database without single points of failure)
- Chukwa (a data collection system to manage large-scale distributed systems)
- Base (a scalable, distributed database, which supports structured data storage for large tables)
- Hive (a data warehouse infrastructure to provide data summarization and ad hoc querying)
- Mahout (a scalable data mining and machine learning library)
- Pig (a high-level dataflow language and execution framework for parallel computation)
- Spark (a general and fast computing engine for Hadoop data)
- Tez (a generalized dataflow programming framework. It has been adopted by Hive, Pig, and other frameworks in the Hadoop ecosystem as well as by other commercial software, such as ETL tools)
- ZooKeeper (a high-performance coordination service for distributed applications)[5]

[3]http://hadoop.apache.org/

[4]http://hadoop.apache.org/

[5]http://hadoop.apache.org/

Hadoop is a an effective tool for managing large amounts of unstructured data. However, it has also some shortcomings in terms of running applications on analytics, in particularly the ones which integrate unstructured and structured data.[6] Conversely, SQL has a long and successful history of enabling heterogeneous data sources to be accessed with almost identical calls.

Besides all socioeconomic and industrial benefits, IoT technologies pose a number of effects including individual level (changes skills) and organizational (changes motivation and resources) and technological capabilities (access, concurrent user licenses, data traffic). Another impact would be coming with changing business styles and lifestyles. In other words, with the IoTs (and of course ubiquitous computing), it is possible to manage the effects of events or phenomena around us more easily. Objects have the potential to directly affect many areas in the ecosystem of the Internet and connected devices, including the information system architecture and indirectly the programming language used in software technology. Technologies that allow working on top of structured data have triggered the development of new applications that allow instant use and analysis of unstructured data with the IoTs. The transition from using SQL to Hadoop is one of the most striking examples that can be shown in this context. The modification of the information system architecture, the software languages, and the database structures has also brought with it the necessity of the connected world of synchronous data flow and use, which has a great potential for the meaninglessness of the data which is not structured according to predetermined procedures. This structure affects a lot of fields from education to health, as well as defense at the forefront of the most affected areas. This technology makes it easy to get to know about the context, and it makes it possible to internalize the processes and make the service planning so much more convenient and personalized. Unlike the "one size fits all" era, which leads to Henry Ford's "Any customer can have a car painted any color that he wants so long as it is black," in today's world personality and customized services have come to the forefront, and convergence has come to foreground in every field. In parallel to this widespread and customized use, the present resources used are reaching to the critical point. One of the most striking examples of potential dangers is the limits of the Internet protocol. It is evident that the number of IP deployed using the IPv4 infrastructure is likely to increase in the near future and that the IPv4 table will soon be consumed. The most basic necessity of starting to use the IPv6 table for this situation is growing (IPv6 $= 3.4 \times 10^{38}$ aka 340 trillion IPs).

Like every aspects of life, the IoTs has impacts for the defense industry too. There are multidimensional impacts; some of these create new threats, whereas others provide new opportunities for defense and military. These are discussed in the next section.

[6]https://internetofthingsagenda.techtarget.com/feature/Focus-on-wide-data-not-just-big-data-in-analytics-systems

14.3 Digitalization of War Theaters

Wars have always been in human life from the ancient times to the present. Motivations, shapes, and sizes of wars have changed drastically over time (Burmaoglu and Saritas 2017). An earlier definition from von Clausewitz (von Clausewitz 1968) highlights some key components of a war such as "opponent," "violence," and "will," which refer to the "nature of war," which is "an act of violence to compel our opponent to fulfill our will" (p.2). Although the concept of "violence" has remained the same, the "means" and "ends" of warfare have changed dramatically in time. For instance, in a more recent definition, Kaldor (2010:272) outlines the key characteristics of present wars: "War is an act of violence involving two or more organized groups framed in political terms." This definition indicates the increasing number of actors involved in warfare.

From a transformative perspective, it is claimed that when the history of warfare is analyzed, three generations can be distinguished (Lind et al. 1989; Hammes 2005). First generation can be described as the "tactics of line and column," and it reflects its age with the calculation of number of barrels. Quantity was equal to power at that time, and keeping the line meant maximizing the firepower. Second generation's distinction came with the usage of technology, mobilization, and power of indirect fires (Artillery). The change of power from manpower to mass power differentiated these first two generations. The third generation of warfare is characterized by Blitzkrieg. In contrast to second generation's technology-driven aspect, Lind et al. (1989) state that the main motivation of the third generation was "ideas," where Germany's superiority in tactics was considered as a superiority. Lind et al. (1989) expressed this superiority from offensive and defensive viewpoints. From an offensive viewpoint, this was as an "attack relied on infiltration to bypass and collapse the enemy's combat forces rather than seeking to close with and destroy them" (p.23). On the other hand, the defensive viewpoint considered this superiority as "the defense was in depth and often invited penetration, which set the enemy up for a counterattack." It is noteworthy that the distinctions between generations were made based on the dominance of military concepts and technologies (Burmaoglu and Saritas 2017).

The aforementioned generations are mainly concerned with the historical evolution of wars, but how about future? The changes and transformations observed today are broader to include overall changes in society, technology, economy, environment, politics, and values (STEEPV). Although being criticized, the "fourth-generation warfare" suggested by Lind et al. (1989) and Hammes (2005) takes into account these broader changes and considers warfare as a twilight zone—between war and peace, between civilian and military, and between tactics and strategy.

Another interpretation of the transformations regarding today's wars comes from Umberto Eco. Narrated by Lucas (Lucas 2010), after two world wars and the Cold War, by contrast, Eco believed war could no longer be defined with Clausewitzian fashion, in terms of the straightforward linear vectors of force operating between clearly defined rival centers of power. In contrast, Eco quoted that "power is no longer monolithic and monocephalous: it is diffused, packeted, made of the continuous agglomeration and breaking down of consensus." War, based on Eco's interpretation, may be characterized more than "two opposing states." There are a number of proxy forces on the war theater as well as the controlling governments of states versus their own internal, opposition parties and religious factions, the media, and financial sector. This interpretation explains the war with an analogy as parallel processing not serial computing. Finally, Eco states that war is no longer a simple serial sequence of events but all sorts of events going on at once.

Another interpretation regarding the future of war comes from Hoffman (2006), who considers warfare as a world of asymmetric and ethno-political phenomena, where machetes and Microsoft merge with apocalyptic extremes wearing Reeboks and Ray Bans dream of acquiring mass destruction. Moreover, these adversaries are considered to be beyond low-tech. Opponents will be capable of undertaking "advanced irregular warfare," with access to encrypted command systems and other modern lethal systems, such as man-portable air defense missiles. In these structures, they will not need formal networks and will make use of cellular structures with greater autonomy. In brief, what he proposed may be considered as "complex irregular warfare," which consists of organizations with distributed network structures.

Irregular warfare is also termed as "asymmetric warfare" (Grange 2000), where the opponent is not a nation and they have limited capabilities. Thus an asymmetry emerged between the sides involved in the warfare. Arreguín-Toft (2001) considers asymmetric warfare, as where the weak wins wars. Because of the involvement of large number of forces with headquarters, bases, and troops and their distributed organization, operations in asymmetric warfare demand increased communication and coordination as well as greater demand for flexibility, mobility, and networking of distributed forces.

Besides changing concepts, the evolution of technologies has also affected the nature of warfare (Aydogdu et al. 2017). Evolution of military technology was examined by van Creveld (2010) with four stages. The first period from 2000 BC to 1500 AC emphasizes that most military technology utilized its energy from muscles of men and animals. Second stage is covering from 1500 to 1830, and this stage is called as "the age of the machines." During this period, the military operations were characterized by mobilization, coordination, and communication, which raised the need for synchronization and energy dramatically (van Creveld 2010). Third stage is called as "the age of systems" and emphasizes the integration of technology into complex networks. Moreover, using tanks, railways, and highways and improving means of logistics made this stage more complex with increasing integration. Hence, it became more important than before to supply energy to military units and share the intelligence between units online at that time for sustaining the on-going operations (Saritas and Burmaoglu 2016).

As of today, rapid technological progress and innovation have increased the information intensity for running military units and making decisions as well as carrying out missions and undertaking operations. Collection, processing, and synthesis of this vast amount of information require higher level of "digitalization," "computerization," and a "network structure." Moreover, there is a greater demand

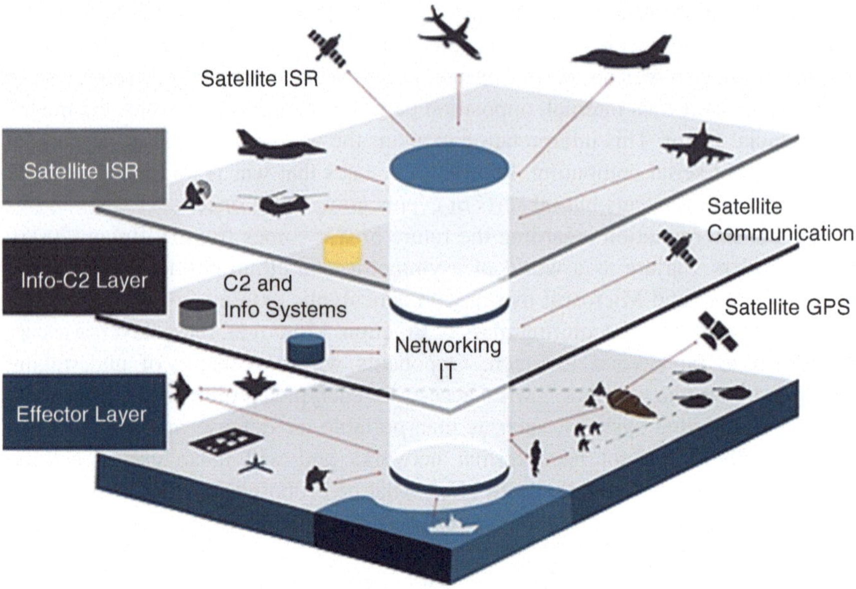

Fig. 14.1 Network centric warfare demonstration (http://mil-embedded.com/articles/the-internet-things-the-intelligence-community/ Access Date: 01.06.2016)g. Source: Author

for the seamless flow and diffusion of information and data between military forces and other actors in war theater.

Combining the conceptual and technological transformations observed in warfare, it can be asserted that future wars may be characterized by involving non-state actors, distributed and cellular type forms, asymmetric nature, and high-tech with the help of information systems and power of social media. Recent years have seen the emergence of the network-centric warfare concept. This concept was created with the increasing need for command and coordination. Such a system constitutes an information grid that connects the elements of all combat soldiers, weapons, military equipment, and so on by utilizing computers, sensors, and wired/wireless networks. In this concept the war power is increased by intelligence superiority through information sharing and integration as connecting intelligence collection systems, command and control system, and strike system (Yang et al. 2015).

Figure 14.1 demonstrated the layers of the network structure.

As can be seen in the figure, a central network body is constructed to get the data from various sources and after analyzing the data transmit it to the theater based on their levels. This layered structure of network becomes more complicated with adding more and more data points. Traditional approaches to military doctrine have been transformed by the network-centric warfare with expanding communication gateways connecting battlefield assets and headquarters and increasing data sharing between legacy assets and new deployments.

An important application of network-centric warfare should be considered as Command, Control, Communications, Computers, Intelligence, Surveillance, and Reconnaissance (C4ISR) systems. The electronic understanding of war theater may be exemplified by the Command, Control, Communication, Intelligence, and Electronic Warfare (C3IEW) concept, which was first introduced by the Defense Science Board in the USA in the early 1990s and aimed to ensure interoperable and cost-effective military systems for the Department of Defense. C4ISR is the improved version of this concept with its first version issued in 1996, and the second version is in 1998 (http://cecomhistorian.armylive.dodlive.mil/2013/03/11/history-of-c4isr/).

In network-centric concept, IoTs may find potential application areas some of which have been applied. For example, IoT can be applied to surveillance and reconnaissance systems, soldier combat uniform, medical services, logistics services, precise guided munitions, combat platforms, and so on.

Credencys Solutions Inc. proposed that IoTs may help military with six ways. These are (1) providing battlefield situational awareness, (2) proactive equipment maintenance, (3) monitoring warfighter's health, (4) remote training, (5) real-time fleet management, and (6) efficient inventory management.[7] From these six ways, it can be asserted that intelligence, maintenance, logistics, fleet management, training, and medical services are the potential sub-working areas for exploiting IoTs in defense.

Meanwhile, it is clear that application of IoTs with today's technology level may cause important security threats such as illegal remote control, information leakage, false information insert, and signal disturbance. As an issue of vulnerability, cyberthreats have also become more common and sophisticated. Four areas concern the security of the IoT environment for military including (1) making sure information is reliable and systems are resilient, (2) keeping pace with technology, (3) focusing on the insider threat, and (4) embracing (big and community) data analytics to minimize cyberthreats (Daly 2016).

Besides vulnerability of information systems, according to retired Marine Corps General James Cartwright (2015), culture is the most important obstacle for not applying IoT in military. Based on his speech at CSIS institution, not only the culture is an obstacle, but also "man-machine partnering" should be focused for implementing IoT. Last statement should be considered as another discussion point, because IoT is mostly designed for machine-to-machine communication; however for war theater, man-machine interaction is another level to be achieved.

Man-unmanned teaming (MUM-T) project may be a special example for man-machine interaction. Based on Colonel Eschenbach (2016)'s article in Army Aviation Magazine, the history of this concept can be traced back to the World War II with the designation of BQ-8 robots in B-17 Flying Fortresses for remote piloting. Moreover, the concept gained importance in the last decade with controlling AH-64D Apache and its payload utilizing an add-on system designated MUM-T2.

[7]https://www.credencys.com/blog/internet-of-things-the-agent-of-change-for-the-defense-system/. Access Date: 16.04.2016.

According to Colonel Eschenbach, these unmanned systems can be used for situational awareness and enrooting air transportation forces. He envisioned that future organizational force structure of Army Aviation has been shaped by MUM-T advantages. Iriarte (2016) has been supporting Eschenbach's ideas and narrated the Unmanned Aircraft Systems (UAS) usage interoperable with AH-64 Apache. Based on her article, the future of this concept will be with increasing autonomy, reducing workload, and manpower.

From the healthcare point of view, it can be seen that IoTs have the potential to change the dynamic of healthcare itself. By networking different devices together on the battlefield and in garrison, information sharing may be more convenient based on the findings of Military Health System Communications Office (2015). Moreover, automated alerts for medical staff are produced by theater mobile computing applications which can increase the quality of medical decisions. These applications should help improve the readiness of warfighters and perhaps increase their chances of survivability. IoTs has potential for transformative change in human health. Using connected devices and wearables, to continuously monitor patients as they live their lives particularly those with chronic conditions like diabetes, it can improve patient adherence to prescribed therapies (as procedures that already planned), avoid hospitalizations (and post-hospitalization complications), and improve the quality of life for hundreds of millions of patients. McKinsey report predicted this could have an economic impact of $170 billion to $1.6 trillion per year in 2025.

Finally, it can be said that IoTs has many potential areas in defense sector, but the obstacles are very hard to overcome in a short time. Hence, it can be interpreted that the adoption and dissemination of IoT applications need various stages. In the next section, alternative scenarios are presented to provide alternative narratives on the context and adoption of IoTs in military.

14.4 Possible Future Scenarios for IoT in Military

As discussed in the previous section, IoT has many advantages and disadvantages for military forces. However, it should be admitted that for some of the sub-military fields such as reconnaissance and logistics, IoT applications will have a great contribution. These contributions may also be called as force multipliers. In this section, the future of war contexts and potential usage of IoTs in these contexts are represented by using alternative scenario trajectories.

There is a consensus on the existing literature that future wars will be characterized by their distributed, synchronous, and complex nature. The key uncertainties are about the visibility of war theaters as well as their boundaries in terms of where they begin and where they end. New technological developments, which may be called as intelligent systems, work autonomously with varying controlled and supervision degrees and collect information by sensing, communicating, and collaborating each other in a war theater. They are enabled by machine learning, perception, and reasoning. They can process information, undertake defensive actions, and unleash a variety of effects on the adversary. These intelligent systems would be used for carrying equipment, shielding troops, and sensing fields at first. Hence they can be produced with different physical entities ranging from insect-sized to mid-size and can move over the ground or air. However, especially for the physical engagement at war, it should be a requirement to authorize humans for asking responsibility. The physical entities and virtual counterparts would be the other side of this environment. These virtual self-operating intelligent systems may be called as "cyber robots." They can protect C4ISR systems and energy grids and warn and prevent about incoming cyber threats. Moreover, these systems should be thought systemically, and all other physical or virtual entities should be considered as parts of this big collaborative network. The battlefield of the future will be populated by fewer humans. But these warriors will be physically and cognitively augmented. They can interact with the autonomous intelligent systems by using improved man-machine interaction algorithms.

However, from the human side, not only the interaction but also the visibility of war theaters is an important issue. Disinformation and deception will be essential to survive and operate on the battlefield of the future. The war theater of the future is expected to feature a crowded and synchronous battlefield. However, physical entities will play minor role and will be replaced by cyber entities, which will be even more difficult to detect and track. Therefore, quantity and quality of information and availability of communications, as well as rapid decision-making, will be crucial. Conventional concepts and approaches would not work because of their hierarchical structure of decision-making environment, which makes the process slower and inefficient. In this new environment, decision-making will be distributed as well as the structure of network. Hence, for future wars, military leadership and the competencies of leaders would become more important than the present. They can self-organize and collaborate within these dense networks with the aid of human-machine teams. Within this context, it is considered that that the key scenario variations would be defined by the different levels and types of human-machine interaction. The main question here is to what extent human supervision should be maintained? Two scenarios are proposed to consider the trajectories of "human-supervised" and "autonomous" systems. The first scenario is more predictable considering the current trends, where the second one, as a probable scenario, also becomes more and more feasible as the speed of technological development increases and leads to the real Defense 4.0.

14.4.1 Predictable Scenario: Human-Supervised War Environment

One of the key features of the changing nature of warfare is that the future wars will take place in contexts, where it will be difficult to distinguish friends and enemy. Instead of battlefields, future wars are expected to take place much closer to human settlements and cities and, therefore, will become increasingly scattered in multiple locations. In these cases, operations will increasingly involve UAVs and thus will be more efficient with the engagement of smaller but more efficient security personnel. Within this context, seamless connection, continuous data flow, and synchronization will be crucial. Armies in human-supervised war environment will set up centers to manage military operations. As there is no need to set up such centers physically on site, where the operations take place, they can be established in more secure sites in a more flexible project-based structure instead of distributed and self-sufficient nodes. What is important here is to provide continuous energy supply both for headquarters and operational centers.

Besides energy supply, it is considered that cloud systems will play an important role to enable and facilitate this process. Current cloud technologies are based on a single centralized logic. It is currently a technological challenge to set up a more scattered cloud system for distributed and different types of operations. During the transition time, a hierarchical cloud system can be designed in-line with the constraints of the operation sites. Thus, a war theater cloud system can be created by considering the structure operation command.

At the individual level, the hierarchy in the military leaders will evolve from a rigid hierarchy to a more functional hierarchy. This will enable a joint decision-making among the military leaders and shared responsibility. However, it can be said that the leaders, which are not involved and interact with the war environment, will be less sympathetic and more relentless in decision-making. In the literature, this situation is called the "trolley dilemma" (Foot 1967).

The trolley dilemma is an experiment applied in ethics. The general form of the problem is the following:

> There is a runaway trolley barreling down the railway tracks. Ahead, on the tracks, there are five people tied up and unable to move. The trolley is headed straight for them. You are standing some distance off in the train yard, next to a lever. If you pull this lever, the trolley will switch to a different set of tracks. However, you notice that there is one person on the side track. You have two options: (1) Do nothing, and the trolley kills the five people on the main track. (2) Pull the lever, diverting the trolley onto the side track where it will kill one person. Which is the most ethical choice? And other scenario is a trolley is hurtling down a track towards five people. You are on a bridge under which it will pass, and you can stop the train by putting something very heavy in front of it. As it happens, there is a very fat man next to you—your only way to stop the trolley is to push him over the bridge and onto the track, killing him to save five. Should you proceed? (Eagleman 2015:126–128)

For the first ethical scenario, most of the people are willing to pull the lever. However, in the second one, no one wants to push a man over the bridge. Even if the result is the same, personal interaction changes the behavior.

In order to overcome this problem, training programs should be designed and implemented for the military leaders and users of weapon systems to make them capable of developing empathy with the opponents and civilians. Virtual reality or augmented reality systems are considered to be immensely useful for this purpose.

Consequently, it should be born in mind that challenges related to human-supervised war environments will exist at all levels, from systems to individuals. Therefore, both infrastructures, including energy, the ICT, and cloud systems, and future military personnel and leaders should be developed considering the context and requirements of this new context of warfare.

Overall, this scenario is considered as an initial phase toward the probable scenario, "autonomous war environment," which will be described in the next section.

14.4.2 Probable Scenario: Autonomous War Environment

Autonomous war environment can be also called as a new machine era, where the Defense 4.0 concept will reach its true meaning. In this context, operations, logistics, and intelligence units will talk to each other and plan actions through artificial intelligence devices and algorithms. Protocols for authority and responsibility must be clearly defined in this system, particularly when intelligence is received and operations activated. Not inferential-based but evidence-based operations should be authorized. Among the biggest challenges for this scenario are continuous energy supply, human-machine interaction limitations, and development of cloud computing and supervised artificial intelligence algorithms.

From the social and human point of view, this can also be considered as a catastrophic scenario. Although there are a number of benefits, questions regarding the reliability of systems and the possibility of losing their control create doubts about the development of this capability.

At first, autonomous systems are expected to be used for the purposes of intelligence and logistics. While cost reductions and efficiency increases will be observed in logistics, systems like drone swarms will help to collect intelligence from a wider range of geographical areas. The use of biomimetic colonies in operations within built up areas will reduce the loss and damage of personnel and vehicles.

From the organizational point of view, information will be collected and processed of undertaken in data centers. Therefore, it can be said that this scenario also suggests a degree of centralization in decision-making. This will be particularly observed in logistics and intelligence units. At the individual level, the processes of human-machine interaction need to be well considered and planned during operations.

An important aspect in autonomous war is the role of military leadership. New generation military leaders should be equipped with virtual management and abstract thinking skills. The assessment of information from autonomous devices and developing an operation plan through human-operated systems require new capabilities beyond conventional defense leadership perspectives.

Overall, access to information and cloud systems, accessibility and synchronization, and assessment and decision-making will be among the key characteristics of this scenario. Without doubt, energy supply will play a crucial role for the seamless operation of the autonomous systems.

14.5 Conclusions and Discussion

Digitalization and its exploitation may have many facets. One of them is the use of sensors and sensor technologies. Sensors are increasingly becoming an important part of our daily life and are used for acquiring online data from technological or biologic subjects that take part in the war theater. In peace times they are used for training troops, optimizing weapons, vehicles or maintenance systems, and so on. The other crucial technology is social media. It is mostly used for human intelligence. One of the successful examples of using social media is UCINET, a social network analysis software, which was developed by Harvard scholars to find Saddam Hussein. Finally, it can be briefly summarized that these technologies are collecting data or gathering it, and after analysis in appropriate software, the results are used even in intelligence, operations, logistics, or whatever else. However, using IoT in military is one-step forward. There are some similarities, but in the case of IoT, sensors can talk to each other and can act according to their manufacturing objective. In both scenarios, it can be asserted that the main challenge that military authorities face is energy. After solving energy sustainability, human-supervised war environment may be activated easily, because on-going operations will be strengthened by benefiting technological capabilities. However, autonomous war environment is one-step forward from human-supervised war environment.

Both scenarios will have same bottlenecks as ethics and social acceptance. Even these developments may reveal benefit for community; they may eventually destroy privacy and freedom of choice. For instance, continuous surveillance of the society online and offline reminds George Orwell's world-famous novel *1984* which may raise concerns about "privacy." A fine balance will still be required when collecting information for intelligence while caring for the private lives of citizens, especially in autonomous war environment.

Acknowledgments The contributions by Professor Ozcan Saritas in this study were prepared within the framework of the Basic Research Program at the National Research University Higher School of Economics and supported within the framework of the subsidy by the Russian Academic Excellence Project '5-100'. The contributions by Asst. Professor Haydar Yalcin in this study was supported by Scientific and Technological Research Council of Turkey Postdoctoral Research Programme (TUBITAK BIDEP 2219) [1059B191700840].

References

Arreguín-Toft I (2001) How the weak win wars: a theory of asymmetric conflict. Int Secur 26 (1):93–128

Ashton K (2009) That 'internet of things' thing. RFiD J 22(7):97–114

Aydogdu A, Burmaoglu S, Saritas O, Cakir S (2017) A nanotechnology roadmapping study for the Turkish defense industry. Foresight 19(4):354–375. https://doi.org/10.1108/FS-06-2017-0020

Bandyopadhyay S, Balamuralidhar P, Pal A (2013) Interoperation among IoT standards. J ICT Stand 1(2):253–270

Burmaoglu S, Saritas O (2017) Changing characteristics of warfare and the future of Military R&D. Technol Forecast Soc Change 116:151–161
Cartwright J (2015.) www.defensesystems.com/articles/2015/11/12/internet-of-thing-dod-cartwright-csis.aspx. Accessed 18 Mar 2016
Daim TU, Chiavetta D, Porter AL, Saritas O (eds) (2016) Anticipating future innovation pathways through large data analysis. Springer, Berlin
Daly M (2016.) http://www.darkreading.com/mobile/internet-of-things-4-security-tips-from-the-military/a/d-id/1297546. Accessed 18 Mar 2016
Ding C, Wu X, Lv Z (2009) Design and implementation of the Zigbee-based body sensor network system. In: WiCom'09, 5th International Conference on Wireless Communications, Networking and Mobile Computing. IEEE, pp 1–4
Eagleman D (2015) The brain: the story of you. Pantheon Books, New York
Foot P (1967) The problem of abortion and the doctrine of the double effect. Oxf Rev 5:1. http://pitt.edu/~mthompso/readings/foot.pdf. Accessed 09 Jun 2017
Grange DL (2000) Asymmetric warfare: old method, new concern. Natl Strategy Forum Rev. Winter
Hammes TX (2005) War evolves into the fourth generation. Contemp Secur Policy 26(2):189–221
Health System Communication Office (2015.) http://science.dodlive.mil/2015/07/19/defense-officials-use-internet-of-things-to-monitor-health/. Accessed 18 Mar 2016
Hoffman FG (2006) Complex irregular warfare: the next revolution in military affairs. Orbis 50:395–411
Iriarte M (2016) MUM-T operations on the US Army's UAS roadmap. Military Embedded Systems. http://mil-embedded.com/articles/mum-t-armys-uas-roadmap/. Accessed 08 Jun 2016
Kaldor M (2010) Inconclusive wars: is Clausewitz still relevant in these global times? Glob Policy 1 (3):271–281
Lind W, Nightengale K, Schmitt J, Sutton J, Wilson GI (1989) The changing face of war: into the fourth generation. Marine Corps Gazette, October 22–26
Lucas GR Jr (2010) Postmodern war. J Milit Ethics 9(4):289–298. https://doi.org/10.1080/15027570.2010.536399
Mantyika J, Chui M, Dobbs R, Bisson P, Marrs A (2013) Disruptive technologies: advances that will transform life, business, and the global economy. McKinsey Global Institute Report, May
Miles I, Saritas O, Sokolov A (2016) Foresight for science, technology and innovation. Springer, Berlin
Rathmell A (1997) Cyber-terrorism: the shape of future conflict? RUSI J 142(5):40–45
Saritas O, Burmaoglu S (2016) Future of sustainable military operations under emerging energy and security considerations. Technol Forecast Soc Change 102:331–343
Van Creveld M (2010) Technology and war: from 2000 BC to the present. Simon and Schuster
von Clausewitz GC (1968) On war, translated by Colonel JJ Graham. Routledge & Kegan Paul, London
von Eschenbach T (2016) Unmanned aircraft systems-manned-unmanned teaming. http://armyaviationmagazine.com/index.php/archive/not-so-current/589-unmanned-aircraft-systems-manned-unmanned-teaming. Accessed 08 Jun 2016
Yang JS, Lee HJ, Park MW, Eom JH (2015) Security threats on National Defense ICT based on IoT. Adv Sci Technol Lett 97:94–98

Serhat Burmaoglu is Associate Professor on Quantitative Decision Making at Izmir Katip Celebi University and Visiting Scholar in Science, Technology and Innovation Policy Studies, School of Public Policy, Georgia Institute of Technology. His research interest is investigating the relationship between economic growth, productivity, competitiveness, innovation and knowledge economy in macroeconomic level by using multivariate statistical analysis and data mining applications for extracting usable patterns to direct technology policy. Dr. Burmaoglu performs bibliometric and patent analyses on emerging technology to prepare scenario-based foresight studies. He published many book chapters and articles in different national and international journals.

Ozcan Saritas is Professor of Innovation and Strategy at the HSE and editor-in-chief of "Foresight"—the journal of future studies, strategic thinking and policy. He worked as Senior Research Fellow at the Manchester Institute of Innovation Research, The University of Manchester. His research focuses on innovation and policy research with particular emphasis on socio-economic and technological Foresight. With a PhD from the "Foresight and Prospective Studies Program," he introduced the "Systemic Foresight Methodology", and has produced a number of publications on the topic. Dr. Saritas has extensive work experience with the international organizations, including the United Nations, OECD, and the European Commission. He has been involved in large scale national, multinational and corporate research and consultancy projects on sectors including energy, climate change, agriculture, food, water, transportation, and ICT among others; published a number of articles in respected journals; and have delivered keynote speeches in more than 50 countries across the world. Besides his research and publication activities, he designs and delivers academic and executive education courses on Foresight and Strategic Planning. He has recently co-authored a book, entitled "Foresight for Science, Technology and Innovation" published by Springer, which has become one of the key readings in the field.

Haydar Yalcin is Assistant Professor on Information Management and Technology. His research interest is measuring scientific productivity using big data methods and data mining applications for extracting usable patterns to direct information and technology policy. Dr. Yalcin performs scientometric analysis, bibliometric and social network analysis on emerging technologies to prepare scenario-based foresight studies. He published many articles and book chapters in different national and international journals.

Part V

Challenges to STI Policy

How to Stimulate Convergence and Emergence of Technologies?

15

Dirk Meissner, Leonid Gokhberg, and Ozcan Saritas

Until the last century, technology development and application, e.g., manufacturing and commercial application, were often centered in reasonable proximity which led to the thinking that regional industry absorbs regionally developed technology at reasonable pace. While this view is certainly true, it neglects that also industry is developing routines over the years which are often hard to overcome, e.g., although technology and skills are available locally, industry might postpone modernization and skills upgrading due to delayed decision-making and limited willingness to leave the old routines as David (1990) describes for the advance of the "New Economy."

From the technology point of view, accelerated diffusion speed brings positive effects when it comes to the sophistication of technologies and the search for complementary application fields and so on. But from the point of view of regional and local development, this becomes even more questionable because the absorptive capacity of existing industries in close proximity of the technology's place of origin is not necessarily sufficient to take advantage of the next technological wave. Moreover, it appears that building new industrial zones for economic development is more comfortable for companies, namely, large companies, than upgrading existing facilities to master new technologies well. Among the reasons for preferring, establishing new facilities is the fact that technology intensive manufacturing is frequently capital intensive which for accounting and controlling reasons might tempt companies to favor new facilities over existing ones due to the depreciation of existing assets and also for reasons of public subsidies, loans, grants, or the like which is often offered to companies investing at green field. It therefore cannot be assumed any longer that the supporting technology development will provide local

D. Meissner (✉) · L. Gokhberg · O. Saritas
National Research University, Higher School of Economics, Moscow, Russia
e-mail: dmeissner@hse.ru; lgokhberg@hse.ru; osaritas@hse.ru

D. Meissner et al. (eds.), *Emerging Technologies for Economic Development*, Science, Technology and Innovation Studies,
https://doi.org/10.1007/978-3-030-04370-4_15

and regional spillovers for the benefit of respective economic and social value with close proximity to the place of origin. In this respect one might argue that local and regional development should change the existing paradigm of "investment into research and technology development creates lasting regional economic advantage" toward a broader understanding of the resulting spillovers. These spillovers are not necessarily bound to appear in a region but might appear elsewhere in the world which lets the question arise why investment was done in one region but the impact achieved in another region.

Accordingly governments are challenged to justify why public intervention is actually required to stimulate the emergence and convergence of technology. At first sight any government intervention seems doubtful in light of the market economy thinking which postulates that the market will balance demand and supply. The failure in this thinking however is that technologies in many cases don't generate new demand but replace existing solutions, e.g., making established technologies obsolete. Assessing the eventual impact resulting from these replacements on the respective economy appears difficult for several reasons. First of all challenges arise in assessing the actual technology-induced economic but also societal impact. Economic impact is frequently measured by job creation and economic returns of a technology but still in most cases the impact which a technology has on revenue generation remains uncertain. For example, frequently products, services, and processes involve several technologies which is why solid economic assessment is only partially possible. Similar is true for the societal impact which evolves. It's not always clear and predictable which impact a technology or bundle of technologies might have on society in advance and also the causality between the emerging technology and societal impact is difficult to assess. Having said this we find that impact assessment is still characterized by the chicken and egg problem, but for government interventions, it's almost essential to know the possible impact. Against this background numerous reasons for governmental intervention in technology emergence and convergence appear:

- Among the governments' ambitions for technology development is the explicit aim to enable companies to develop new products and improve processes at broader scale for manufacturing which eventually results in economic growth (Cooke and Leydesdorff 2006). This is an obvious intention by policy makers; however firms' competitive advantages result from their presence in the value chain (Krugman 1995; Porter 1990, 1998), e.g., especially in course of open markets and global value chains, companies tend to diversify their value generation activities. Value chain-related activities are typically spread across different locations, e.g., countries and regions. But this does not imply that each step or activity of the value chain is established in one region only. Quite on the contrary, companies, especially large companies, operate different facilities in different regions for various reasons. Among the motivations are the end consumer proximity as well as the regional supply chain for selected products and service features, the servicezation of products for which regional proximity is a plus but also the human resource dimension in all facets and in some cases national regional regulations which requires this (Miozzo et al. 2016; Miles 2016; Miles and Miozzo 2015).

- Whereas in the early stages of the technology life cycle only few knowledgeable individuals and entities are involved and competent to apply and further develop the technology, the number of parties increases considerably with growing application diffusion. Typically, there is uncertainty about the eventual possible applications an initial invention might bring over the lifetime which is due to the small and often closed community of individuals who are engaged. Frequently these individuals are primarily scientists and dedicated engineers who aim at the perfect scientific and/or engineering solution with little or limited attention to application fields. Thus discussions about the technologies are mainly limited to community internal debates at least for a while. Following the involvement of a broader public audience, namely, a scientific and engineering audience, more views and perspectives on the technology under consideration appear which provides the ground for very early adoption (Helpman and Trajtenberg 1994). It follows that information about the existence of the invention, hence technology, diffuses and more actors are becoming aware of the potentials. Increasing awareness of the inventions outside the initial communities also brings spillover effects which cross regions and countries in addition to the initial technology field. Therefore, an invention might show potential to stimulate activities in other regions which for some reason provide better framework conditions for the invention to unfold its economic and application potential than locally only.
- Technologies typically originate at a dedicated location but show limited application potential and economic impact, respectively, in the very early stages. The challenge arising is that technologies diffuse fast within and beyond communities, and therefore locations and regions of origin might experience that the merits of applying and exploiting technologies are grasped at another destination. This observation is by no means a twenty-first century phenomenon but has been described in the 1980s already (see, e.g., Cooke et al. 1984). Creating measurable economic value from technology is hardly an issue for scientific and engineering work and related competencies only but is extended to the common business, operations, and maintenance skills which are essential for manufacturing and production hence for economic value. Economic value here involves local and regional employment creation but also tax revenues to the local and regional public budgets. In addition value to society arises through more indirect value, e.g., by means of employment which in turn contributes to social welfare and less social tensions among many others. For these reasons technology creation is often seen by the policy community as a means of long-term value creation. However, in many cases respective support measures are aimed at initial technology development in the first instance but less on timely providing adequate labor force competences, which is companies' absorptive capacities in the broadest sense. Beyond scientists, engineers, and other related highly qualified, labor industry demands lower skilled—but still flexible—manpower for manufacturing. Again this is hardly a new observation but has been described by Saxenian (1981) already. What has changed since then is the even faster

diffusion of information thus technology and obviously technology's complexity and interdisciplinary nature.

As in any technology development case, completion is always uncertain both with regard to completion in time and budgeted resources which is why it appears to be little constructive to talk about emerging technologies if there is no certainty about completion. Furthermore, users' technology acceptance might be indicated, but this again is highly risky and uncertain because competing technologies are potentially also in the development stages, and the actual technology performance remains rather vague with much of the potential value being assigned to expectations and enthusiasm. The latter is also a potential barrier for acceptance and diffusion if the technology under development eventually doesn't provide reasonable value to users to replace existing solutions or to enter new grounds base on the new technology. Therefore, the psychological dimension is crucial to consider. Against these arguments the authors understand emerging technologies as:

> Solutions of which basic principles and modes of action have been developed and demonstrated successfully. Initial applications of the technologies are known and at least partially understood but there are additional yet unexplored application fields. Emerging technologies are in principle platform technologies in their early technology life cycle stages. A technology might be entitled 'emerging' if it is in operation for demonstration purposes at least and multiple application fields are possible.

Among other features emerging technologies are characterized by their potential to initiate new discoveries and inventions which are based on their initial invention, e.g., the level and degree of their multiple usage potential (David 1990; Youtie et al. 2008). This includes that emerging technologies aren't diffusing a single application field only but provide the basis for complementary technologies which in turn form significant parts of new technological solutions in other fields. From a technological point of view, this is closely related to complementary technologies, in other words platform technologies, which share a common main principle stemming from an earlier invention.

Like any technology emerging technologies are frequently challenged by the prevailing uncertainty about potential side effects and less favorable impacts on society, namely, in the environment, health, and safety (EHS) context. Thus in order to generate economic benefits, thorough assessments of the technologies are required which are aimed at society and related impacts in the first instance. These assessments also request a dedicated media and information campaign targeted at informing society and raising awareness. Experience with various technologies provides evidence that the emergence of technologies and related economic effects might suffer from societal resistance if no early-stage awareness and information campaigns are in place. In such cases technological development progresses at high speed, but knowledge and information about the technology are trapped in a rather closed community with selected information pieces being made available to a broader audience. Consequently, media and interested communities use the available information pieces to communicate among society but run danger of drawing misleading pictures of technological impacts which might influence societal opinion about the technology at large thus finally determine acceptance or resistance. Accordingly, investments into further technology and application development are at stake as long as investors are confronted with uncertainty about society attitude toward technologies.

Emerging technologies are characterized by numerous uncertainties including technological development (achieving reliability and operationality under time and budget constraints), competitive technologies development, market and user development, standards, state regulations and certifications, among others. All these features evolve and develop over time with different impact on the technology itself, and also they influence each other to some extent. A suitable instrument to visualize and understand these developments and effects are roadmaps, i.e., roadmaps dedicated to emerging technologies and economic development. However, such roadmaps are only impactful if building on a series of mini roadmaps under a common umbrella which enable regional actors at least to build their own targeted strategies and roadmaps (Walsh 2004).

Another important challenge for achieving economic value from technology development, namely, the diffusion of emerging technologies, lies with human resources. It's frequently argued that labor is mobile but still causing considerable cost to employers which was postulated 35 years ago by Dorfman (1983). The labor mobility-related cost has increased considerably over the last years when numerous countries, regions, and locations (municipalities) have decided to develop local hubs for technology development at the leading edge. Often these initiatives are challenged by a shortcoming in available talent with respective competences and the necessary integration into the appropriate communities and networks. Thus demand for such talent has grown, while supply remained at the similar level. Furthermore, it cannot be expected that supply of such talent increases at the same speed as demand develops.

In many cases technology is non-rival, e.g., it provides multiple application possibilities which can be developed at marginal costs (Fu et al. 2011). Even the digital (knowledge) economy technologies field remain featured by a significant share of tacit—hence non-codified—knowledge which provides advantages for regional and local innovation (technology) ecosystems as postulated by Jaffe et al. (1993). Such advantage is mainly found in a time advantage which the research (viz., research institutions and universities) and also the innovation (namely companies) communities in a region enjoy over other actors outside these regional communities. Yet these advantages are hardly of long-lasting nature since reverse engineering and international labor mobility enable competitors to copy or invent other related solutions. In this regard labor mobility is especially important as this affects the tacit knowledge which becomes accessible if skilled labor is moving to other places and occupations.

Diffusion paths of technologies, including emerging technologies, take a broad range of shapes, among which are trade of goods and capital by means of inward and outward foreign direct investment, mobility of people, cross border R&D and innovation collaboration, media and social network communication, and, last but not least, the global value chains (Pietrobelli 1996). Global value chains and labor mobility are becoming more and more central to diffusion as these channels include the physical transfer of technology (GVC) and the tacit knowledge which is necessary to operate technologies. Information-related channels (social networks and media) more likely have an awareness and less sophisticated information function.

Regional proximity of actors is an important driver for technologies to diffuse in the application sphere. It is often found that face-to-face communication of actors is supportive for technologies to emerge and diffuse at higher speed (Ku et al. 2005). The rationale behind this observation is that face-to-face communication allows the actors involved exchanging tacit knowledge, e.g., direct verbal communication about the technology under question takes different forms than in more structured and documented communications. Closer regional proximity of communities demonstrates positive effects on the social relationships of actors which results in free discussion of ideas leading to positive externalities—i.e., information and knowledge dissemination—and building of trust among individuals. Social relationships and resulting trust development provide a clear contribution to the absorptive capacities of actors, namely, firms, which also contribute to technology diffusion and adoption speed (Fu 2008). The latter offers strong potentials for technological development and thus economic development but at the same time inherits reasonable threats for entities, namely, commercial entities when it comes to labor mobility between the actors involved (Ku et al. 2005).

In this respect emerging technologies clearly provide strong opportunities for generating economic impact at company and at regional level, e.g., at micro- and macro-level. However, in order to leverage emerging technologies' economic potential for the advantage of regional economic development, a much broader approach to science, technology, and innovation (STI) policy is required with dedicated features:

- Standard (common practice) STI policy measures targeting at supporting technology development but should provide more room for creativity in the design of projects and application fields.
- In order to establish lasting economic impact and provide the respective framework conducive to sustainable technology-based leadership, policies need to look beyond the initial technology horizon. This implies the active support of related regional innovation milieus and ecosystems by means of developing and keeping human resources which are at the front of the technological dimension but which also possess a broader experience and related soft skills.
- While there is a reasonable amount of scholarly works done on soft skills, the key messages haven't diffused to the national STI policy-making communities. Related policy measures share the common understanding so that featured ecosystems evolve in clusters and platform or by attracting talent without any additional support. But this is only a part of the truth. Clusters, platforms, and the like certainly play a role and might act as nucleus but hardly involve the potential to influence individual's attitudes which on the other hand is key.

STI policy is hence required to take these challenges into account if it aims at finding ways to enhance technology emergence and convergence. Obviously measures responding to the described challenges are hardly found in the narrow understanding of STI policy but beyond the common measures. It requires unorthodox approaches which provide reasonable space for interpretation of societal attitudes, legal issues, and other related regulations. Although this might in some cases contradict the standard rules and procedures of public spending—which is in almost all countries worldwide strongly regulated—policy should develop models which allow a more creative and pragmatic support. This involves especially:

- Supporting individuals aiming at extending and broadening their horizon in fields other than their current or previous education but always in line with the clear ambition to use the experiences gathered for the advantage of regional development. This is easily said but difficult to implement. Some might argue that such public support schemes take the form of scholarships which can hardly provide guarantees of receivers' return and impact generated. Right on the contrary, if a reasonable effect is expected, regulations should be as flexible as possible without putting much administrative burden on the receiver's side and leaving aside attempts to assess and quantify the resulting impact.
- It's appropriate to establish schemes which take the shape of "play money" being spent and invested with uncertain return.
- Further this requires that aims and goals of public support are formulated in a more flexible form as currently practiced and no definite fixed indicators and deliverables are described. In doing so governments are asked to obtain a more entrepreneurial attitude which is not expressed in standardized public announcements.
- Establish a system innovation thinking which incorporates user understanding, e.g., STI policy measures need to account for the requirements and perceptions of technology and innovation adopters from the initial support phases.

Summing up, we find that emerging technologies provide significant opportunities for companies and research institutions. However, the widespread expectation that emerging technologies deliver significant regional economic impact is often fulfilled partially only due to economic constraints and global spillovers. In order to leverage the economic impact in favor of the region of origin, policy makers need to look beyond the existing policy measures. This said means especially concerted—e.g., consistent and coherent—STI policy approaches are required. It is a common policy maker dilemma to develop new STI policy measures which aim at supporting emerging technologies, but in very few rare cases, the existing STI policy mix is rethought fully. For technology developers and applicants, however, it is much more important to experience a seamless and consistent sustainable policy mix; however, the respective actors are often critical and skeptical against considering changing framework conditions.

The editors wish to express their gratefulness to all authors who contributed to this volume. The book proves how emerging technologies are identified, selected, and evaluated against economic value and impact. Furthermore, the chapters provide clear strategic intelligence for exploiting emerging technologies in different fields.

Acknowledgments The book chapter and the whole book were prepared within the framework of the Basic Research Program at the National Research University Higher School of Economics and supported within the framework of the subsidy by the Russian Academic Excellence Project '5-100'.

References

Cooke P, Leydesdorff L (2006) Regional development in the knowledge-based economy: the construction of advantage. J Technol Transfer 31:5–15

Cooke P, Morgan K, Jackson D (1984) New technology and regional development in austerity Britain: the case of the semiconductor industry. Reg Stud 18(4):277–289. https://doi.org/10.1080/09595238400185281

David P (1990) The dynamo and the computer: an historical perspective on the modern productivity paradox. Am Econ Rev 80(2):355–361

Dorfman NS (1983) Route 128: The development of a regional high technology economy. Res Policy 12:299–316

Fu X (2008) Foreign direct investment, absorptive capacity and regional innovation capabilities: evidence from China. Oxf Dev Stud 36(1):89–110

Fu X, Pietrobelli C, Soete L (2011) Role of foreign technology and indigenous innovation in the emerging economies: technological change and catching-up. World Dev 39(7):1204–1212

Helpman E, Trajtenberg M (1994) A time to sow and a time to reap: growth based on general purpose technologies. NBER Working Paper No. 4854. National Bureau of Economic Research, Cambridge, MA

Jaffe AB, Trajtenberg M, Henderson R (1993) Geographic localization of knowledge spillovers as evidenced by patent citations. Q J Econ 108(3):577–598

Krugman P (1995) Development, geography and economic theory. MIT Press, Cambridge, MA

Ku YL, Liau S-J, Hsing W-C (2005) The high-tech milieu and innovation-oriented development. Technovation 25:145–153

Miles I (2016) Twenty years of service innovation research. In: Service innovation. Springer, Japan, pp 3–34

Miles I, Miozzo M (2015) The globalization of knowledge intensive services. In: The handbook of global science, technology, and innovation. Wiley, pp 171–190

Miozzo M, Desyllas P, Lee HF, Miles I (2016) Innovation collaboration and appropriability by knowledge-intensive business services firms. Res Policy 45(7):1337–1351

Pietrobelli C (1996) Emerging forms of technological cooperation: the case for technology partnerships–inner logic, examples and enabling environment. In: Science and technology issues. UNCTAD-United Nations, Geneva

Porter M (1990) The competitive advantage of nations. The Free Press, New York

Porter M (1998) On competition. Harvard Business School Press, Boston, MA

Saxenian A (1981) Silicon chips and spatial structure: the industrial base of urbanization in Santa Clara County. California Working Paper 345. Institute of Urban and Regional Development, University of California, Berkeley, March

Walsh ST (2004) Roadmapping a disruptive technology: a case study. The emerging microsystems and top-down nanosystems industry. Technol Forecast Soc Change 71:161–185

Youtie J, Iacopetta M, Graham S (2008) Assessing the nature of nanotechnology: can we uncover an emerging general purpose technology? J Technol Transfer 33:315–329

Dirk Meissner is Deputy Head of the Laboratory for Economics of Innovation at HSE ISSEK and Academic Director of the Master Program "Governance for STI". Dr. Meissner has 20 years experience in research and teaching technology and innovation management and policy. He has strong background in policy making and industrial management for STI with special focus on Foresight and roadmapping, funding of research and priority setting. Prior to joining HSE Dr. Meissner was responsible for technology and innovation policy at the presidential office of the Swiss Science and Technology Council. Previously he was management consultant for technology and innovation management with Arthur D. Little. He is member of OECD Working Party on Technology and Innovation Policy. He is Associate Editor of Technological Forecasting and Social Change, IEEE Transactions on Engineering Management, Journal of the Knowledge Economy, member of Editorial Review Board at Small Business Economics and Journal of Knowledge Management. He guest edited Special Issues in Industry and Innovation journal, Journal of Engineering and Technology Management, Technological Analysis and Strategic Management among others.

Leonid Gokhberg is First Vice-Rector of the HSE and also Director of HSE ISSEK. His area of expertise is statistics and indicators on STI as well as foresight and policy studies in this area. He has authored over 400 publications in Russian and international peer-reviewed journals, monographs, and university textbooks. Prof. Gokhberg has coordinated dozens of national and international projects funded by public agencies, businesses, and international organizations. He has served as a consultant of the OECD, Eurostat, UNESCO, and other international and national agencies. Leonid is also a member of the Global Innovation Index Advisory Board, the OECD Government Foresight Network, and OECD and Eurostat working groups and task forces on indicators for S&T and, as well as steering committees of various prestigious international and national initiatives. Prof. Gokhberg is Editor-in-Chief of the Scopus-indexed scientific journal Foresight and STI Governance and editor of the Springer academic book series Science, Technology, and Innovation Studies, and participates on the editorial boards of several other influential journals. He holds PhD and Dr. of Sc. degrees in Economics.

Ozcan Saritas is Professor of Innovation and Strategy at the HSE and editor-in-chief of "Foresight"—the journal of future studies, strategic thinking and policy. He worked as Senior Research Fellow at the Manchester Institute of Innovation Research, The University of Manchester. His research focuses on innovation and policy research with particular emphasis on socio-economic and technological Foresight. With a PhD from the "Foresight and Prospective Studies Program," he introduced the "Systemic Foresight Methodology", and has produced a number of publications on the topic. Dr. Saritas has extensive work experience with the international organizations, including the United Nations, OECD, and the European Commission. He has been involved in large scale national, multinational and corporate research and consultancy projects on sectors including energy, climate change, agriculture, food, water, transportation, and ICT among others; published a number of articles in respected journals; and have delivered keynote speeches in more than 50 countries across the world. Besides his research and publication activities, he designs and delivers academic and executive education courses on Foresight and Strategic Planning. He has recently co-authored a book, entitled "Foresight for Science, Technology and Innovation" published by Springer, which has become one of the key readings in the field.

The manufacturer's authorised representative in the EU is Springer Nature Customer Service Centre GmbH, Europaplatz 3, 69115 Heidelberg, Germany. If you have any concerns regarding our products, please contact ProductSafety@springernature.com

Printed and bound by CPI Group (UK) Ltd, Croydon, CR0 4YY

30/09/2026

02225651-0010